AF182853

G.G. Delaini • T. Skřička • G. Colucci
Editors

Intestinal Polyps and Polyposis

From Genetics to Treatment and Follow-up

Foreword by Prof. John Nicholls

Springer

Editors
Gian Gaetano Delaini
Department of Surgery and Gastroenterology
University of Verona
Verona, Italy

Gianluca Colucci
Department of Surgery and Gastroenterology
University of Verona
Verona, Italy

Tomáš Skřička
Department of Surgery
Masaryk Memorial Cancer Institute
Brno, Czech Republic

Library of Congress Control Number: 2008938916

ISBN 978-88-470-1568-5 Springer Milan Berlin Heidelberg New York
e-ISBN 978-88-470-1124-3

Springer is a part of Springer Science+Business Media
springer.com

Softcover reprint of the hardcover 1st edition 2009

Cover: Simona Colombo, Milan, Italy
Typesetting: Compostudio, Cernusco s/N (Milan), Italy
Printing and binding: Printer Trento S.r.l., Trento, Italy

Springer-Verlag Italia S.r.l., Via Decembrio 28, I-20137 Milan, Italy

To Rosa,
for her courage
and love for life

Foreword

The malignant potential and familial association of polyps in the intestine have been recognized since the 19th century. There has been a huge amount of subsequent work classifying polyps, identifying polyposis syndromes, and establishing surveillance programs for cancer prevention. Advances in imaging and endoscopy have led to the accurate diagnosis and identification of malignant invasion, and therapeutic colonoscopy has completely transformed management. This has had the effect of reducing the need for conventional surgery which still, however, has a vital role in well-defined circumstances. Technical advances such as laparoscopy, transanal endoscopic microsurgery, and endoscopic mucosal resection have, in important instances, reduced the need for conventional open surgery. All these developments have been paralleled by greater genetic understanding, resulting in a degree of refinement of the assessment of cancer risk, particularly in polyposis syndromes.

Intestinal Polyps and Polyposis, includes an all-embracing account of this important group of diseases. It deals thoroughly with aspects ranging from genetics and pathology, through diagnosis and treatment, to follow-up and surveillance. It consolidates present knowledge but also looks to the future. Each of the 23 chapters deals with a specific topic. There is useful information on quality of life, psychological aspects, and the role of patient associations towards the end of the book. Surgical and endoscopic techniques are fully covered by chapters which contain much wisdom and many points of practical importance. Unusual situations such as residual polyps and uncommon polyposis variants receive full attention. Advice on establishing a surveillance service will be very useful to doctors, nurses, and other healthcare professionals, which will lead to real benefit to patients.

London, October 2008

John Nicholls *MA (Cantab), M Chir (Cantab), hon FRCP (Lond), FRCS (Eng), hon FRCSE, hon FRCS (Glasg), hon fellow ASCRS, EBSQ (Coloproctology)*
Emeritus Consultant Surgeon
St Mark's Hospital London
Visiting Professor Imperial College London

Contents

Contributors

Giampaolo Angelini MD Director of Unit of Digestive Endoscopy, Institute of Gastroenterology, University of Verona, Verona, Italy

Luigi Benini MD Associate Professor, Gastroenterology Unit, Department of Biomedical and Surgical Sciences, University of Verona, Verona, Italy

Laura Bernardoni MD Unit of Digestive Endoscopy, Institute of Gastroenterology, University of Verona, Verona, Italy

Silvia Biasin MD Department of Public Health and Medicine, Division of Hygiene and Preventive Environmental and Occupational Medicine, University of Verona, Verona, Italy

Fabrizio Bonfante MD Endoscopy Assistant, Department of Internal Medicine and Digestive Endoscopic Unit, Desenzano Hospital, Desenzano del Garda (BS), Verona, Italy

Paola Capelli MD Department of Pathology, University of Verona, Verona, Italy

Gianpaolo Cengia MD Associate Professor of Surgery, University of Brescia, Digestive Endoscopy Unit, A.O. Spedali Civili, Brescia, Italy

Renzo Cestari MD Professor of Surgery, University of Brescia, Digestive Endoscopy Unit, A.O. Spedali Civili, Brescia, Italy

Andrea Chimetto MD Department of Surgery and Gastroenterology, University of Verona, Verona, Italy

Gianluca Colucci MD Department of Surgery and Gastroenterology, University of Verona, Verona, Italy

Michele Comberlato MD Department of Gastroentrology and Digestive Endoscopy, Central Hospital, Bolzano, Italy

Gian Gaetano Delaini MD Department of Surgery and Gastroenterology, University of Verona, Verona, Italy

Domenico Della Casa MD Surgeon, University of Brescia, Digestive Endoscopy Unit, A.O. Spedali Civili, Brescia, Italy

Annalisa de Leone MD Gastroenterologist, Unit of Gastroenterology, "Magrassi-Lanzara" Department of Clinical and Experimental Medicine, 2nd University of Naples, Naples, Italy

Giuseppe Di Falco MD Pelvic Floor Unit, Section of Anal Physiology and Ultrasound, Colproctology Service, I° Department of Surgery, Regional Hospital, Treviso, Italy

Ilaria Esposito MD Director, Unit of Gastroenterology, "Magrassi-Lanzara" Department of Clinical and Experimental Medicine, 2nd University of Naples, Naples, Italy

Pavel Fabian MD PhD Surgical Pathologist, Department of Oncological and Experimental Pathology, Masaryk Memorial Cancer Institute, Brno, Czech Republic

Paola Facchin MD, PhD Associate Professor of Pediatrics, Coordinating Center for Rare Diseases, Veneto Region, Padua, Italy

Niccoló Faccioli MD Department of Radiology, G.B. Rossi University Hospital, Verona, Italy

Patrizia Ferroni MD PhD Chief of Thrombosis and Haemostasis Laboratory, Department of Laboratory Medicine and Advanced Biotechnologies, IRCCS San Raffaele Pisana, Rome, Italy

Lenka Foretová MD PhD Clinical Geneticist and Head of Department of Cancer Epidemiology and Genetics, Masaryk Memorial Cancer Institute, Brno, Czech Republic

Luca Frulloni MD Associate Professor, Gastroenterology Unit, Department of Biomedical and Surgical Sciences, University of Verona, Verona, Italy

Oliviana Gelasio MD Coordinating Center for Rare Diseases, Veneto Region, Padua, Italy

Michele Ghedi MD Surgeon, University of Brescia, Digestive Endoscopy Unit, A.O. Spedali Civili, Brescia, Italy

Giuseppe Gizzi MD Department of Medicine and Gastroenterology, University of Bologna, Bologna, Italy

Fiorella Guadagni MD PhD Head of Department of Laboratory Medicine and Advanced Biotechnologies, IRCCS San Raffaele Pisana, Rome, Italy

Gian Paolo Lancini MD Surgeon, University of Brescia, Digestive Endoscopy Unit, A.O.Spedali Civili, Brescia, Italy

Marco Lo Muzio MD Department of Surgery and Gastroenterology, University of Verona, Verona, Italy

Sandro Magrini MD Division of Surgery, Vittorio Veneto Hospital, Vittorio Veneto, Italy

Tania Magro MD Department of Surgery and Gastroenterology, University of Verona, Verona, Italy

Maurizio Mainente MD Department of Surgery and Gastroenterology, University of Verona, Verona, Italy

Silvia Manea MD PhD Coordinating Center for Rare Diseases, Veneto Region, Padua, Italy

Riccardo Manfredi MD MBA Professor, Department of Radiology, University of Verona, Verona, Italy

William Mantovani MD Department of Public Health and Medicine, Division of Hygiene and Preventive Environmental and Occupational Medicine, University of Verona, Verona, Italy

Federico Martin MD Chief of the Department of Surgery, Central Hospital, Bolzano, Italy

Mick J. Mason Founder/Secretary, FAPGene, Melton Mowbray, UK

Monica Mazzucato MD PhD Epidemiologist, Coordinating Center for Rare Diseases, Veneto Region, Padua, Italy

Fabio Menestrina Professor, Department of Pathology, University of Verona, Verona, Italy

Luigi Minelli MD Surgeon, University of Brescia, Digestive Endoscopy Unit, A.O. Spedali Civili, Brescia, Italy

Cinzia Minichiello PharmD Pharmacologist, Veneto Region Coordinating Center for Rare Diseases, Padua, Italy

Guido Missale MD Associate Professor of Surgery, Department of Surgery, University of Brescia and Digestive Endoscopy Unit, A.O. Spedali Civili, Brescia, Italy

Dario Moneghini MD Surgeon, University of Brescia, Digestive Endoscopy Unit, A.O. Spedali Civili, Brescia, Italy

Filippo Nifosì MD Department of Surgery and Gastroenterology, University of Verona, Verona, Italy

Raffaele Palmirotta MD PhD Chief of Molecular Diagnostic Laboratory, Department of Laboratory Medicine and Advanced Biotechnologies, IRCCS San Raffaele Pisana, Rome, Italy

Sara Pecori MD Department of Pathology, University of Verona, Verona, Italy

Luciano Pellegrini MD Service of Gastrointestinal Endoscopy, MF Toniolo Hospital, Bologna, Italy

Laura Peraro MD Gastroenterology Unit, Department of Biomedical and Surgical Sciences, University of Verona, Verona, Italy

Johann Pfeifer MD Associate Professor of Surgery, Department of Surgery, Medical University of Graz, Graz, Austria

Walter Piubello MD Head of the Department of Internal Medicine and Digestive Endoscopic Unit, Desenzano Hospital, Desenzano del Garda (BS), Italy

Albino Poli PhD Professor, Department of Public Health and Medicine, Division of Hygiene and Preventive Environmental and Occupational Medicine, University of Verona, Verona, Italy

Gabriele Riegler MD Gastroenterologist, Unit of Gastroenterology, "Magrassi-Lanzara" Department of Clinical and Experimental Medicine, 2nd University of Naples, Naples, Italy

Mauro Risio MD Director, Unit of Pathology, Institute for Cancer Research and Treatment (IRCC), Turin, Italy

John D. Roberts Chairman, FAPGene, Derby, UK

Mario Roselli MD Head of Medical Oncology, Department of Internal Medicine, Medical Oncology, University of Rome "Tor Vergata", Policlinico Tor Vergata, Rome, Italy

Anna Rostello MD Gastroenterology Unit, Department of Biomedical and Surgical Sciences, University of Verona, Verona, Italy

Giulio Aniello Santoro, MD PhD Head of the Pelvic Floor Unit, Section of Anal Physiology and Ultrasound, Coloproctology Service, I° Department of Surgery, Regional Hospital, Treviso, Italy

Chiara Scattolini MD, PhD Gastroenterologist and Medical Assistant, Gastroenterology Unit, Department of Biomedical and Surgical Sciences, University of Verona, Verona, Italy

Tomáš Skřička MD PhD Associate Professor, Department of Surgery, Masaryk Memorial Cancer Institute, Brno, Czech Republic

Stefano Tardivo MD Professor, Department of Public Health and Medicine, Division of Hygiene and Preventive Environmental and Occupational Medicine, University of Verona, Verona, Italy

Morena Tebaldi MD Endoscopy Assistant, Department of Internal Medicine and Digestive Endoscopic Unit, Desenzano Hospital, Desenzano del Garda (BS), Italy

Francesco Tonelli MD Director of Surgery of the Digestive Tract, Department of Clinical Pathophysiology, University of Florence, Florence, Italy

Rosa Valanzano MD Associate Professor of Surgery, and Dean of Medical Curriculum, Department of Clinical Pathophysiology, Surgery Unit, University of Florence, Florence, Italy

Italo Vantini MD Professor, Chief of Gastroenterology Unit, Department of Biomedical and Surgical Sciences, University of Verona, Verona, Italy

Tiziana Venesio PhD Assistant in Anatomy and Molecular Geneticist, Unit of Pathology, Institute for Cancer Research and Treatment (IRCC), Turin, Italy

Marco Vergine MD Department of Pathology, University of Verona, Verona, Italy

Irene Zagni MD Endoscopy Assistant, Department of Internal Medicine and Digestive Endoscopic Unit, Desenzano Hospital, Desenzano del Garda (BS), Italy

Chiara Zugni MD Department of Surgery and Gastroenterology, University of Verona, Verona, Italy

The Impact of Bowel Cancer Screening

Are We Going to Observe a Different Disease?

1

Stefano Tardivo, Silvia Biasin, William Mantovani and Albino Poli

Abstract Colorectal cancer (CRC), the third most prevalent cancer worldwide, imposes a significant economic and humanitarian burden on patients and society. Nearly 10% of all cancer incidence worldwide is CRC, and it is the only major malignancy with a similar prevalence in men and women. Since CRC is generally a disease of the elderly, its economic burden is expected to grow in the near future, mainly due to population aging. Ample evidence shows that screening for CRC with any of several available strategies significantly decreases CRC mortality by allowing detection at an early stage, and even prevention by removal of possible precursors like adenomas. Changing trends in incidence rates, stage, mortality, and change from left-sided to right-sided CRC have been seen in the last two decades. However, many people who would benefit from CRC screening do not receive it. Future interventions should focus on reducing modifiable barriers, making follow-up testing more convenient and accessible, and increasing understanding of the benefit of screening and follow-up.

Keywords Bowel cancer screening • Burden of disease • Colorectal cancer epidemiology • Colonoscopy • Computed tomography colonography • Double-contrast barium enema • Fecal occult blood testing • Flexible sigmoidoscopy • Guidelines for screening • Screening test • Stool DNA test

1.1 Epidemiology

Colorectal cancer (CRC) is the second most common cancer and the second leading cause of death from cancer in developed countries. In terms of global incidence there are 1.02 million new cases of cancer per year, and the number of deaths per year includes about half of all new cases (529,000) [1] (Table 1.1). Nearly 10% of all cancer worldwide is CRC, and it is the only major malignancy with a similar prevalence in men and women [2]. The American Cancer Society estimates that in the United States, 148,810 subjects will be diagnosed with CRC and 49,960 will die from this disease in 2008 [3]. In Europe, about 376,000 new cases are diagnosed each year, accounting for 13% of all malignant tumors in adults (Table 1.2), and it represents 11.9% of all cancer deaths (Table 1.3) [4].

In Italy in 2003 there were 38,643 new cases of CRC, and in 2002 there were 29,734 deaths [5].

S. Tardivo (✉)
Department of Public Health and Medicine, Division of Hygiene and Preventive Environmental and Occupational Medicine, University of Verona, Verona, Italy

Table 1.1 Incidence and mortality by sex and cancer site worldwide, 2002. Reproduced from [1]

	Incidence						Mortality					
	Males			Females			Males			Females		
	Cases	ASR (World)	Cumulative risk (age 0-64)	Cases	ASR (World)	Cumulative risk (age 0-64)	Deaths	ASR (World)	Cumulative risk (age 0-64)	Deaths	ASR (World)	Cumulative risk (age 0-64)
Oral cavity	175,916	6.3	0.4	98,373	3.2	0.2	80,736	2.9	0.2	46,723	1.5	0.1
Nasopharynx	55,796	1.9	0.1	24,247	0.8	0.1	34,913	1.2	0.1	15,419	0.5	0.0
Other pharynx	106,219	3.8	0.3	24,077	0.8	0.1	67,964	2.5	0.2	16,029	0.5	0.0
Esophagus	315,394	11.5	0.6	146,723	4.7	0.3	261,162	9.6	0.5	124,730	3.9	0.2
Stomach	603,419	22	1.2	330,518	10.3	0.5	446,052	16.3	0.8	254,297	7.9	0.4
Colon/rectum	550,465	20.1	0.9	472,687	14.6	0.7	278,446	10.2	0.4	250,632	7.6	0.3
Liver	442,119	15.7	1.0	184,043	5.8	0.3	416,882	14.9	0.9	181,439	5.7	0.3
Pancreas	124,841	4.6	0.2	107,466	3.3	0.1	119,544	4.4	0.2	107,479	3.3	0.1
Larynx	139,230	5.1	0.3	20,011	0.6	0	78,629	2.9	0.2	11,327	0.4	0.0
Lung	965,241	35.5	1.7	386,891	12.1	0.6	848,132	31.2	1.4	330,766	10.3	0.5
Melanoma of skin	79,043	2.8	0.2	81,134	2.6	0.2	21,952	0.8	0.0	18,829	0.6	0.0
Kaposi sarcoma*												
Breast				1,151,298	37.4	2.6				410,712	13.2	0.9
Cervix uteri				493,243	16.2	1.3				273,505	9.0	0.7
Corpus uteri				198,783	6.5	0.4				50,327	1.6	0.1
Ovary				204,499	6.6	0.5				124,860	4.0	0.2
Prostate	679,023	25.3	0.8				221,002	8.2	0.1			
Testis	48,613	1,5	0.1				8,878	0.3	0.0			
Kidney	129,223	4.7	0.3	79,257	2.5	0.1	62,696	2.3	0.1	39,199	1.2	0.1
Bladder	273,858	10.1	0.4	82,699	2.5	0.1	108,310	4.0	0.1	36,699	1.1	0.0
Brain, nervous system	108,221	3.7	0.2	81,264	2.6	0.2	80,034	2.8	0.2	61,616	2.0	0.1
Thyroid	37,424	1.3	0.1	103,589	3.3	0.2	11,297	0.4	0.0	24,078	0.8	0.0
Non-Hodgkin lymphoma	175,123	6.1	0.3	125,448	3.9	0.2	98,865	3.5	0.2	72,955	2.3	0.1
Hodgkin disease	38,218	1.2	0.1	24,111	0.8	0.1	14,460	0.5	0.0	8,352	0.3	0.0
Multiple myeloma	46,512	1.7	0.1	39,192	1.2	0.1	32,696	1.2	0.1	29,839	0.9	0.0
Leukemia	171,037	6.9	0.3	129,485	4.1	0.2	125,142	4.3	0.2	97,364	3.1	0.2
All sites but skin	5,801,839	209.6	10.3	5,060,657	161.5	9.5	3,795,991	137.7	6.4	2,927,896	92.1	4.9

*Africa only.

Table 1.2 Estimates of numbers of incident cases of cancer in Europe, both sexes combined (2004) (in thousands). Reproduced from [4], with permission from Oxford University Press

Site	Cases	%
All sites except non-melanoma skin	2886.8	100.0
Lung	381.5	13.2
Colon and rectum	376.4	13.0
Breast	370.1	12.8
Prostate	237.8	8.2
Stomach	171.0	5.9
Uterus	133.8	4.6
Lymphomas	121.2	4.2
Oral cavity and pharynx	97.8	3.4
Leukaemia	75.6	2.6
Larynx	46.1	1.6
Oesophagus	43.7	1.5

Table 1.3 Estimates of numbers of cancer deaths in Europe, both sexes combined (2004) (in thousands). Reproduced from [4], with permission from Oxford University Press

Site	Death	%
All sites except non-melanoma skin	1711.0	100.0
Lung	341.8	20.0
Colon and rectum	203.7	11.9
Stomach	137.9	8.1
Breast	129.9	7.6
Prostate	85.2	5.0
Lymphomas	65.2	3.8
Leukaemia	52.6	3.1
Uterus	49.3	2.9
Oral cavity and pharynx	40.1	2.3
Oesophagus	39.5	2.3
Larynx	24.5	1.4

The overall incidence of CRC has been declining in the United States over the past two decades, (from 66.3 cases per 100,000 population in 1985 to 48.2 in 2004) [3]. The decline has been more steep in the most recent period (2.3% per year from 1998 to 2004), partly due to an increase in screening which can result in the detection and removal of colorectal polyps before they progress to cancer. Mortality rates have also declined in the same period, due to declining incidence rates and improvements in early detection and treatment. (Figs. 1.1, 1.2) [2,3,6]. In Europe the same trend has been observed: from 1997 to 2002 appreciable declines were registered in mortality in both men (–1.6% per year to reach 18.8/100,000) and women (–2.5%). In contrast, in low-risk CRC countries, the incidence and mortality have recently been increasing. In Japan, for example, the number of new cases of CRC among men and women has been predicted to increase 9.5 and 7.5 times by 2005, and 12.3 and 10.5 times by 2020, respectively, from the 1975 baseline [7,8]. The incidence of CRC in Japan is rising dramatically, probably due to Western influence particularly in diet. There is a direct correlation between CRC and diets that are high in red meat, animal fats, and alcohol, and low in fibre, with sedentary lifestyle and excess body weight. Many studies reveal that in groups of migrants from low-risk to high-risk countries, the incidence of CRC tends to increase to the rates of the host country within the first or the second generation, or sometimes within the migrating generation itself [9,10].

The exact causes of CRC are unknown, but several factors have been linked to increased risk of CRC: age above 50 years, personal history of CRC, colorectal polyps or inflammatory bowel disease (ulcerative colitis and Crohn's disease), family history of CRC, race, familial adenomatous polyposis (FAP) and hereditary non-polyposis colon cancer (Lynch syndrome), sedentary lifestyle, and, as previously mentioned, a diet that is high in fat and low in fibre [11].

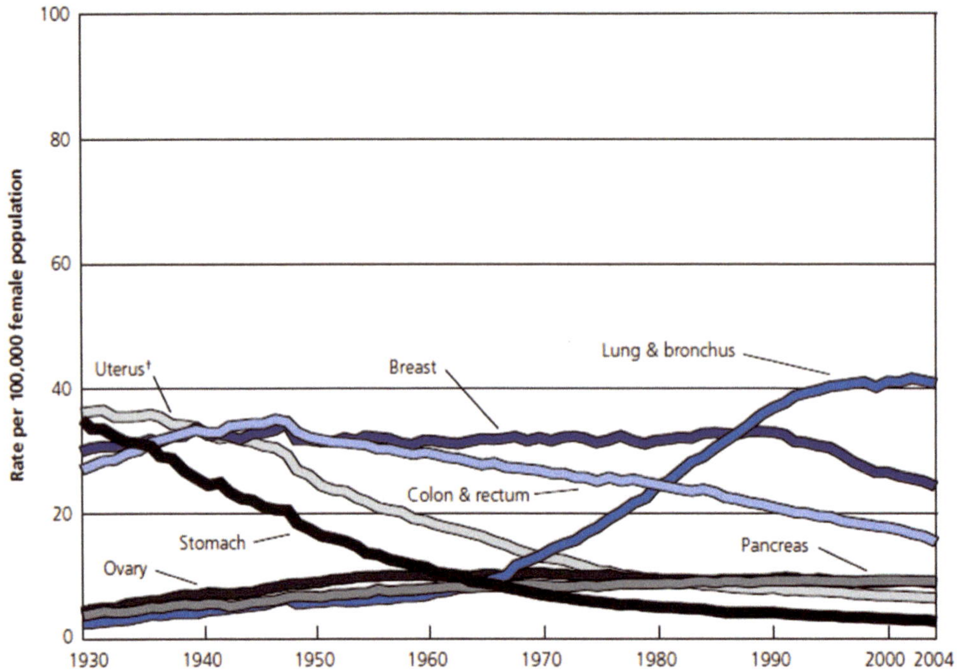

* Per 100,000, age-adjusted to the 2000 US standard population. † Uterus cancer death rates are for uterine cervix and uterine corpus combined. Notes: Due to changes in ICD coding, numerator information had changed over time. Rates for cancer of the lung and bronchus, colon and sectum, and ovary are affected by these coding changes.

Sources: US Mortality Data 1960 to 2004, US Mortality Volumes 1930 to 1959, National Center for Heath Statistics, Centers for Disease Control and Prevention, 2006.

American Cancer Society, Survellance Research, 2008

Fig. 1.1 Age-adjusted cancer death rates,* females by site, US, 1930–2004. Reproduced from [3], with permission from Wolters Kluwer Health

Table 1.4 shows that for both men and women, the incidence of CRC begins to rise around the age of 40 years [3]. Incidence sharply increase at age 50 years: 92% of CRCs are diagnosed in persons aged 50 years or older. People in their 80s continue to be at risk for CRC, with 12.5% of cases diagnosed after the age of 85 years [12]. Due to the aging population and population growth, the expected number of CRC diagnoses will increase in forthcoming years. Thus, prevention and early detection have immense public health importance.

Another important CRC risk factor is a family history of the disease. First-degree relatives of patients with CRC have a two- to threefold increased risk of developing the disease compared with the general population. Risk also depends on the age at which the neoplasm is detected, the number of relatives affected, and the degree of kinship [13]. Familial clustering of sporadic CRC is well recognized. Depending on the number of affected first-degree relatives, the relative risk of CRC varies from 1.85 for one relative, to 8.52 for at least three affected relatives; taking age into account, the risk varies from 2.18 for relatives aged 50 years and older, to 3.55 for affected relatives who are younger than 50 years [14].

High-risk groups with genetic syndromes such as FAP, hereditary non-polyposis colorectal cancer (HNPCC) or Lynch syndrome, and the hamartomatous polyposis syndromes only amount to a small proportion of the many cancers. Lynch syndrome is the most common form of hereditary colorectal cancer, accounting for approximately 1–6 % of all colorectal malignancies. This disorder is characterized by early onset of colorectal cancer and other adenocarcinomas, including endometrial, ovarian, gastric, and urinary tract cancers, as a result of defects in the mismatch

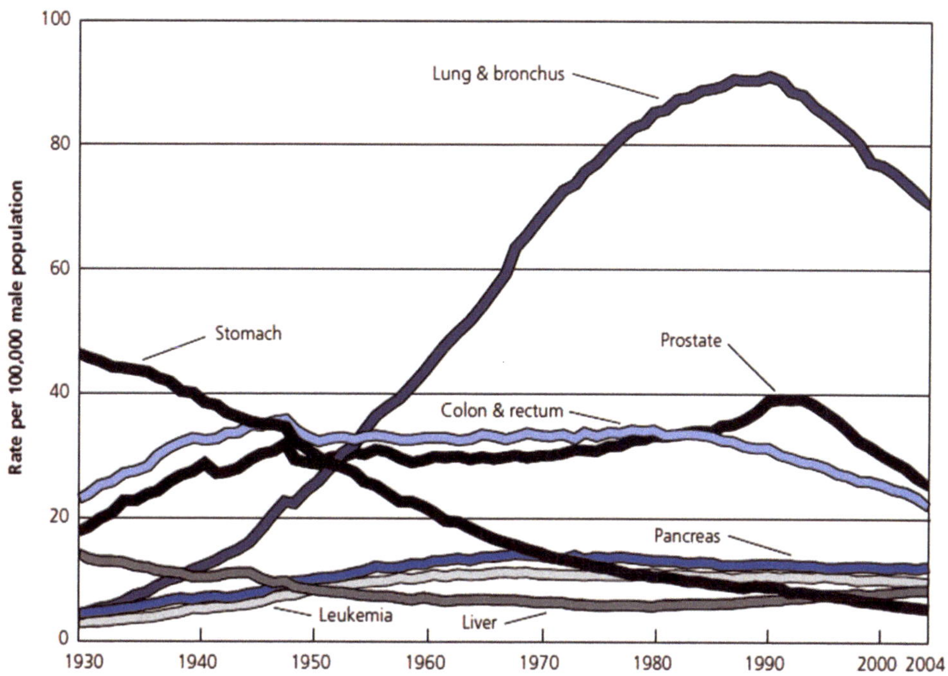

* Per 100,000, age-adjusted to the 2000 US standard population.
Notes: Due to changes in ICD coding, numerator information hat changed over time. Rates for cancer of the liver, lung and bronchus, and colon and rectum are affected by these coding changes.
Sources: US Mortality Data 1960 to 2004, US Mortality Volumes 1930 to 1959, National Center for Heath Statistics, Centers for Disease Control and Prevention, 2006.
American Cancer Society, Survellance Research, 2008

Fig. 1.2 Age-adjusted cancer death rates,* males by site, US, 1930–2004. Reproduced from [3], with permission from Wolters Kluwer Health

repair genes. Individuals have an 80% lifetime risk of colorectal cancer. In FAP, mutation in the adenomatous polyposis coli (*APC*) tumor suppressor gene gives rise to hundreds or thousands of colorectal polyps, some of which will inevitably progress to cancer. FAP affects 1 in 8,000–10,000 individuals, and accounts for <1 % of all colorectal cancers. The hamartomatous polyposis syndromes are uncommon but distinctive disorders in which multiple hamartomatous polyps develop at a young age. Although these polyps were previously considered non-neoplastic, a predisposition for intestinal malignancies is now recognized [15–17].

Incidence and mortality vary greatly by race and ethnicity; for example, the age-adjusted death rate of CRC from 2000 to 2003 was 27.3 per 100,000 African-Americans, and 19.3 per 100,000 for Caucasians in the SEER (Surveillance Epidemiology and End Results) Study areas, while the relative 5-year survival rate was 54.7% for African-Americans and 65.1% for Caucasians in the same study population [18]. Several studies have indicated that the increased mortality in African-Americans with colon cancer can be attributed in part to a more aggressive tumor and to a more advanced stage at diagnosis, and in part to differences in treatment, screening, and postdiagnosis surveillance. Poor socio-economic status and low educational level are associated with poorer health outcomes and increased mortality. Identifying groups at high risk of death from cancer by educational level as well as by race may be useful in targeting interventions and tracking cancer disparities [19–21].

Approximately 70–90% of CRCs arise from adenomatous polyps. About 30% of all polyps are hyperplastic with no malignant potential. Others are adenomatous and are considered premalignant. Adenomatous polyps are very common in adults over the age of

Table 1.4 Probability of developing invasive cancers over selected age intervals by sex, US, 2002–2004*. Reproduced from [3], with permission from Wolters Kluwer Health

		Birth to 39 (%)	40 to 59(%)	60 to 69(%)	70 and Older (%)	Birth to Death (%)
All sites†	Male	1.42 (1 in 70)	8.58 (1 in 12)	16.25 (1 in 6)	38.96 (1 in 3)	44.94 (1 in 2)
	Female	2.04 (1 in 49)	8.97 (1 in 11)	10.36 (1 in 10)	26.31 (1 in 4)	37.52 (1 in 3)
Urinay bladder‡	Male	0.02 (1 in 4,477)	0.41 (1 in 244)	0.96 (1 in 104)	3.50 (1 in 29)	3.70 (1 in 27)
	Female	0.01 (1 in 9,462)	0.13 (1 in 790)	0.26 (1 in 384)	0.99 (1 in 101)	1.17 (1 in 85)
Breast	Female	0.48 (1 in 210)	3.86 (1 in 26)	3.51 (1 in 28)	6.95 (1 in 15)	12.28 (1 in 8)
Colon & rectum	Male	0.08 (1 in 1,329)	0.92 (1 in 109)	1.60 (1 in 63)	4.78 (1 in 21)	5.65 (1 in 18)
	Female	0.07 (1 in 1,394)	0.72 (1 in 138)	1.12 (1 in 89)	4.30 (1 in 23)	5.23 (1 in 19)
Leukemia	Male	0.16 (1 in 624)	0.21 (1 in 468)	0.35 (1 in 288)	1.18 (1 in 85)	1.50 (1 in 67)
	Female	0.12 (1 in 837)	0.14 (1 in 705)	0.20 (1 in 496)	0.76 (1 in 131)	1.06 (1 in 95)
Lung & bronchus	Male	0.03 (1 in 3,357)	1.03 (1 in 97)	2.52 (1 in 40)	6.74 (1 in 15)	7.91 (1 in 13)
	Female	0.03 (1 in 2,964)	0.82 (1 in 121)	1.81 (1 in 55)	4.61 (1 in 22)	6.18 (1 in 16)
Melanoma of the skin	Male	0.15 (1 in 656)	0.61 (1 in 164)	0.66 (1 in 151)	1.56 (1 in 64)	2.42 (1 in 41)
	Female	0.26 (1 in 389)	0.50 (1 in 200)	0.34 (1 in 297)	0.71 (1 in 140)	1.63 (1 in 61)
Non-Hodgkin lymphoma	Male	0.13 (1 in 760)	0.45 (1 in 222)	0.57 (1 in 174)	1.61 (1 in 62)	2.19 (1 in 46)
	Female	0.08 (1 in 1,212)	0.32 (1 in 312)	0.45 (1 in 221)	1.33 (1 in 75)	1.87 (1 in 53)
Prostate	Male	0.01 (1 in 10,553)	2.54 (1 in 39)	6.83 (1 in 15)	13.36 (1 in 7)	16.72 (1 in 6)
Uterine cervix	Female	0.16 (1 in 638)	0.28 (1 in 359)	0.13 (1 in 750)	0.19 (1 in 523)	0.70 (1 in 142)
Uterine corpus	Female	0.06 (1 in 1,569)	0.71 (1 in 142)	0.79 (1 in 126)	1.23 (1 in 81)	2.45 (1 in 41)

* For people free of cancer at beginning of age interval. †All sites exclusive basal and squamous cell skin cancers and in situ cancers except urinary bladder.

‡ Includes invasive and in situ cancer cases.

Source: DevCan: Probability of Developing or Dying of Cancer Software, Version 6.2.1. Statistical Research and Applications Branch, National Cancer Institute, 2007. www.srab.cancer.gov/devcan

American Cancer Society, Surveillance Research, 2008

50 years, but the majority will not develop into adenocarcinoma. Histology and size determine their clinical importance. The characteristics of high-risk polyps are: larger than 1 cm, villous lesion or high-grade dysplasia on histological examination, three or more in number. Polyps larger than 2 cm in diameter have a 50% chance of becoming malignant, about 10% if they are 1–2 cm in diameter, and only 1% if they are smaller than 1 cm in diameter. CRC arises as genetic alterations that cause abnormal cellular proliferation, resulting in progression from normal colonic mucosa to adenoma, or adenomatous polyp to adenocarcinoma. This progression can be induced by a series of mutations involving oncogenes and tumor suppressor genes. The sequence of molecular events is not linear but rather a collection of events that occur over time [22]. When the transformation occurs (2.5/1000 adenomas/year), the progression from adenoma to cancer usually takes several years (5–10 years), and detection and removal of adenomas during this premalignant phase markedly decreases the incidence of colorectal cancer [23]. Recent studies show that a first-degree relative with large adenomas increases an individual risk for colorectal cancer. This risk is more pronounced if the affected family member is younger than 60 years at the time of polyp detection [14].

1.2 Screening

The primary purpose of screening for cancer is to reduce mortality from the disease screened for. Screening also has effects other than those on survival, notably on economic cost and quality of life.

Screening is a public health service in which members of a defined population, who do not necessarily perceive they are at risk of, or are already affected by, a disease or its complications, are asked a question or offered a test, to identify those individuals who are more likely to be helped than harmed by further tests or treatment to reduce the risk of a disease or its complications [24]. The National screening Committee criteria for appraising the viability, effectiveness, and appropriateness of a screening program are based on the criteria developed by Wilson in 1968 [25], and address the condition, the test, the treatment and the screening program:

1. the disease should be an important public health problem, as measured by incidence, mortality, and other measures of disease burden
2. the disease should have a detectable preclinical phase
3. treatment of disease detected before the onset of clinical symptoms should offer benefits compared with treatment after the onset of symptoms
4. the screening test should meet acceptable levels of accuracy and cost
6. the screening test and follow-up requirements should be acceptable to individuals at risk, and to their healthcare providers
7. treatment or intervention that improves survival or quality of life (compared with no screening) should be available for patients with recognized disease
8. adequate staffing and facilities for recruitment, testing, diagnosis, and follow-up, treatment and program management should be available.

The resources allocated to the screening program (including testing, diagnosis, and treatment of diagnosed patients) should be economically balanced in relation to other healthcare priorities.

CRC fulfils most of the criteria for applying screening: the natural history is well known compared with many other cancers, and it may be cured by detection at an early stage, and even prevented by removal of possible precursors such as adenomas [26]. CRC survival is closely related to the clinical and pathological stage at diagnosis. High-quality evidence shows that survival is improved when CRC is treated at early stages. The 5-year survival rate is 90% when cancer is limited to the bowel wall, 68% when lymph nodes are involved, and only 10% if metastasis has occurred by the time of diagnosis (Table 1.5) [3]. CRC detected before lymph node involvement can often be effectively treated without radiation or chemotherapy. Early detection of colorectal neoplasm improves patient outcomes not only by reducing disease-associated

Table 1.5 Five-year relative survival rates*, by stage at diagnosis, 1996–2003. Reproduced from [3], with permission from Wolters Kluwer Health

Site	All stages %	Local %	Regional %	Distant %	Site	All stages %	Local %	Regional %	Distant %
Breast (female)	88.6	98.0	83.5	26.7	Ovary§	44.9	92.4	71.4	29.8
Colon & rectum	64.0	89.8	67.7	10.3	Pancreas	5.0	20.3	8.0	1.7
Esophagus	15.6	33.7	16.9	2.9	Prostate¶	98.4	100.0	–	31.9
Kidney†	65.5	89.6	60.8	9.5	Stomach	24.3	61.1	23.7	3.4
Larynx	62.9	81.1	50.0	23.9	Testis	95.4	99.3	95.8	70.0
Liver‡	10.8	22.3	7.3	2.8	Thyroid	96.7	99.7	96.9	56.0
Lung & bronchus	15.0	49.1	15.2	3.0	Urinary bladder	79.5	92.1	44.6	6.4
Melanoma of the skin	91.1	98.5	65.2	15.3	Uterine cervix	71.6	92.0	55.7	16.5
Oral cavity & pharynx	59.1	81.8	52.1	26.5	Uterine corpus	82.9	95.3	67.4	23.1

*Rates are adjusted for normal life expectancy and are based on cases diagnosed in the SEER 17 areas from 1996-2003, followed through 2004. †Includes renal pelvis. ‡Includes intrahepatic bile duct. § Recent changes in classification of ovarian cancer, specifically excluding borderline tumors, has affected survival rates. ¶The rate for local stage represents local and regional stages combined. *Local*: an invasive malignant cancer confined entirely to the organ of origin. *Regional*: a malignant cancer that 1) has extended beyond the limits of the organ of origin directly into surrounding organs or tissues; 2) involves regional lymph nodes by way of the lymphatic system; or 3) has both regional extension and involvement of regional lymph nodes. *Distant*: a malignant cancer that has spread to parts of the boody remote from the primary tumor either by direct extension or by discontinuous metastasis to distant organs, tissues, or via the lymphatic system to distant lymph nodes.

Source: Ries LAG, Melbert D, Krapcho M, et al. (eds) SEER Cancer Statistics Review, 1975-2004, National Institute, Bethesda, MD, www.seer.cancer.gov/cst/1975_2004/, 2007.

American Cancer Society, Surveillance Research, 2008

morbidity and mortality, but also by preventing cancer occurrence by removal of precancerous polyps. The rationale for screening is comprehensible, and many clinical trials have been performed and are ongoing to detect adenomas and CRC in a favorable stage, but the ideal instrument for that purpose has not been yet identified: unlike other types of cancer, there are several options for screening for CRC. The population can also be divided into three risk groups for developing CRC: average risk, increased risk, and high risk. Different screening strategies are recommended for different each of these three groups.

1.3 Screening Tests

Screening tests for CRC falls into two categories. In one category are the fecal tests: fecal occult blood test (FOBT) – with a guaiac-based test (gFOBT) or with a fecal immunochemical test (FIT), and stool DNA test (sDNA), which are tests that are primarily effective at identifying CRC. Some premalignant adenomatous polyps may be detected, providing an opportunity for polypectomy and the prevention of CRC, but the opportunity for prevention is both limited and incidental, and is not the primary goal of CRC screening with these tests. In the second category are the partial or full structural examinations: flexible sigmoidoscopy (FSIG), colonoscopy, double-contrast barium enema (DCBE), and computed tomography colonography (CTC). These are tests that are effective at detecting cancer and premalignant adenomatous polyps, differ in complexity and accuracy for the detection of CRC, require bowel preparation and an office or hospital visit, and have various level of risk to patients. Significant positive findings on FSIG, DCBE, CTC but also on the fecal tests require follow-up colonoscopy [27].

1.3.1 *Fecal Occult Blood Test*

As already mentioned, FOBTs fall into two primary categories based on the detected analyte: gFOBTs and FITs.

gFOBTs are the most common stool blood tests in use for CRC screening and the only screening tests for which there is evidence of efficacy from prospective, randomized controlled trials (RCTs). These tests are based on peroxidase activity in all hemoglobin, heme and myoglobin, and non-heme peroxidases. The usual home-test kit requires collection of two samples from each of three consecutive bowel movements. Avoidance of red meat, poultry, fish, some raw vegetables, vitamin C, aspirin, and other non-steroidal anti-inflammatory drugs before and during the test is required because diet test interaction with these agents can increase the risk of false-positive and false-negative results. Moreover, dietary restriction may make compliance and follow-up poor.

Screening average-risk individuals over the age of 50 years with annual or biennial gFOBT has been shown in four randomized trials to reduce colorectal cancer incidence and mortality rates by between 15% and 33%. Moreover, it has been demonstrated that screened patients have cancer detected at an early and more curable stage than unscreened patients, and that annual screening results in a greater reduction in mortality rate than biennial screening [28].

A large long-duration trial in the United States (Minnesota) randomly assigned 46,551 volunteers aged 50–80 years to 5 years of screening with either annual gFOBT, biennial gFOBT, or usual care. The cumulative 18-year CRC mortality was 33% lower in the annual group than in the control group (risk ratio (RR) 0.67, 95% confidence interval (CI) 0.51–0.83), whereas the biennial group had a 21% lower colorectal cancer mortality rate than the control group (RR 0.79, 95% CI 0.62–0.97) [29,30] (Table 1.6).

A Danish trial (Funen) randomly assigned 61,933 people aged 45–75 years to usual care or screening with an initial annual followed by biennial gFOBT. At 13 years, for seven rounds of screening, the CRC mortality was 18% lower in those screened than among controls (RR 0.82, 95% CI 0.69–0.97) [31]. After nine screening rounds, at 17 years of follow-up, mortality was 16% lower for annual screening versus controls (RR 0.84, 95% CI 0.73–0.96) and 11% lower for biennial screening versus controls, including deaths attributed to complications from treatment (RR 0.89, 95% CI 0.78–1.01) [30,32] (Table 1.6).

A population-based trial in the United Kingdom (Nottingham) randomly assigned 152,850 people aged 45–74 years to either control or biennial gFOBT. It reported a 13% reduction in CRC mortality for biennial screening after 11 years of follow-up (RR 0.87, 95% CI 0.78–0.97) [30,33] (Table 1.6).

A population-based trial in Sweden (Göteborg) randomly assigned 68,308 people aged 60–64 years to

Table 1.6 Number of CRC deaths, mortality, incidence ratio and mortality reduction for the included trials. Reproduced from [30], with permission from Blackwell Publishing

Study	No. of CRC deaths		Incidence ratio		Mortality reduction (%)
	Screening group	Control group	Screening group (py)	Control group (py)	
Funen	363/30,967	431/30,966	0.84/1,000	1.00/1,000	16
Gotebord	252/34,144	300/34,146	NR	NR	16
Minnesota (A)	121/15,570	177/15,394	0.67/1,000	1.00/1,000	33
Minnesota (B)	148/15,5887	(as above)	0.79/1,000	(as above)	21
Nottingham	684/76,466	684/76,384	0.70/1,000	0.81/1,000	13

A, annual screening; *B*, biennal screening; *NR*, not reported; *py*, persons years

Table 1.7 Rehydratation of slides, positivity rates, sensitivity, and positive predictive value (PPV) for CRC and adenomas. Reproduced from [30], with permission from Blackwell Publishing

Study	Rehydration	Positivity Rate (%)	Sensitivity (%)	PPV (CRC) (%)	PPV (Adenoma) (%)
Funen	No	0.8–3.8	55.0	5.2–18.7	14.6–38.3
Goteborg	Yes	1.7–14.3	82.0	NR	NR
	No	1.9	NR	NR	NR
Minnesota	Yes	3.9–15.4	92.2	0.9–6.1	6.0–11.0
	No	1.4–5.3	80.8	5.6	NR
Nottingham	No	1.2–2.7	57.2	9.9–17.1	42.8–54.5

NR, not reported

control or two rounds of gFOBT screening at baseline and then at 16–24 months. Mortality data have not been published, but were made available for a meta-analysis of the four gFOBT trials, which showed a 16% reduction in CRC death (RR 0.84, 95% CI 0.78–0.90); when adjusted for screening attendance in the individual studies, there was a 23% reduction in CRC mortality [34]. A recent Cochrane systematic review includes seven new publications and unpublished data concerning CRC screening using FOBT, and confirms previous research demonstrating that FOBT screening reduces the risk of CRC mortality (16% RR for CRC mortality and 25% when adjusted for screening attendance). The results also indicate that there is no difference in all-cause mortality between the screened and non-screened population [30] (Table 1.6).

The sensitivity and specificity of a gFOBT varies, based on the brand or variant of the test, specimen-collection technique, number of samples collected, whether or not the stool specimen is rehydrated, and variation in interpretation, screening interval, and other factors. The sensitivity of the Hemoccult test used in the four RCTs, defined as the proportion of all CRCs detected during screening, varied from 55% to 57% for the non-rehydrated slides, and from 82% to 92% for the rehydrated slides (Table 1.7) [30]. Sensitivity for CRC is relatively low at 30–50%, and even lower for adenomas <20% [35]; if we consider that the test is used as part of an annual screening program, sensitivity achieved 90% (Table 1.8) [28]. The majority of trials reported that the positive predictive value (PPV) of Hemoccult for CRC was fairly low, suggesting that over 80% of all positive tests were false-positives. Investigations of these false-positive participants may have resulted in some negative psychosocial consequences and a small chance of significant adverse consequences from the diagnostic tests [30].

FITs are immunochemically based tests that use a reaction to human globin. A change of diet before and during the test is not necessary, and sampling is less demanding. The spectrum of benefits, limitations, and harms is similar to a gFOBT with high sensitivity. Recent studies show that sensitivity and specificity of FITs tends to be higher for all distant and advanced neoplasia, but the sensitive gFOBT shows superior performance for advanced adenomas.

Annual screening with FOBT (both FIT and gFOBT) have been shown to detect a majority of prevalent CRC in an asymptomatic population, and the tests are an acceptable option for colorectal screening in average-risk adults aged 50 years and older. Any positive test should be followed up with colonoscopy (Table 1.9) [27].

Table 1.8 Operating characteristics for colorectal cancer screening tests*. Reproduced from [28], with permission

Test	Sensitivity, %	Specificity, %	Notes
Fecal occult blood test	~50	>90	The 50% sensitivity figure is for a 1-time test, but the test is 90% sensitive when used as part of an annual screening program
Flexible sigmoidoscopy with biopsy	88–98 for large, distal adenomas or cancer	92–94 for large, distal adenomas and 92–96 for distal cancers	Only evaluates distal colon and rectum, should not be used alone to evaluate symptoms or signs, especially if a patient is over age 40
Colonoscopy with biopsy	90–97	>98	Preferred evaluation for positive screening tests and suggestive symptoms or signs, colonoscopy is considered the "gold standard" for both screening and evaluation of the colon
Double-contrast barium enema	~80	~80	Can be used if colonoscopy is not available or contraindicated
Virtual colonoscopy (CT colonography)	lesion ≤5 mm; 4 lesion 6–9 mm: 33 lesion 10 mm: 82	90	Awaits further study before clinical application can be generally recommended

CT, computed tomography

Table 1.9 Testing options for early detection of colorectal cancer and adenomatous polyps for asymptomatic adults aged 50 years and older. Reproduced from [27], with permission from Elsevier

Tests that detect adenomatous polyps and cancer
FSIG every 5 years, or
CSPY every 10 years, or
DCBE every 5 years, or
CTC every 5 years
Tests that primarily detect cancer
Annual gFOBT with high test sensitivity for cancer, or
Annual FIT with high test sensitivity for cancer, or
sDNA, with high sensitivity for cancer, interval uncertain

FSIG, flexible sigmoidoscopy; *CSPY*, colonoscopy; *DCBE*, double-contrast barium enema; *CTC*, computed tomograhy colonography

1.3.2 Stool DNA Test

This new method of CRC screening is based on the presence of known DNA alteration in the adenoma–carcinoma sequence, and on a continuous shedding of cells that contain altered DNA into the bowel lumen and consequently in the feces. More gene mutations are present in cells of adenomas or CRCs, so a multitarget DNA stool assay is required to achieve adequate sensitivity. Only a single stool collection is required, because shedding of cells is not intermittent and non-specific like occult bleeding. Several studies have been published on sDNA: sensitivity for CRC in these studies ranged from 52% to 91%, with specificity ranging from 93% to 97% [27]. Imperiale et al highlighted better sensitivity compared to Hemoccult II for CRC, high-grade dysplasia, and all advanced adenomas [36]. There are, however, insufficient data at present to include sDNA as an acceptable option for CRC screening; moreover, these methods are expensive and labour intensive [26].

1.3.3 Flexible Sigmoidoscopy

FSIG is an endoscopic procedure that examines the lower half of the colon lumen and rectum, where 70% of CRCs and adenomas are located. It carries a small risk of perforation but sedation is not required and it can be performed with a more-simple bowel preparation than is required for standard colonoscopy. The use of FSIG for CRC screening is supported from case-control and cohort studies. It is associated with a 60–80% reduction in CRC mortality for the area of the colon within its reach, and this protective effect appears to persist for ten years or more [37]. More-definitive data are awaited from ongoing trials. Additional evidence supporting FSIG derives from colonoscopy studies: it is 60–70% sensitive for advanced adenomas and CRC, compared with colonoscopy [38]. Differences in the distal and proximal lesions based on age, gender, and ethnicity (proximal CRCs are more common after age 65 years, in women than in men and in African-Americans than Whites), and the benefits and limitations of CRC screening with FSIG among these different groups remain important areas of continued investigation [27]. FSIG can be performed alone or in combination with FOBT; in this second option more CRCs can be detected, and mortality rates can be reduced more than by using either modality alone. In one study, FOBT detected 23.9% of patients with advanced neoplasia, FSIG detected 70.3%, and the tests combined detected 75.8% [39].

It is recommended that FSIG is performed for screening of an average population every 5 years. It can be performed alone or in combination with annual FOBT. Positive tests findings will need to be followed by colonoscopy (Table 1.9) [27].

1.3.4 Colonoscopy

Colonoscopy permits visual examination of the entire colon and rectum and removal of lesion at the same time as screening, rather than requiring referral for a second test. In addition to polypectomy, it allows biopsy of other lesions to be performed, as well as other procedures, such as cauterization of a bleeding lesion, dilatation of strictures, or injection of dye, to localize a tumor for subsequent surgical removal. Numerous studies have shown increased detection of adenomas and CRCs with colonoscopy compared with FOBT or FSIG, but no RCTs have been published. The National Polyp Study and other cohort studies suggest that colonoscopy with polypectomy reduces the incidence of CRCs by 76–90% [40,41]. Because colonoscopy is commonly used as the criterion standard examination, it is difficult to calculate its sensitivity. Rex and colleagues found single test sensitivity to be 90% for large adenomas and 75% for small adenomas (<1 cm); sensitivity for cancer probably exceeds 90% [42]. The specificity of colonoscopy with biopsy is generally reported to be 99% or 100%, but this assumes that all detected adenomas represent true positive results. Most detected adenomas, especially small ones, will never develop into cancer; if detection of an adenoma like this is considered a false-positive result that subjects a patient to risk without benefit, then the actual specificity of colonoscopy would be much lower [43].

Colonoscopy has several limitations: a colonoscopic preparation is required, sedation may be necessary, and although the risks of colonoscopy and polypectomy are small, the procedure may result in bleeding, perforation, and other complications. Risk of perforation with colonoscopy is 1/1000 compared with 0.02/1000 for FSIG [35]. A recent prospective study of 502 asymptomatic patients who had a colonoscopy for screening, surveillance, or follow-up of another positive screening test, found that although 34% reported mild complications (bloating and pain), only six patients (0.01%) had unexpected hospitalizations or emergency department visits within 30 days following colonoscopy, and 94% of patients lost two or fewer days from normal activities [44].

Colonoscopy every 10 years is an acceptable options for CRC screening in average-risk adults beginning at the age of 50 years (Table 1.9) [27].

1.3.5 Double-Contrast Barium Enema

DCBE evaluates the entire colon, by coating the mucosal surface with high-density barium and distending the colon with air introduced through a flexible catheter that is inserted into the rectum.

Multiple radiographs are acquired while varying the patient position during direct fluoroscopic evaluation. For colonic preparation, a 24-hour dietary and laxative regimen is usually necessary, and patients may experience mild to moderate discomfort during and after the procedure. There are no randomized trials that examine the effectiveness of DCBE in reducing incidence or death from CRC. The National Polyp Study showed that the sensitivity of DCBE was 32% (95% CI 25–39%) for polyps smaller than 0.5 cm, 53% (95% CI 40–66%) for polyps 0.6 to 1 cm, and 48% (95% CI 24–67%) for polyps larger than 1 cm, including two cases of cancerous polyps; specificity was 85% (95% CI 82–88%) [45]. Important complications occurred in 1 in 10,000 examinations, and perforation in 1 in 25,000 examination. At present, DCBE remains an option for direct imaging of the entire colon, particularly when colonoscopy is contraindicated, but any abnormal result must be followed by endoscopy. DCBE every 5 years is an acceptable option for CRC screening in average-risk adults aged 50 years and older (Table 1.9) [27].

1.3.6 Computed Tomography Colonography

Colonography, also known as virtual colonoscopy has been developed as a minimally invasive imaging examination of the entire colon and rectum, as an alternative to conventional colonoscopy. In 2005, two meta-analyses reviewed the cumulative published CTC performance data including both high-risk and screening cohorts, with one analysis representing 33 studies on 6393 patients [46,47]. Sensitivity for large polyps (>1 cm) was found to be 85–93%, and specificity 97%. For detection of small polyps (6–9 mm), sensitivity was found to be 70–86%, and specificity was 86–93%. The accuracy of CTC is influenced by lesion size, and sensitivity and specificity improves with polyp size. Although evidence supports the effectiveness of CTC in detecting colonic neoplasm, there are no studies of the effectiveness as a screening test in reducing mortality from CRC, and it is not yet among the tests recommended for CRC screening. A number of CTC trials are currently in progress within the United States and Europe. The technique still requires bowel preparation and colonic insufflation with air, and some patients find the procedure uncomfortable; moreover it subjects patients to ionizing radiation. Abnormal findings require referral to colonoscopy. In addition, approximately 11% of patients will have new extracolonic abnormalities identified during CTC and these may require investigation or intervention. A recent study suggests that CTC with no reporting of diminutive lesions (<6 mm) could be the most cost-effective and safest screening option [48]. However, non-reporting is unethical, cannot permit retrospective examination of the appropriateness of this approach, and does not permit any survey of a "not-reported" lesion [49]. Standardization of the evolving technology and consensus related to the reporting of findings will be essential for effective implementation of CTC screening. However, in terms of detection of colon cancer and advanced neoplasia, recent data suggest CTC is comparable to colonoscopy. The American Cancer Society, based on the accumulation of evidence, includes CTC as an acceptable option for CRC screening of an average-risk population from the age 50 years. The interval for repeat examinations after a negative CTC is uncertain, but if current studies confirm the previously reported high sensitivity for detection of CRC and of polyps ≥6 mm, it would be reasonable to repeat examinations every 5 years (Table 1.9) [27].

1.4 Guidelines for Screening

In describing screening tests we have focused on screening in an average-risk population: American Cancer Society screening recommendations are summarized in Table 1.9 [27]. Several available screening options seem to be effective, but the single best screening approach cannot yet be determined.

Persons within an average-risk population should initiate screening at 50 years of age. This is also confirmed by findings from a study based on colonoscopic screening: the prevalence of advanced neoplasia in an average-risk population in the 50–59-year age group may be higher than that in the 40–49-year age group, although the prevalence of total adenomas is similar in both groups [50]. Experts recommend that African-Americans begin screening at the age of 45 years, because of the lower survival rates and delay in making the diagnosis in this population. Colonoscopy is the preferred screening test because of the propensity for lesions in the upper (proximal) colon in this group. The age at which to stop CRC screening is not known with certainty, but depends on life expectancy and the anticipated benefit of screening.

The increased-risk population can be divided into three groups: patients with a history of polyps at prior colonoscopy, patients with CRC, and patients with family history of CRC. Patients with small rectal hyperplastic polyps have the same screening recommendation as average-risk individuals. For patients with one or two small tubular adenomas with low-grade dysplasia, colonoscopy is recommended but the precise timing (from 5 to 10 years after the initial polypectomy) should be based on other clinical factors such as prior colonoscopy findings, family history, judgment of the physician, and the preferences of the patients. For patients with three to ten adenomas or one adenoma >1 cm, or any adenoma with villous features or high-grade dysplasia, colonoscopy is recommended 3 years after the initial polypectomy. For patients with more than ten adenomas on a single examination, the possibility of an underlying familial syndrome should be considered, and colonoscopy is recommended within 3 years after the initial polypectomy. Moreover, for patients with sessile adenomas that are removed piecemeal, colonoscopy is recommended after 2 to 6 months, to verify complete removal [27,51]. If the cancer is non-obstructing, patients should undergo a preoperative colonoscopy to view the entire colon and remove other polyps if present. If the cancer is obstructing and no unresectable metastases are found during surgery, colonoscopy should be performed 3–6 months after the cancer resection; alternatively colonoscopy can be performed intraoperatively. For people who have had colon or rectal cancer removed by surgery, colonoscopy is recommended within a year after cancer resection; if the results are normal, the examination should be repeated every 5 years, and if it is again normal, after a further 5 years [52]. For patients with either CRC or adenomatous polyposis in a first-degree relative before the age of 60 years, or in two or more first-degree relatives at any age, colonoscopy is recommended every 5 years beginning at age 40 years or 10 years before the youngest case. For patients with either CRC or adenomatous polyps in a first-degree relative aged 60 years or older, or in two second-degree relatives, screening should be at an earlier age (40 years), but individuals may choose to be screened with any recommended form of testing [27].

High-risk groups are: genetic syndromes (the most frequent are: FAP and HNPCC) and inflammatory bowel disease (chronic ulcerative colitis and Crohn's disease).

In HNPCC, screening should be started at the age of 20–25 years with full colonoscopy, or ten years younger than the youngest age of the person diagnosed in the family, and recommended every one to two years [15,53,54] (Table 1.10). In FAP, FSIG annually or semi-annually should begin at puberty (10–12 years) and continue until age 35 years, at which time the interval can be reduced to 3 years if no polyps have been detected. Once polyps emerge, yearly colonoscopy is required, although prophylactic colectomy should closely follow this occurrence. Interval surveillance for the remaining rectal mucosa in patients who undergo a prophylactic subtotal colectomy is required [15] (Table 1.10). Cancer risk for inflammatory bowel disease begins to be significant at 8 years after the onset of pancolitis, or 12–15 years after the onset of left-sided colitis. Colonoscopy every one to two years with biopsies for dysplasia is recommended.

1.5 Burden of Disease

CRC, the third most prevalent cancer worldwide, imposes a significant economic and humanitarian burden on patients and society. Precise quantification of the costs of CRC is required to improve management and to evaluate the cost-effectiveness of screening. One study conservatively estimates the annual expenditures for CRC to be approximately 5.3 billion dollars in the US in 2000, including both direct and indirect costs [55]. No worldwide data have been published, but assuming that the US represents 25–40% of the total expenditure in oncology, as seen for breast and lung cancers, a rough estimate for CRC would be in the range of US $14–22 billion [55]. A recent study in France on the cost of management of CRC and the effects of age, stage at diagnosis, health-care pattern, and level of comorbidities, confirms the major economic burden of CRC and that total costs depend mainly on the stage at diagnosis. The mean cost for the first year of management is estimated at 24,966€ in 2004. Costs increase significantly with cancer progression, from 17,596€ for stage I to 35,059€ for stage IV. Mass screening could contribute to decreasing the cost of managing CRC by enabling diagnosis at an earlier stage [56]. Since CRC is generally a disease of the elderly, its economic burden is expected to grow in the near future, mainly due to population aging. A projection of the costs associated with CRC care in the US by the year 2020 estimates that among individuals aged 65 years and older, there will be increases of 53% and 89%, in a fixed scenario and in the current trend scenario respectively [57].

Screening for CRC has a cost per life-year saved that is similar to other nationally recommended screening programs [28]. A 2002 systematic review of

Table 1.10 Screening recommendations. Reproduced from [15]

Disorder	Gene(s)	CRC Risk/penetrance	Age of onset	GI manifestations	Extra intestinal manifestations
Nonpolyposis disorders					
HNPCC	DNA mismatch repair genes: hMSH2, hMLH1, hPMS1, hPMS2, hMSH6	~80%	Polyps 20–30 years CRC 30–40 years Endometrial cancer 30–40 years	Discrete, often multiple, adenomatous colorectal polyps; early onset and multiple colorectal cancers	Early onset endometrial ovarian, transitional cell, stomach, small intestine, hepatobiliary cancers; multiple cancers in the same individual
Muir–Torre	hMSH2, hMLH1				As in HNPPC plus sebacous gland and breast carcinomas
Polyposis disorders					
FAP	Tumour suppressor gene: APC	~100%	Polyps 10–20 years CRC 20–40 years	100–5000 adenomatous colorectal polyps; fundic gland polyps; ileal and jejunal polyps; duodenal/ampullary adenomas/cancer	Papillary thyroid cancer; hepatoblastoma; adrenal hyperplasia/carcinoma
Gardner's syndrome	APC	~100%	As in FAP	As in FAP	Desmoid tumors; soft-tissue tumors; osteomas; dental abnormalities; CHRPE; plus those tumors seen in FAP
Turcot's syndrome	APC (70%) hMLH1, hPMS2 (30%)	~100%	As in FAP	As in FAP	Central nervous system tumors – medulblastomas, astrocytomas, espendynomas
AAPC	APC (proximal or distal ends)	Very high but not 100%	Polyps 20–30 years CRC 20–50 years	< 100 polyps; flat polyps; proximal colon distributions; abundant upper GI polyps	
Hamartomatous disorders					
Peutz-Jeghers	Tumor suppressor: LKB1	~40%	Polyps early chilhood CRC 30–40 years	5–100's of hamartomatous polyps throughout GI tract; polyps demonstrate smooth muscle pseudoinvasion	Mucocutaneous pigmentation; cancer of the breast (bilateral), cervix, gonads, thyroid and pancreas;
Juvenile Polyposis	Tumor suppressors – SMAD4 and BMPR1A; PTEN	10–40%	Polyps 5–15 years CRC 15–40 years	50–200 polyps throughout GI tract, colonic predominance; polyps have abundant lamina propria; intussussception/ prolapse of polyps; anemia; bleeding; protein-losing enteropathy	Congenital abnormalities, including cardiac, craniofacial and bowel rotations Cowden's syndrome
Cowden's syndrome	Protein tyrosine phosphatese gene: PTEN	No increased risk		Hamartomatous polyps throughout the GI tract	Verrucous skin lesions on face and extremities; skin, breast and thyroid cancer; macrocephaly; Lhermitte-Duclos disease
Ruvalcaba–Myhre–Smith	Protein tyrosine phosphatase gene: PTEN			Hamartomatous polyps throughout the GI tract	Unusual facies, macrocephaly, developmental delay, penile pigmented papules; thyroiditis; skeletal abnormalities
Heroditaty mixed polyposis		~30%	Polyps 20–40 years CRC 30–50 years	1–15 atypical juvenile polyps; a mix of adenomamatous, hyperplastic and hamar-tomatous polyps	No extraintestinal manifestations

GI, gastrointestinal; *HNPCC*, hereditary non-polyposis colorectal cancer; *FAP*, familial adenomatous polyposis; *ACP*, adenomatous polyposis coli; *PTEN*, phosphate and tensin; *CHRPE*, congenital hypertrophy of the retinal pigment epithelium

seven cost-effectiveness analyses found that the cost-effectiveness of the commonly used screening modalities was between $10,000 and $25,000 per year of life saved, compared with no CRC screening. This review found that no single strategy consistently had the best cost-effectiveness ratio, and that additional analyses are necessary to determine the optimal ages of initiation end cessation [58]. A recent study estimated the clinical preventable burden if a birth cohort of 4 million individuals were offered screening at recommended intervals; 31,500 deaths would be prevented and 338,000 years of life would be gained over the lifetime of the birth cohort. In 2000, the cost-effectiveness of offering patients aged 50 years and older a choice of CRC screening options was $11,900 per year of life gained [59].

Every 7 seconds, someone turns 50 years old; every 3.5 minutes, someone is diagnosed with CRC; every 9 minutes, someone dies from CRC; and every 5 seconds, someone who should be screened for CRC is not [12]. Compliance with screening is low even in the US, the country with the highest awareness and compliance in the world, with 35% of the general population aged over 40 years, and 60% of the high-risk population [55]. Other studies show that the adherence rate to screening programs in high-risk persons is only 38% [13]. Lack of physician recommendation for both the average-risk and the high-risk population is the most commonly endorsed barrier to adoption of CRC screening [60]. Also, low socio-economic status, low level of education, and non-Caucasian race, particularly African-American, are well know to negatively influence CRC mortality, mainly due to a later stage at diagnosis and less-aggressive treatment [21,61]. Later stage at diagnosis largely depends on a low adherence rate to screening programs or follow-up. Future interventions should focus on reducing modifiable barriers, making follow-up testing more convenient and accessible, promoting the acceptance of complete diagnostic evaluations, and educating the public regarding the risk factors of CRC and increasing understanding of the benefit of screening and follow-up [62].

1.6 Are We Going to Observe a Different Disease?

A population-based study by Gupta et al summarized changing trends in CRC incidence rates, stage, mortality, and change from left-sided to right-sided CRC seen in the last two decades [2]. The study showed a significant decline in CRC incidence during the study period, from 60 to 46.4 per 100,000 per year, with a 40% reduction in left-sided CRC, where rates fell from 32.4 to 19.5 per 100,000. This corresponds with what has happened in most developed countries in the last 20 years, and one potential explanation is the increase in CRC screening as well as an improvement in treatment (Figs.1.1 and 1.2).

Cancer of the left and right colon shows different prevalence at varying ages, in high- and low-incidence nations and in men and women. In particular, proximal CRC is most common in elderly people and women, and in developed countries. The progressive increase of right-sided colon cancers in proportion to left-sided cancers over the past 40 years has been seen in multiple studies. Differences of epidemiology of proximal and distal CRC could derive from different procarcinogenic factors in the ascending colon compared with the descending colon; nevertheless, potential differences in tumor biology might also play a part [63]. However, the proximal CRC shift over time may not be attributed to a real increase in the incidence of right-sided CRC, but to a decrease in the incidence of distal CRC coupled with the aging of the population [64]. Another hypothesis is that a greater likelihood of prior polypectomy, and thus cancer prevention, occurs more on the left side than on the right [65]. These factors are important in evaluating potential strategies for instituting advances in diagnosis and prevention. Elderly people are an age group most affected by right-sided CRC, so they would be the most adversely affected by CRC screening methods that do not asses the total colon [64].

The population-based study by Gupta et al [2] showed a significant and favorable stage shift over time, and earlier cancer stage at diagnosis: the proportion of stage A doubled, rising from 7% to 13%; conversely, there was a decrease in stage D which fell from 13% to 7% (the proportions of stages B and C did not change significantly). Moreover, the screen-detected cases had a survival advantage with significantly higher 5- and 10-year survival rates with an increased proportion of screen-detected cancer.

Early CRC detection has an important role in reducing the transition from preclinical Dukes' stages A and B to preclinical stages C and D, and consequently improving the prognosis of CRC. Five-year survival rate is 90% when cancer is limited to the bowel wall, 68% when lymph nodes are involved, and only 10% if metastasis has occurred by the time of diagnosis (Table 1.5) [3]. The progression rates of CRC by Dukes' stage in a high-risk group were estimated and applied to evaluate the efficacy of different screening regimens using colonoscopy. The predicted reductions of Dukes' stage C and D achieved by annual, biennial, 3-yearly, and 6-yearly screening regimes against the control group were 60%, 49%, 40%, and 25% respectively. The corresponding predicted mortality reductions were 39%, 33%, 28%, and 18% respectively [66]. This suggest that selective screening with colonoscopy for a high-risk group is important for reducing Dukes' stages C and D disease, which in turn leads to a reduction in CRC mortality. Mass screening contributes to decreasing the cost of managing CRC and reduces mortality by improving the stage at diagnosis and by identification and removal of the precursor lesion – the adenomatous polyp [67].

References

1. Parkin DM, Bray F, Ferlay J, Pisani P (2005) Global cancer statistics 2002. CA Cancer J Clin 55:74–108.
2. Gupta AK, Melton LJ, Petersen GM, et al (2005) Changing trends in the incidence, stage, survival and screen-detection of colorectal cancer: a population-based study. Clin Gastroenterol Hepatol 3:150–158
3. Jemal A, Siegel R, Ward E, et al (2008) Cancer statistics, 2008. CA Cancer J Clin 58:71–96.
4. Boyle P, Ferlay J (2005) Cancer incidence and mortality in Europe, 2004. Ann Oncol 16:481–488.
5. Sessa A, Abbate R, Di Giuseppe, et al (2008) Knowledge, attitudes, and preventive practices about colorectal cancer among adults in an area of Southern Italy. BMC Cancer 8:1171.
6. Golfinopoulos V, Salanti G, Pavlidis N, Ioannidis JP (2007) Survival and disease-progression benefits with treatment regimens for advanced colorectal cancer: a meta-analysis. Lancet Oncol 8:898–911.
7. Kukiri K, Tajima K (2006) The increasing incidence of colorectal cancer and the preventive strategy in Japan. Asian Pac J Cancer Prev 7:495–501.
8. Lee KJ, Inonue M, Otani T, et al (2007) Colorectal cancer screening using fecal occult blood test and subsequent risk of colorectal cancer: a prospective cohort study in Japan. Cancer Detect Prev 31:3–11.
9. Flood DM, Weiss NS, Cook LS, et al (2000) Colorectal cancer incidence in asian migrants to United States and they descendants. Cancer Causes Control 11:403–411.
10. Marchland LL (1999) Combined influence of genetic and dietary factors on colorectal cancer incidance in Japanese Americans. J Natl Cancer Inst Monogr 26:101–105.
11. Brenemann AE (2008) Clinical watch, colon cancer: an update on screening. JAAPA 3:17–18. http://jaapa.com/issues/j20080301/articles/watch0308.htm (accessed 27 August 2008).
12. Benson AB (2007) Epidemiology, disease progression and economic burden of colorectal cancer. J Manag Care Pharm 13:5–18.
13. Bujanda L, Sarasqueta C, Zubiaurre L, et al (2007) Low adherence to colonoscopy in the screening of first-degree relatives of patients with colorectal cancer. Gut 56:1714–1718.
14. Cottet V, Pariente A, Nalet B, et al (2007) Colonscopic screening of first-degree relatives of patients with large adenomas: increased risk of colorectal tumors. Gastroenterology 133:1086–1092.
15. Strate LL, Syngal S (2005) Hereditary colorectal cancer syndromes. Cancer Causes Control 16:201–213.
16. Lipton L, Tomlinson I (2006) The genetics of FAP and FAP-like syndromes. Fam Cancer 5:221–226.
17. Jass GR (2008) Colorectal polyposes: from phenotype to diagnosis. Pathol Res Pract 204:431–447.
18. Du XL, Meyer TE, Franzini L (2007) Meta-analysis of racial disparities in survival in association with socioeconomic status among men and women with colon cancer. Cancer 11:2161–2170.
19. Mayberry RM, Coates RJ, Hill HA, et al (1995) determinants of black/white differences in colon cancer survival. J Natl Cancer Inst 22:1686–1693.
20. Du XL, Fang S, Vernon SW, et al (2007) Racial disparities and socioeconomic status in association with survival in a large population-based cohort of elderly patient with colon cancer. Cancer 3:160–169.
21. Albano JD, Ward E, Jemal A, et al (2007) Cancer mortality in the United States by education level and race. J Natl Cancer Inst 19:1384–1394.
22. Vogelstein B, Fearon ER, Hamilton SR, et al (1988) Genetic alteration during colorectal tumor development. N Engl J Med 319:525–532.
23. Brooks DD, Winawer SJ, Rex DK, et al (2008) Colonscopy surveillance after polypectomy and colorectal cancer resection. Am Fam Physician 77:995–1002.
24. UK National Screening Committee. Definition of screening. www.nsc.nhs.uk/whatscreening/whatscreen_ind.htm (accessed 25 September 2008).
25. Wilson JM, Junger G (1968) Principles and practice of screening for disease. World Health Organization, Geneva.
26. Hakama M, Hoff G, Kronborg O, Pahlman L (2005) Screening for colorectal cancer. Acta Oncol 44:425–439.
27. Levin B, Lieberman DA, McFarland B, et al (2008) Screening and surveillance for the early detection of colorectal cancer and adenomatous polyps, 2008: a joint guideline from the American Cancer Society, the US Multy-Society Task Force on colorectal cancer, and the American College of Radiology. Gastroenterology 134:1570–1595.
28. Weinberg DS (2008) In the clinic. Colorectal cancer screening. Ann Intern Med 148(3):ITC-2-1–ITC-2-16.
29. Mandel JS, Church TR, Ederer F, Bond JH (1999)

Colorectal cancer mortality: effectiveness of biennal screening for fecal occult blood. J Natl Cancer Inst 91:434–437.
30. Hewitson P, Glasziou P, Watson E, et al (2008) Cochrane systematic review of colorectal cancer screening using the fecal occult blood test (hemoccult): an update. Am J Gastroenterol 103:1541–1549.
31. Jorgensen OD, Kronborg O, Fenger C (2002) A randomised study of screening for colorectal cancer using fecal occult blood testing: results after 13 years and seven biennial screening rounds. Gut 50:29–32.
32. Kronborg O, Jorgensen OD, Fenger G, Rasmussen N (2004) Randomized study of biennal screening with a fecal occult blood test: results after nine screening rounds. Scand J Gastroenterol 37:846–851.
33. Scholefield JH, Moss F, Sufi F, et al (2002) Effect of faecal occult blood screening on mortality from colorectal cancer: results from a randomised controlled trial. Gut 50:840–844.
34. Towler BP, Irwing L, Glasziou P, et al (2000) Screening for colorectal cancer using the fecal occult blood test, hemoccult. Cochrane Database Syst Rev 1:CD001216.
35. Atkin W (2003) Options for screening for colorectal cancer. Scand J Gastroenterol 237(suppl):13–16.
36. Imperiale TF, Ranshoff DF, Itzkowitz SH, et al, Colorectal Cancer Study Group (2004) Fecal DNA versus fecal occult blood for colorectal cancer screening in an average-risk population. N Engl J Med 351:2704–2714.
37. Newcomb PA, Noorfleet RG, Storer BE, et al (1992) Screening sigmoidoscopy and colorectal cancer mortality. J Natl Cancer Inst 84:1572–1575.
38. Imperiale TF, Wagner DR, Lin Y, et al (2000) risk of advanced proximal neoplasms in asymptomatic adults according to the distal colorectal findings. N Engl J Med 343:169–174.
39. Lieberman DA, Weiss DG, Veterans Affairs Cooperative Study Group (2001) One-time screening for colorectal cancer with combined fecal occult blood testing and examination of the distal colon. N Engl J Med 345:555–560.
40. Winawer SJ, Zauber AG, Ho MN, et al (1993) Prevention of colorectal cancer by colonscopic polypectomy. The National Polyp Study Workgroup N Engl J Med 329:1977–1981.
41. Thiis-Evensen E, Hoff GS, Sauar J, et al (1999) Population based surveillance by colonoscopy: effect on the incidence of colorectal cancer. Telemark Polyp Study I. Scand J Gastroenterol 34:414–430.
42. Rex DK, Coutler CS, Lemmel GT, et al (1997) Colonoscopic miss rate of adenomas detemined by back-to-back colonoscopies. Gastroenterology 112:24–28.
43. Pignone M, Rich M, Teutsch SM, et al (2002) Screening for colorectal cancer in adults at average risk: a summary of the evidence for the US Preventive Services Task Force. Ann Intern Med 137:132–141.
44. Ko CW, Riffe S, Shapiro JA, et al (2007) Incidence of minor complications and time lost from normal activities after screening or surveillance colonoscopy. Gastrointest Endosc 65:648–656.
45. Winawer SJ, Stewart ET, Zauber AG, et al (2000) A comparison of colonoscopy and double contrast barium enema for surveillance after polypectomy. National Polyp Study Work Group. N Engl J Med 342:1766–1772.
46. Halligan S, Altman DG, Taylor SA, et al (2005) CT colonography in the detection of colorectal polyps and cancer: systematic review, meta-analysis, and proposed minimum data set for study level reporting. Radiology 237:893–904.
47. Mulhall BP, Veerappan GR, Jackson JL (2005) Metaanalisis: computed tomographic colonogrephy. Ann Intern Med 142:635–650.
48. Pickhardt PJ, Hassan C, Laghi A, et al (2007) Cost effectiveness of colorectal cancer screening with computer tomography colonography. Cancer 109(11):2213–2221.
49. Romagnuolo J (2008) Cost effectiveness of colorectal cancer screening with computer tomography colonography: the impact of not reporting diminutive lesions. Cancer 112:222.
50. Rundle AG, Lebwohl B, Vogel R, et al (2008) Colonoscopic screening in average-risk individuals ages 40 to 49 vs 50 to 59 years. Gastroenterology 134:1311–1315.
51. Winawer SJ, Zauber AG, Fletcher RH, et al (2006) Guideline for colonoscopy surveillance after polipectomy:a consensus update by the US Multy Society Task Force on colorectal cancer and the American Cancer Society. CA Cancer J Clin 56:143–159.
52. Rex DK, Kahi CJ, Levin B, et al (2006) Guidelines for colonoscopy after cancer resection: a consensus update by the American Cancer Society and the US Multy Society Task Force on Colorectal Cancer. Gastroenterology 130:1865–1871.
53. Mecklin JP, Aarnio M, Laara A, et al (2007) Development of colorectal tumors in colonoscopic surveillance in Lynch Syndrome. Gastroenterology 133:1093–1098.
54. Lindor NM, Peterson GM, Hadley DW, et al (2006) Recommendations for the care of individuals with inherited predisposition to Lynch Syndrome. JAMA 296:1507–1517.
55. Redaelli A, Cranor CW (2003) Screening, prevention and socioeconomic costs associated with the treatment of colorectal cancer. Pharmacoeconomics 21:1213–1238.
56. Clerc L, Jooste V, Lejeune C, et al (2007) Cost of care of colorectal cancers according to health care patterns and stage at diagnosis in France. Eur J Health Econ Nov 21, epub ahead of print.
57. Yabroff KR, Mariotto AB, Feuer E, Brown ML (2007) projections of the costs associated with colorectal cancer care in the United States 2000–2020. Health Econ 17:947–959.
58. Pignone M, Saha S, Hoerger T, Mandelblatt J (2002) Cost-effectiveness analyses of colorectal cancer screening: a systematyc review for the US Preventive Services Task force. Ann Intern Med 137:96–104.
59. Maciosek MV, Solberg LI, Coffield AB, et al (2006) Colorectal cancer screening. Health impact and cost effectiveness. Am J Prev Med 31:80–89.
60. Rawl SM, Menon U, Champion VL, et al (2005) Do benefit and barriers differ by stage of adoption for colorectal cancer screening? Health Educ Res 20:137–148.
61. Byers TE, Wolf HJ, Bauer KR, et al (2008) The impact of socioeconomic status on survival after cancer in the United

States: findings from the National Program of Cancer Registries Patterns of Care Study. Cancer 113:582–591.
62. Zheng YF, Saito T, Takahashi M, et al (2006) Factors associated with intentions to adherence to colorectal cancer screening follow.up exams. BMC Public Health 6:272.
63. Glebov OK, Rodriguez LM, Nakahara K, et al (2003) Distinguishing right from left colon by the pattern of gene expression. Cancer Epidemiol Biomarkers Prev 12:755–762.
64. Rabenek L, Davila JA, El-Serag HB (2003) Is there a true shift to the right colon in the incidance of colorectal cancer? Am J Gastroenterol 98:1400–1409.
65. Saltzstein SL, Behing CA (2007) Age and time as factors in the left-to-right shift of the subsite of colorectal adenocarcinoma: a study of 213,383 cases from the California Cancer Registry. J Clin Gastroenterol 41:173–177.
66. Wong JM, Yen MF, Lai MS, et al (2004) Progression rates of colorectal cancer by Dukes' stages in a high risk group: analisys of selective colorectal cancer screening. Cancer Journal 10:160–169.
67. Rae LC, Gibberd RW (2000) survival of patients with colorectal cancer detected by a community screening program. Med J Aust 172:13–15.

Pathological Features of Sporadic Colonic Adenoma

2

Sara Pecori, Paola Capelli, Marco Vergine and Fabio Menestrina

Abstract Colorectal cancer is one of the most common neoplasms of industrialized nations. Most colorectal cancer develops from adenomas. There are four categories of adenoma: tubular, villous, tubulo-villous, and flat-depressed, and the histological features of adenomas may be defined as low- or high-grade dysplasia.

Morphological features that determine the malignant potential of an adenoma are size, growth pattern, and grade of dysplasia. Colorectal adenoma containing invasive carcinoma corresponds to a carcinoma invading the submucosa, and represents the earliest form of clinically relevant colon cancer. Improved prognostic power may derive from advancements in histopathological evaluation. The pathological features that are crucial for evaluating risk of adverse outcome include histological grade, completeness of resection margin, vasoinvasiveness, tumor budding, and level of invasion of the submucosa. These pathological parameters define two groups of early colorectal cancer with different risk of nodal and/or local recurrence: low- and high-risk early colorectal cancer.

Phenotypic characteristics seen on histopathological examination are essential to planning patient management and should continue to be the major focus of pathologists' efforts.

Keywords Adenoma • Early colorectal cancer • Malignant polyp

2.1 Introduction

Colorectal cancer is one of the most common neoplasms of industrialized nations, and accounts for approximately 9% of all cancer [1]. It is the second leading cause of cancer-related death in the Western world [2].

Most colorectal cancer develops from adenomas, the precursor lesions [2–5]. Adenomas are benign neoplasms with malignant potential; they may harbor an invasive carcinoma. Adenomas occur sporadically or as part of a polyposis syndrome. Hereditary polyposes account for approximately 1% of all colorectal carcinomas; hereditary non-polyposis colorectal cancer (HNPCC) accounts for approximately another 5%, and perhaps 30% or more of sporadic carcinomas may be inherited [6]. In addition to its clinical relevance as a precancerous lesion, the adenoma provides a model of early neoplastic change that has contributed to our understanding of the mechanisms of colorectal carcinogenesis [7].

Colorectal cancer is highly curable if diagnosed in

S. Pecori (✉)
Department of Pathology, University of Verona, Verona, Italy

the early stages [8], and malignant polyps constitute the precursors of early colorectal cancer. The pathologist plays a critical role in the management of the patient with endoscopically removed polyps, especially malignant polyps, because the histopathological interpretation is the most important consideration for subsequent management [9].

2.2 Adenomas

Adenoma is a benign intraepithelial neoplasm composed of dysplastic cells. Most colorectal adenomas are present as protuberant masses or polyps. They must be differentiated from other types of epithelial polyps. They are classified according to the pathological process that is believed to underlie their origin [7].

Adenomas, the benign glandular neoplasms that precede colon cancer development, originate from the intestinal epithelium. They occur singly or in multiples. When multiple, the patients may have a genetic syndrome.

2.3 Biological Alterations in Adenomas

Despite their differing structure, there are two common features in adenomas: a dysregulated proliferation and the failure to fully differentiate the epithelium. The dysregulated proliferation is evidenced by an upward shift in the proliferative compartment. Mitotic figures, including abnormal ones, are present throughout the entire length of the hyperchromatic, adenomatous epithelium.

In the normal colon most apoptosis occurs near the luminal surface. Adenomas contain numerous apoptotic cells which often lie at the adenomatous base, a reversal of the normal distribution. This suggests that adenomas exhibit a reversed epithelial cell migration and have an inward growth pattern directed toward the crypt base rather than toward the lumen [10].

Adenomas also tend to show abnormalities in epithelial cell differentiation: adenomatous epithelium resembles the replicating cells normally present in the crypt base. Tall cells with prominent, elongated, hyperchromatic nuclei produce a characteristic "picket fence" pattern as they line the adenomatous glands. The adenomatous epithelium contains incompletely differentiated goblet cells and absorptive cells at all levels of the crypt, including the free surface. Adenomatous glands show no evidence of differentiation toward the luminal surface.

2.4 Adenoma Growth

Small adenomas represent neoplastic clonal populations of colonic epithelial cells, suggesting that they arise from a single abnormal precursor stem cell. Adenomas begin in a single crypt, and then grow by replacing normal epithelium in a centrifugal manner. Unicryptal adenomas are rare and most typically affect patients with adenomatous polyposis syndrome.

The neoplastic cells appear to cluster at the luminal aspect of the mucosa without extending to the base of the glands. Normal-appearing mucosa lies below the adenomatous glands. In 86% of early tubular adenomas, the number of glands opening along the polyp surface is larger than the number of gland bases; this difference increases with polyp size [11]. Gland proliferation is predominant in the upper crypts and along the surface of the lesions.

Early adenomas are present as small growths with a very benign tubular histology. The progression of most small adenomas is slow, and occurs over several years. On average, small adenomas double their diameter in 10 years [12]. Some adenomas ultimately progress to invasive cancers, but not all adenomas progress; some may stay stable and may even regress or disappear while new ones may form [13].

2.5 Incidence

Adenomas are the most commonly biopsied tumors of the large bowel [14]. Incidence rates of adenomas vary considerably throughout the world. Geographic areas exhibiting a high risk for colon cancer also exhibit a high risk for adenoma development, and vice versa.

The incidence in the general population varies from 0% to 69%, depending on the country of origin [15,16] and on how the adenomas are detected [17]. In Western populations, the average prevalence rate for adenomas from flexible sigmoidoscopy screening is 10%, and colonoscopic screening prevalence averages 25% [18]. Adenomas accounted for 68% of all polyps removed by colonoscopy in the National Polyp Study [19].

In the 50- to 59-year age group, population screening studies and autopsy studies show an adenoma prevalence rate of 41.3% to 69% [20], increasing in advancing years up to 88% in centenarians [21]. Arminski and McLean [22] documented a 7.5% increase in adenoma incidence per decade. Adenoma incidence peaks at age 60 to 70 years; it also occurs more frequently in men (61.6%) than in women (38.4%) [19].

Based on endoscopic studies, most sporadic adenomas arise in the rectosigmoid colon (66% to 77%) [23]. Adenomas also occur from a distal to a proximal location as patients age [15–17,24]; thus, left-sided adenomas are found more commonly in younger age groups, and right-sided lesions increase in frequency in individuals older than 65 years of age.

Some adenomas tend to cluster. This means that multiple adenomas tend to occur closer together than would normally be expected from the general distribution of adenomas. This phenomenon occurs in all colonic segments, but is less pronounced in the rectum than in other parts of the large intestine [25].

2.6 Multiple Polyps

Individuals with one adenoma have a 40% to 55% likelihood of having additional synchronous lesions [23,26,27]. The additional adenomas can be detected at the same time as the initial adenoma (synchronous adenomas), or at a different time (metachronous adenomas). The prevalence of multiple adenomas increases with age (about 9% of those under 60 years, and 28% of people older than 75 years have three or more adenomas). The incidence of large intestinal adenomas occurring synchronously with carcinomas is approximately double that of adenomas occurring alone.

A relationship exists between adenoma multiplicity and histological findings. In patients with a single adenoma, 38.8% are villous, whereas those with multiple adenomas have a 60.1% chance of having at least one villous adenoma [28]. Patients with multiple adenomas are also more likely to harbor at least one adenoma that contains high-grade dysplasia (13.8%) versus patients with a single adenoma (7.3%).

The overall recurrence rates for new adenomas are estimated from 20% up to 60%, with average follow-up times of 3 to 10 years after index polypectomy [17,23,29]. Most recurrences occur in the first two years following polypectomy. The estimated time of finding new adenomas is 58 months for patients clear on the first colonoscopy, and 16 months for patients who had adenomas on the first examination [30,31]. Villous tumors, particularly broadly sessile ones, usually have less well-defined borders than tubular adenomas, and therefore have a greater tendency to recur after local resection than smaller, pedunculated adenomas.

Endoscopic follow-up studies to evaluate new adenomas are hampered by the fact that as many as 25% to 27% of adenomas measuring less than 5 mm in diameter, and up to 6% of adenomas measuring 1 cm in diameter are missed during a single endoscopic examination [32,33].

Right-sided adenomas are missed more often (27%) than left-sided adenomas (21%) [32]. Relatives of individuals with colorectal cancer have an adenoma prevalence rate of 39%.

2.7 Clinical Features

Bleeding is the most frequent symptom reported, and occurs more often in left-sided lesions than right-sided adenomas [34]. Small adenomas, ranging up to 1 cm in maximum diameter, usually remain asymptomatic unless they are traumatized by the passage of well-formed, hardened stool. Larger lesions become symptomatic, with the symptoms depending on polyp size and location. The bleeding is seldom severe. The incidence of bleeding increases with increasing adenoma size and once a carcinoma develops within the adenoma. Villous tumors are more likely to bleed than tubular ones, since they tend to be larger [35]. Cecal lesions that block the appendiceal orifice may produce symptoms mimicking acute appendicitis.

2.8 Gross Features

Grossly, adenomas assume one of three major growth patterns: (a) pedunculated, (b) sessile, or (c) flat or depressed. Most sporadic colorectal adenomas appear as exophytic [8]. The categorization of adenomas according to their macroscopic appearance is important, as it may influence surgical treatment.

2.8.1 Pedunculated Adenomas

Pedunculated adenomas appear as exophytic, mucosal protrusions with a lobulated head and a stalk covered by normal mucosa (Fig. 2.1). In pedunculated polyps, the adenomatous epithelium remains confined to the mucosa of the head of the polyp. The stalk consists of normal mucosa, including the muscularis mucosae and submucosal tissue, in continuity with the major part of the bowel wall.

2.8.2 Sessile Adenomas

Sessile adenomas attach to the mucosa by a broad base (Fig. 2.2). Sessile adenomas are often less well circumscribed than pedunculated ones. Because of their ill-defined edges, they are difficult to delineate, and have a greater tendency to recur following local excision.

2.8.3 Flat (Depressed) Adenomas

The terms superficial, flat, and depressed non-polypoid adenoma are used synonymously to describe this entity [8], but have two different macroscopic aspects. The overall prevalence of non-polypoid colorectal neoplasms is variable, and it accounts for from about 35% [36] to 42% of adenomas [37].

Flat adenomas are lesions that lack an exophytic polypoid configuration. They consist of slightly elevated dysplastic mucosal plaques that are never greater than twice the thickness of the surrounding normal colonic mucosa [38] (Fig. 2.3). They constitute a special subgroup of adenomas with a greater potential for malignant transformation, while still being smaller than exophytic adenomas [8].

Depressed adenomas have a collarette of epithelium similar to that seen in a flat adenoma, but with a depression that is usually central. Because flat or depressed adenomas display little or no mucosal elevation, they can be very difficult to see endoscopically and pathologically, especially in the proximal colon [39]. They are often more clearly delineated endoscopically after spraying the mucosa with methylene blue or indigo carmine [40–43]. The failure to recognize these flat lesions may account for the lingering concept of *de novo* colorectal carcinoma [44].

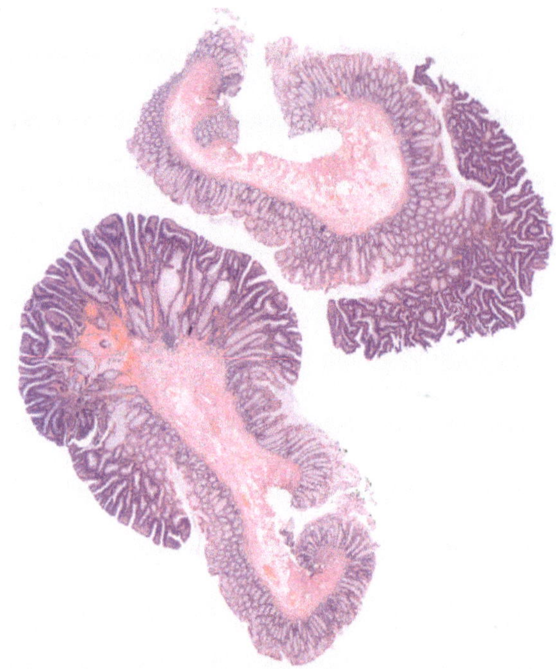

Fig. 2.1 Pedunculated polyp with the typical lobulated head and a stalk covered by normal mucosa

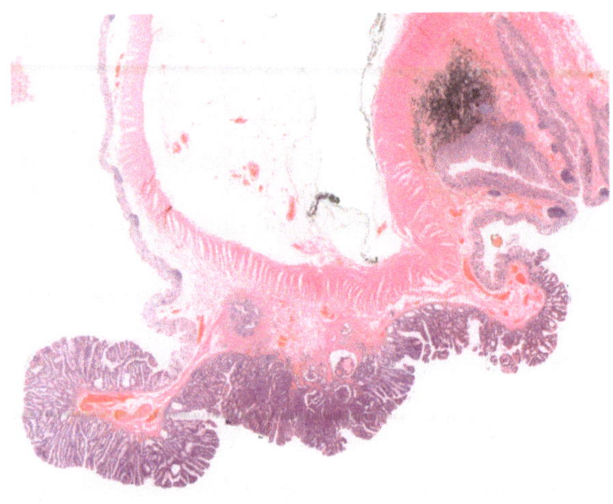

Fig. 2.2 Sessile polyp attached to the mucosa by a broad base. On the right is a colonic tattoo: a collection of black non-degradable pigments in the submucosa

Depressed adenomas tend to arise more commonly in the right colon than elsewhere [44]. They occur in HNPCC syndrome, sporadically, or in patients with familial adenomatous polyposis (FAP) [45]. The frequency of flat adenoma is 50.7% in HNPCC

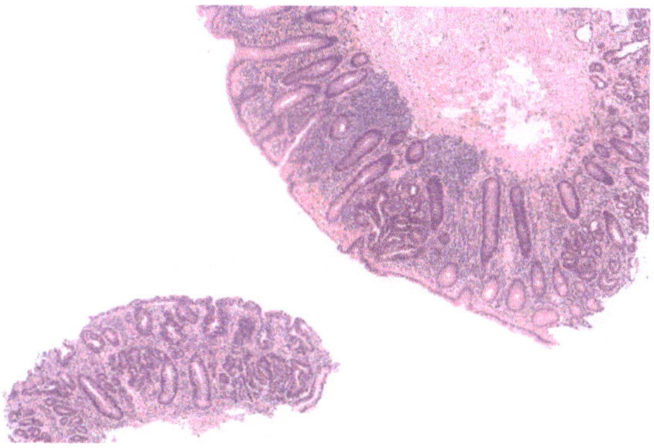

Fig. 2.3 Flat adenoma: low-power photomicrograph demonstrating a flat adenoma with approximately the same thickness of non-neoplastic colonic mucosa and containing crowded glands lined by hyperchromatic and mucin-depleted epithelium concentrated at the surface. In the biopsy on the right there are non-neoplastic glands on each side of the dysplastic epithelium

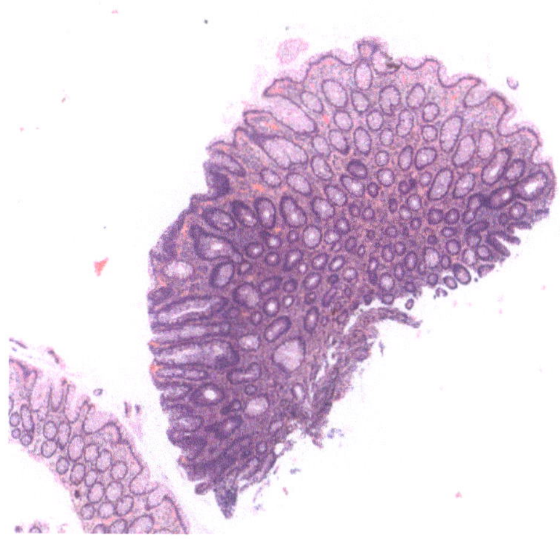

Fig. 2.4 Tubular adenoma maintaining its original crypt architecture, but adenomatous epithelium (darker) replaces the normal colonic epithelium of the crypts

patients. Generally, adenomas appear as grossly homogeneous, soft lesions without induration, ulceration, or fixation. Areas of ulceration, depression, or firmness suggest the possibility of a coexisting carcinoma.

2.9 Histological Features

There are four categories of adenoma: tubular, villous, tubulo-villous and flat-depressed [8]. The factors controlling the growth pattern of adenomas are unknown [7].

2.9.1 Tubular Adenoma

Tubular adenomas maintain the original crypt architecture, but adenomatous epithelium replaces the normal colonic epithelium in lining the crypts (Fig. 2.4). This is the most common type of adenoma (about 68% to 87%) [19,46,47]. Tubular lesions are those that contain greater than 80% of a tubular component. Tubular adenomas consist of closely packed branching tubules separated by varying amounts of lamina propria. The tubule may be relatively regular, or when the adenomatous tubules grow, they may branch and show considerable irregularity. Small tubular adenomas usually have a dysplastic surface epithelium overlying normal epithelium in the crypt base.

2.9.2 Villous Adenoma

Villous adenomas (approximately 20%) have villi with cores of lamina propria covered by a single layer of adenomatous epithelium. Villous lesions are those that contain greater than 80% of a villous component [46] (Fig. 2.5). Villous adenomas fall into three types: (a) flat, carpet-like masses; (b) lobulated, bulky, sessile masses; (c) pedunculated lesions with short, broad pedicles.

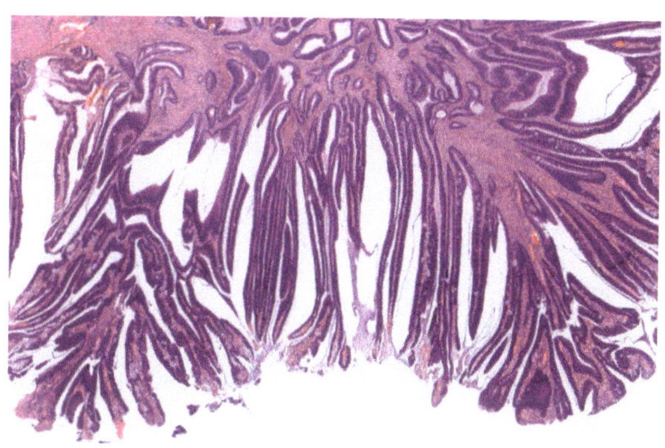

Fig. 2.5 Villous adenoma characterized by long finger-like fronds lined by neoplastic epithelium

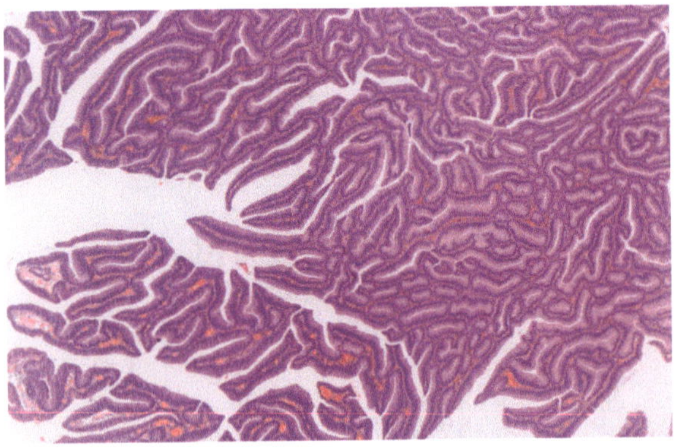

Fig. 2.6 Tubulo-villous adenoma: mixture of tubular and villous architecture – villous fronds and tubular glands

2.9.3 Tubulo-Villous Adenoma

Tubulo-villous adenomas contain a mixture of both tubular and villous patterns, or have broad villi containing short tubular structures. Tubulo-villous lesions are those that contain from 20% to 79% villous components [46]. They tend to be larger than tubular adenomas, with a mean diameter of 19 mm [22] (Fig. 2.6). A villous component is present in 35% to 75% of all adenomas measuring more than 1 cm in largest diameter [48].

2.9.4 Flat-Depressed Adenoma

Flat or depressed adenomas are a variant of tubular adenoma with little or no mucosal elevation. The thickness of the adenomatous mucosa does not exceed twice than that of the normal mucosa [8], and the adenomatous changes concentrate near the luminal surface. Flat adenomas have a high incidence of high-grade dysplasia [38,45], and they are more likely to harbor invasive carcinoma than is typically seen in polypoid counterparts [49]. There is a high association with synchronous and metachronous invasive colorectal carcinomas [8]. Depressed adenomas measuring less than 1 mm in diameter show horizontal growth between the normal adjacent crypts, often leaving normal crypts entrapped as residual islands.

2.10 Diagnosis

The histological features of adenomas may be defined as low- or high-grade dysplasia.

2.10.1 Low-Grade Dysplasia

Low-grade dysplasia consists of stratified dysplastic epithelium that retains its columnar shape. The nuclei are spindle or oval shaped. The stratified nuclei tend to remain in the basal epithelium, extending no more than three-quarters of the height of the epithelium (Fig. 2.7). Minor cytological variations including numerous mitoses, mild nuclear pleomorphisms, and variations in cell size and shape may occur in adenomatous epithelium; however, these features (more common in larger polyps) are insufficient for a diagnosis of high-grade dysplasia.

Sometimes it is difficult distinguish a small tubular

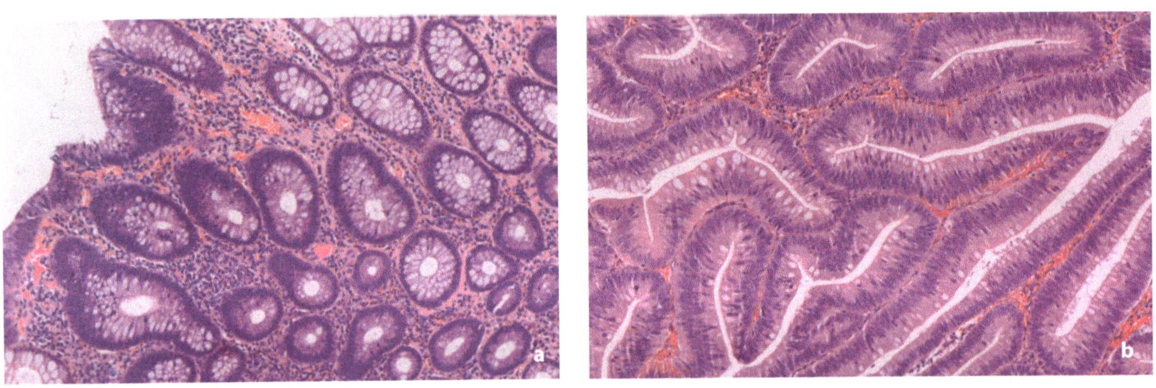

Fig. 2.7a,b Low-grade dysplasia; a, small tubular adenomatous gland with very little atypia; normal colonic glands on the right; b, small tubular adenomatous gland with moderate atypia

adenoma from reactive epithelium present in an inflamed mucosa, because reactive glands appear more basophilic than normal and the nuclei may exhibit pseudostratification. In these cases it is useful examine the degree of differentiation of the epithelium along the length of the tubular crypt. If the entire gland is not replaced by basophilic epithelium, then its restriction to the bottom portion of the crypt serves to identify the epithelium as regenerative. Conversely, in small adenomas, the adenomatous glands appear more basophilic at the surface of the lesion, and non-neoplastic epithelium lies below it [8].

2.10.2 High-Grade Dysplasia

High-grade dysplasia is characterized by the presence of marked cytological atypia, the loss of cellular polarity, stratification of cells to the luminal surface of the glands, and crowding with occasional formation of solid nests of dysplastic cells. The cells show loss of columnar shape with cellular rounding and an increase of nuclear-to-cytoplasmic ratios. Cells remain confined within the basement membrane of the original colonic crypt, or they may extend into the surrounding lamina propria, with a cribriform pattern obliterating the intervening stroma. Glandular density increases. (Fig. 2.8b) The presence of high-grade dysplasia strongly correlates with a contiguous invasive carcinoma.

High-grade dysplasia represents the extreme end of the spectrum of abnormal histological changes, short of invasive carcinoma in the adenoma–carcinoma continuum. Individual adenomas may contain transitions between high-grade and low-grade dysplasia. The percentage of adenomas containing high-grade dysplasia increases significantly with increasing adenoma size, villous architecture, multiplicity of adenomas, and age greater than 60 years [50,51].

High-grade dysplasia encompasses the histological changes called carcinoma *in situ* [9] and intramucosal carcinoma (Fig. 2.8a). The latter is when there is extension of the neoplastic cells through the basement membrane of the crypt into the surrounding lamina propria but not beyond [52,53]; intramucosal carcinoma includes that which involves the muscularis mucosae. Neoplastic glands in and among a splayed muscularis mucosae is not invasive cancer. Only when cancer invades into the submucosa does it have the potential to metastasize [8,9,52].

Neither carcinoma *in situ* nor intramucosal carcinoma have a clinically significant potential for metastasis (if all neoplastic tissue is removed), and the lesions do not require additional treatment [8]. Hence this term should only be used in conjunction with the comment that intramucosal adenocarcinoma lacks the potential for metastases, and if totally removed it has been adequately treated [9].

The pathology report should state the macroscopic description (pedunculated or sessile polyp, and the greatest dimension), the highest degree of dysplasia present in the adenoma, whether or not it has villous features, the completeness of its removal, and the presence or absence of invasive tumor [54].

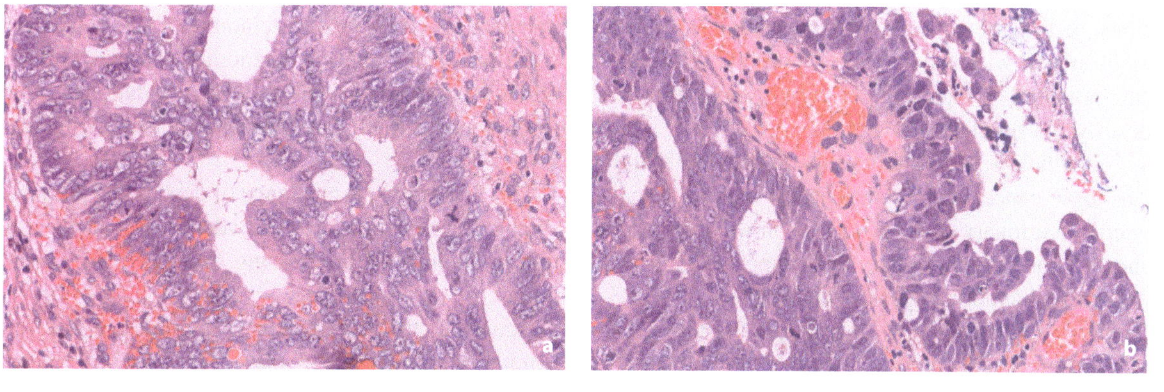

Fig. 2.8a,b High-grade dysplasia; **a**, characterized by cellular disorganization and more marked cytologic atypia, the degree of nuclear pleomorphism is sufficient to call it intramucosal carcinoma; **b**, stratification of cells to the luminal surface of the glands, a feature of high-grade dysplasia

2.11 Reporting Colorectal Adenoma

Gross features:

- Macroscopic growth pattern: pedunculated/sessile/flat polyp.
- Greatest dimension.

Histological features:

- Architecture: tubular/villous/tubulo-villous.
- Grade: low-grade/high-grade dysplasia.
- Status of the resection margin.

2.12 Adenoma–Carcinoma Sequence

Adenoma constitutes the precursor lesion for most colorectal carcinomas [2–5]. Two concepts can explain the understanding that now exists in relation to the evolution of colorectal neoplasia. The first is the model provided by the adenoma–carcinoma concept and is supported by clinical, pathological, and epidemiological data collated over several decades [3]. The second model is related to the hereditary bowel cancer syndromes (FAP and HNPCC) that led to the discovery of important cancer genes [55,56].

The earliest lesions consist of pseudostratified, immature, mildly dysplastic, adenomatous cells. In some cases, one may see a continuous histological spectrum of increasing degrees of dysplasia culminating in the development of an invasive carcinoma [8].

There are publications purporting "*de novo*" carcinomas that are open to various interpretations. It should be recalled that adenomas may, on rare occasions, be flat or even depressed, presenting essentially as dysplasia within flat mucosa [40,57]. "*De novo*" carcinoma may represent an early cancer that has destroyed a small adenoma [58].

Nevertheless, some studies support the view that the "*de novo*" cancer and classical cancer represent *divergent* evolutionary pathways. "*De novo*" carcinoma shows a non-polypoid, superficially spreading, growth pattern, and a more aggressive course [7].

Morphological features that determine the malignant potential of an adenoma are size, growth pattern, and grade of dysplasia [7].

Carcinomas are more likely to arise in larger adenomas than smaller ones. The incidence of carcinoma in an adenoma increases as the size of the adenoma increases. The prevalence of cancer in adenomas under 1 cm is only about 1%, in those between 1 and 2 cm in diameter it is about 10%, and in those over 2 cm there is nearly a 50% malignancy rate.

Adenomas with a villous pattern have a higher malignant potential than those with a tubular pattern. The malignancy rate for tubular adenomas is about 5%, but rises to 40% in villous ones. In tubulo-villous types, the malignancy rate is about 22%.

Although histological type is very important in the assessment of malignant potential, it seems that size is the paramount feature [59].

The malignant potential of an adenoma increases as grading of dysplasia increases, irrespective of histological growth pattern. Both growth pattern and dysplasia grade correlate with adenomatous size.

Usually, small adenomas (those under 1 cm) show low-grade dysplasia and have very low malignant potential. The risk of cancer developing in such adenomas is only 5% after 15 years. The malignancy rate rises to 27% if a high-grade dysplasia is present; however, it is rare in a polyp of this size. A similar relationship is seen in adenomas that are 1 to 2 cm in diameter in relation to the grade of dysplasia. In adenomas over 2 cm in size, the malignancy rate is high but bears little relation to the degree of dysplasia.

Although the trend observed for size and malignant change is considerably greater than the trend for dysplasia and malignancy, there are reasons to suspect that at the biological level of dysplasia is the most selective marker of increased malignant potential [7].

Even though adenomas clearly constitute the precursor lesion for most carcinomas, a vast gap exists in the prevalence rates of adenomas and carcinomas, indicating that some 90% to 95% of adenomas will never become malignant during a person's lifetime [16].

This fact offers the challenge of developing markers for the identification of those adenomas that have a high probability of progressing to an invasive carcinoma.

Actuarial analysis reveals a cumulative risk of developing cancer in adenomas that are not removed at 5, 10, and 20 years of 2.5%, 8%, and 24%, respectively. It is estimated that the conversion rate of adenomas to cancer is 0.25% per year [60].

2.13 Adenomas Containing Carcinoma (Malignant Polyps)

A malignant polyp is an adenoma containing invasive carcinoma. The diagnosis of invasive carcinoma is made when neoplastic glands have invaded and penetrated through the muscularis mucosae into the submucosa of the bowel wall or into the submucosa of the stalk of an adenoma [9,61]. Invasion into, but not through, the muscularis mucosae is still "intramucosal carcinoma". Desmoplasia often surrounds the invading glands, which have irregular, angled contours and show cytological features of malignancy [8]. This feature must be differentiated from "entrapped" (pseudoinvasive) mucosa.

Submucosal invasion is most easily recognized by the intermingling of the malignant glands with normal submucosal structures including medium-sized blood vessels, fat, nerves, ganglia, and large lymphatics [8]. Various degrees of substitution of an adenoma by carcinoma may occur. A polypoid carcinoma is a polyp consisting entirely of cancer with no remaining benign adenoma.

Malignant adenomas represent an early form of colorectal carcinoma.

Approximately 42 to 85% of early colorectal cancers are pedunculated, and 15–58% are sessile [62,63]. Carcinomas arising from pedunculated adenomas cause the biggest clinical questions with regard to further management.

Various opinions exist for managing patients after endoscopic removal of malignant polyps. Some of these lesions require further therapy, others do not. One possibility is that all patients with malignant polyps should undergo standard resection [64]; another opinion is that a conservative approach should be maintained in the absence of cancer at the resection line [65].

The present mainstream opinion, however, is that all malignant polyps removed by endoscopic polypectomy require evaluation of histological parameters that have been demonstrated to be significant prognostic factors related to the risk of adverse outcome (i.e. lymph node metastases or local recurrence from residual malignancy) after polypectomy [54,65–69]. The management of these malignant adenomas depends upon their histological risk factors and the patient's general condition [70].

The dilemma about managing patients after endoscopic removal of malignant polyps is best resolved by a multidisciplinary team involving the surgeon, pathologist, and endoscopist, and taking the patient's condition and wishes into account [70].

The clinician faces the therapeutic decision as to whether or not polypectomy alone is adequate therapy or whether the patient requires a definitive surgical resection; therefore, the metastatic risk must be determined to plan future therapy.

After endoscopic polypectomy, all the histological risk factors need to be simultaneously and carefully evaluated by the pathologist to identify and classify patients into low-risk or a high-risk group associated with an adverse outcome (i.e. lymph node metastasis or local recurrence from residual malignancy) [68,71].

2.14 Prognostic Factors of Metastatic Risk or Residual Disease Present in Malignant Adenomas

Histological parameters have been developed over the years to identify prognostic factors of metastatic risk and reduce the number of unnecessary additional laparotomies, while selecting which adenomas have very little or virtually no risk of nodal metastasis and/or local recurrence [68].

The pathological features that have independent prognostic significance and that are crucial for evaluating risk of adverse outcome (e.g. increased risk of residual disease or lymph node metastases) include histological grade, completeness of resection margin, lymphatic-venous vessel involvement, tumor budding, and level of invasion of the submucosa.

2.14.1 Grade of Differentiation

The grading system is based on gland or tubule formation and the cytological features of adenocarcinoma (how closely it approximates normal epithelium) [9]. The neoplastic components should be divided into well-differentiated (grade 1 – G1), moderately differentiated (grade 2 – G2), poorly differentiated adenocarcinoma (grade 3 – G3), and undifferentiated carcinomas (grade 4 – G4). The World Health Organization (WHO) classifies the neoplastic components into just two categories: low-grade (G1 and G2) and high-grade (G3 and G4) adenocarcinomas [61].

Well-differentiated (G1) adenocarcinoma exhibits glandular structures in more than 95% of the tumor. Moderately differentiated (G2) adenocarcinoma has 50 to 95% glandular structure. Poorly differentiated (G3) adenocarcinoma has 5 to 50% glandular structure. Undifferentiated (G4) carcinoma has less than 5% glandular structure [61]. In order to reduce the degree of inter-observer variability in the grading of adenocarcinoma, and in light of its prognostic value and relative simplicity and reproducibility, a two-tiered grading system for colorectal carcinoma has been recommended: low-grade carcinoma (gland formation greater than or equal to 50%) and high-grade carcinoma (gland formation less than 50%) [72].

The histological grade is assigned according to the least-differentiated area found, even though this may appear to be quantitatively insignificant [8]. Tumor grade is classified as a favorable grade (low-grade adenocarcinoma) or an unfavorable grade (high-grade adenocarcinoma) [68].

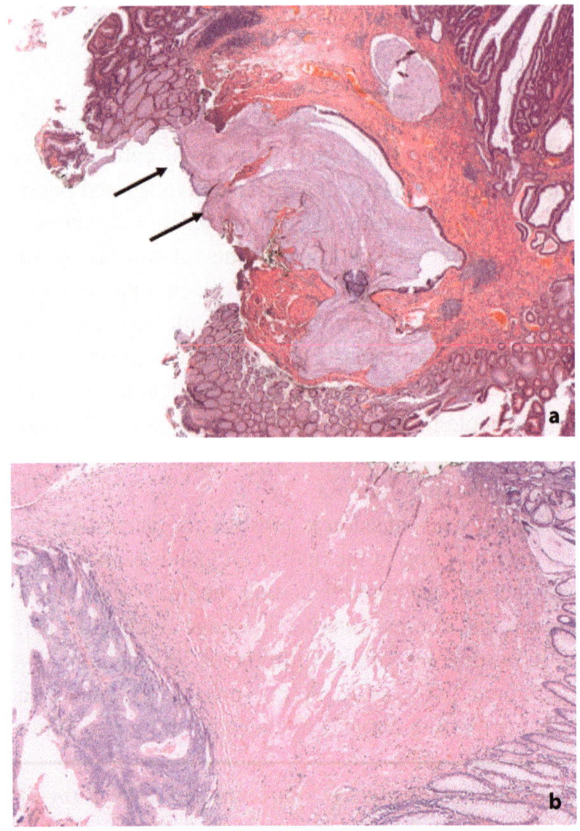

Fig. 2.9a,b Malignant polyps; a, with cancer at the resection margin, arrows indicate resection margin; b, with cancer near (≤1 mm) the resection margin (diathermy effect is evident on the right)

2.14.2 Margins

The margin of resection, or transection point, is defined as the actual free edge of the submucosal connective tissue that contains diathermy change [9]. A tumor at the margin is defined as cancer cells extending up to the actual transected soft tissue margin (Fig. 2.9a) A tumor near the margin is defined as cancer cells less than or equal to 1 mm from the transected margins, cancer within the diathermy change, or within one high-power field of the cautery effect [9] (Fig. 2.9b). The presence of a tumor at or near the resection margin has the same clinical significance and is associated with intramural recurrence after local excision an adverse outcome [9], even in the absence of any other unfavorable parameters [8].

2.14.3 Lymphatic Invasion

The diagnosis of lymphatic invasion requires the presence of cancer cells within endothelium-lined channels [9] (Fig. 2.10). Lymphatic invasion may be confused with retraction artifact. These are most commonly encountered within the invasive tumor itself, rather than in the submucosa away from and surrounding the actual invasive cancer [73]. The retraction artifact is often seen around small clusters of tumor cells, where reactive fibroblasts often surround tumor cells and mimic endothelium-lined channels.

When questionable areas for lymphatic invasion are present, subsequent serial and deeper sections are recommended [9]. Immunohistochemistry studies have not been of great help in establishing or excluding lymphatic invasion [9].

2.14.4 Venous Invasion

Venous invasion is defined as tumor emboli within endothelium-lined channels surrounded by a smooth muscle wall. When one suspects venous invasion, multiple serial or deeper sections (and possible elastic

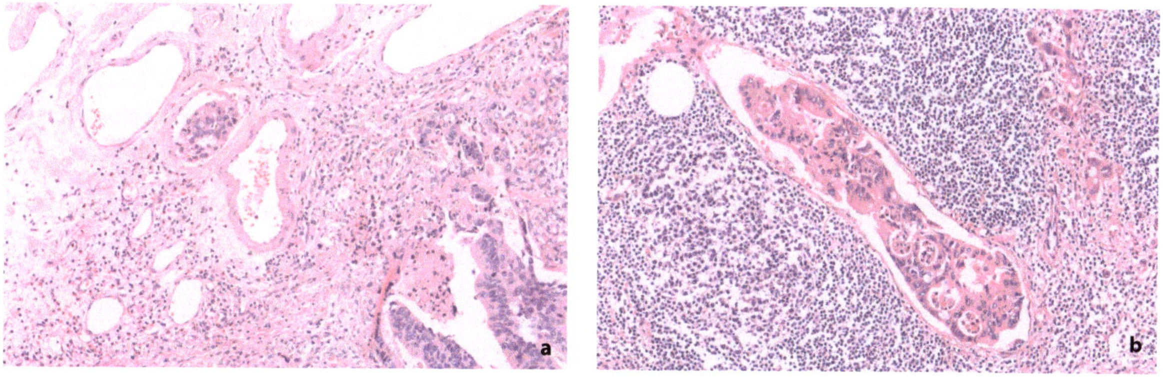

Fig. 2.10a,b Lymphatic invasion encountered in the submucosa (a, magnification 20×; b, magnification 30×)

stains) are quite helpful in deciding whether venous invasion is present [9].

The degree of lymphovascular invasion has been defined by the Japanese Society for Cancer of the Colon and Rectum. Lymphatic (ly) or vascular (v) invasion may be absent (ly0, v0), slight (ly1, v1), moderate (ly2, v2), or massive (ly3, v3) [62].

2.14.5 Tumor Budding

Tumor budding, also known as dedifferentiation, is a recently recognized feature that represents a high-grade, undifferentiated component of a tumor at the leading invasive edge [74] (Fig. 2.11). It is defined as an isolated single cancer cell or a cluster composed of fewer than five cancer cells observed in the stroma of an invasive frontal region [75]. A budding count must be done after choosing one field where budding is the most intensive. in a field measuring 0.785 mm^2, using a 20× objective lens [69]. A field with fewer than five budding foci is viewed as negative [76]; one with five or more buds is viewed as positive [68].

Nonetheless, the intensity of tumor budding also seems to be important [77]. Recent evidence suggests that tumor budding is associated with both lymphatic invasion and nodal metastases [75,78,79]. A number of 0 to 9 foci are classified as a low-grade or low-“intensity” tumor budding, while 10 or more buds are a high-grade or high-“intensity” tumor budding [79]. Higher intensity of tumor budding is significantly associated with higher risk of postoperative recurrence [80]. The disease-free survival and the overall survival rates dramatically decrease in patients with an intensity greater than nine tumor buds [80]. Intensity greater than nine may be considered to be an adverse prognostic indicator in patients with colon carcinoma [80].

Because cell clusters (buds) at the leading invasive edge may be quite small and do not form glands or produce mucin, identification on histopathological examination may be difficult [74] (Fig. 2.12a). A pan-cytokeratin immunostain may be helpful in their identification, especially if accompanied by an inflammatory reaction that obscures their presence on hematoxylin-eosin stain [74,80] (Fig. 12b).

Results indicate that tumour budding is a useful risk factor for predicting lymph node metastases in cases of early colorectal cancer [81].

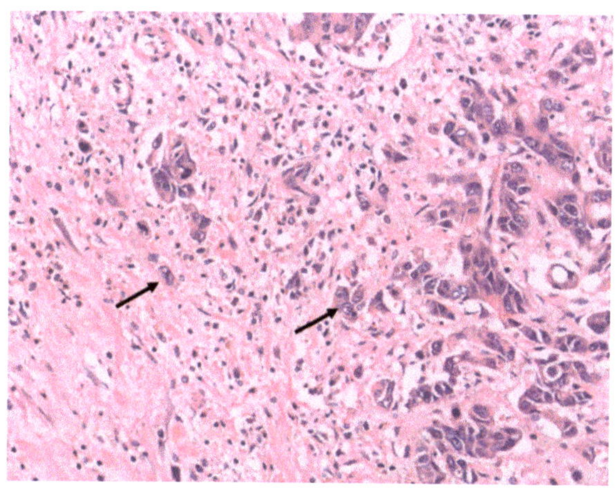

Fig. 2.11 Early colorectal cancer with high degree of tumor budding: isolated single cancer cell or a cluster composed of fewer than five cancer cells is defined as a budding focus. *Arrows*, budding foci

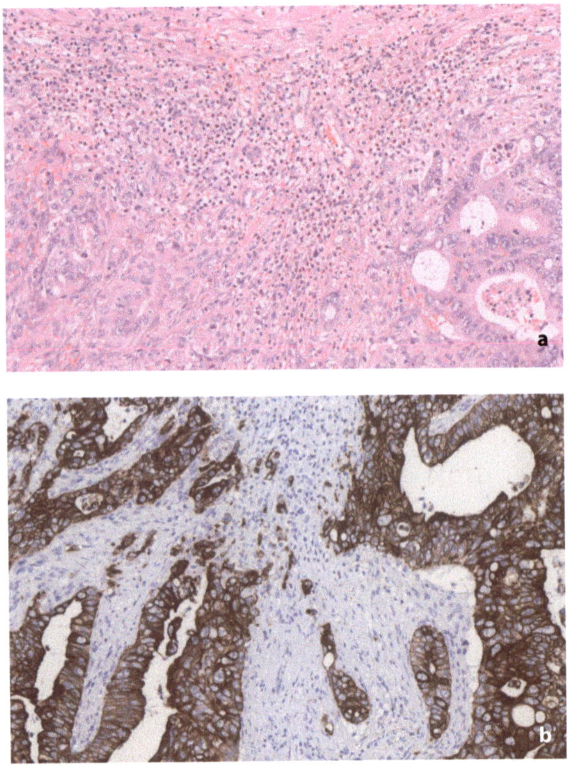

Fig. 2.12a,b Colorectal cancer. a, early colorectal cancer with low degree of tumor budding. The identification of budding foci is difficult because the buds are obscured by inflammatory reaction (magnification 12×); b, tumor budding highlighted by immunohistochemistry: by using a pan-cytokeratin antibody, budding is easily seen (magnification 20×)

2.14.6 Adenocarcinoma in the Submucosa (Microstaging)

If invasive cancer is present, it should be reported the amount of adenocarcinomatous component in terms of the volume of adenoma replaced by the carcinoma, the depth of its invasion, and the width of horizontal spread in the submucosa. This process can be called microstaging, and allows the ability to report both the level and the extent of the infiltration into the submucosal layer.

2.14.6.1 Volume of Adenoma Replaced by the Carcinoma

The volume of adenoma replaced by the carcinoma can be measured. This is a quantitative ratio, expressed as a percentage. Lesions with small foci of invasive carcinoma have lower metastatic capability than polyps that are mostly made of invasive carcinoma [82].

2.14.6.2 Depth or Level of Invasion of the Submucosa

Different staging of invasion into the submucosa has been proposed for pedunculated and sessile polyps. The Haggitt levels are used for carcinoma in pedunculated polyps [53], and the Kikuchi levels [64] are used for carcinoma in sessile polyps [62].

Haggitt level (Fig. 2.13)
The level of invasion in a *pedunculated malignant polyp* is defined within four levels:

- *Level I*: invasion is limited to the head of the polyp (Fig. 2.14).
- *Level II*: invasion into the junction of head and stalk (Fig. 2.15).
- *Level III*: invasion into the stalk.
- *Level IV*: invasion in the submucosa below the stalk.

Kikuchi level (Fig. 2.16)
The level of submucosal (sm) invasion in *sessile malignant polyp* is defined within three levels:

- *Sm1*: slight submucosal invasion from the muscularis mucosae to the depth of 200–300 μm (Fig. 2.17).
- *Sm2*: intermediate invasion.
- *Sm3*: carcinoma invasion near the inner surface of the muscularis propria.

Considering polyp morphology, the sessile type is associated with a unfavorable outcome as compared with that of pedunculated type. Although patients with sessile polyps frequently underwent surgery (85%) [83], their overall mortality remained roughly eight times higher when compared with patients with pedunculated polyps. This seems to be mainly due to a significantly higher prevalence of all the histological risk factors in this group rather than to a predetermined biologically aggressive behaviour [84,85]. In detail, a positive resection margin seemed to be by far the most crucial risk factor in sessile polyps, probably because of an inadequate endoscopic removal of these lesions. This confirms the previous analysis by Haggitt et al [53], in which the level of invasion, but not the sessile morphology, seemed to be an independent risk factor for an adverse outcome [83].

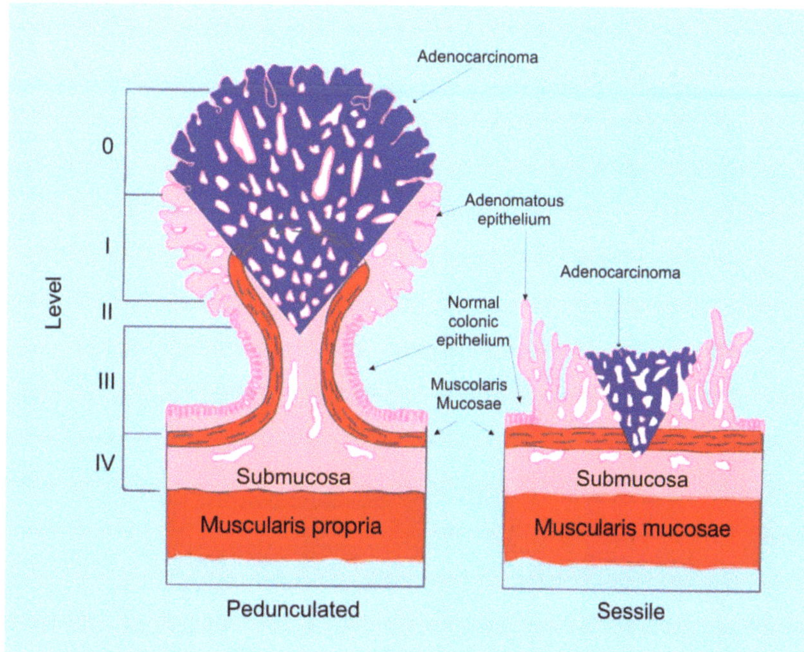

Fig. 2.13 Haggitt's levels. Modified from [53], with permission from Elsevier

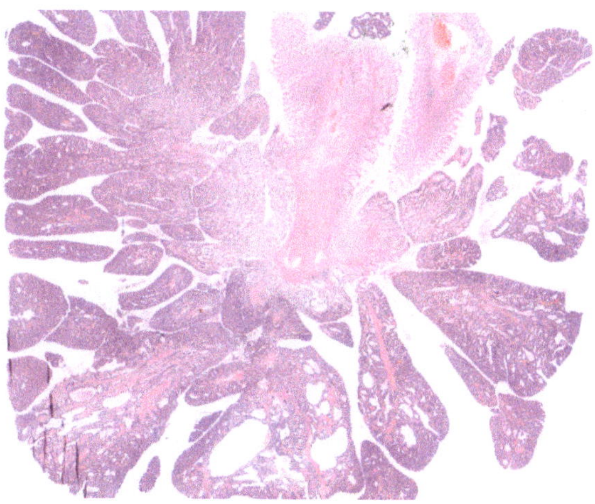

Fig. 2.14 Pedunculated early colorectal cancer: Haggitt's level I with invasion limited to the head of the polyp

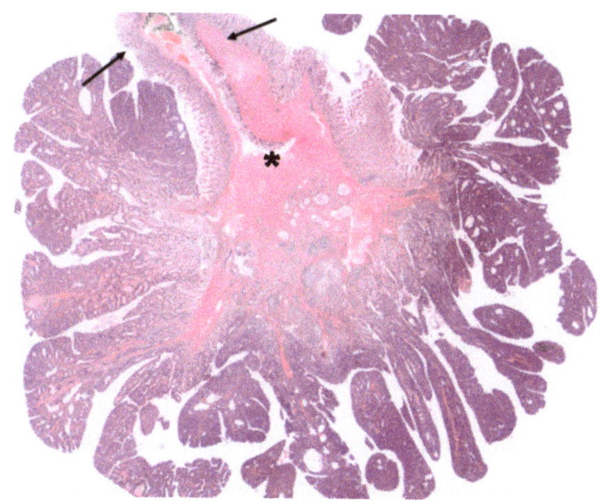

Fig. 2.15 Early colorectal cancer: Haggitt's level II (invasion to the junction of head and stalk). The margin is the cauterized submucosa (*black star*) and not the dangling wings of mucosa (*arrows*)

The Haggitt classification is less useful for sessile tumors. According to these criteria, invasive cancer arising in a pedunculated adenoma could be classified as level I to level IV. Invasive cancer arising in a sessile adenoma is, by definition, a level IV lesion (Fig. 2.17). In sessile and semi-sessile adenomas there will most likely be an invasion into the submucosa of the bowel wall, and the patient will therefore be at higher risk for metastasis compared to early invasive carcinomas arising in pedunculated adenomas [8].

2.14.6.3 Measuring the Level of Submucosal Invasion

Extension into the submucosal layer may be expressed by micrometric measurement of depth and width of

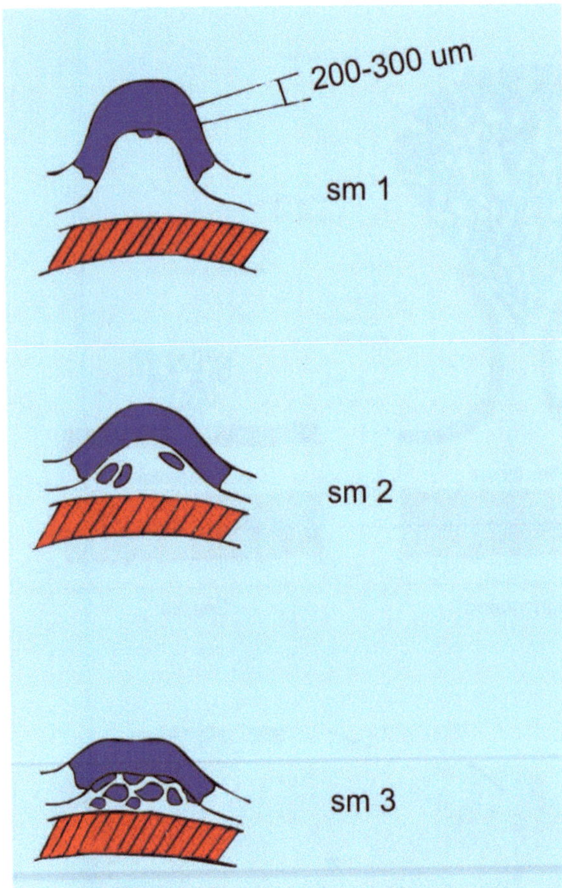

Fig. 2.16 Kikuchi's levels. Modified from [63]

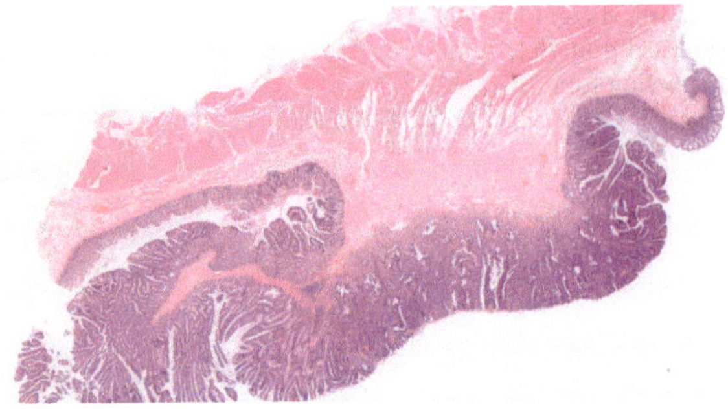

Fig. 2.17 Sessile early colorectal cancer: Haggitt's level IV; Kikuchi's level sm1 with invasion from the muscularis mucosae measuring less than 300 μm deep

submucosal invasion. It is a numerical measurement regarding depth and width of tumor invasion. Depth invasion (vertical distance) is estimated from the lower edge of the muscularis mucosae to the deepest invasive front. When the muscularis mucosae cannot be identified, the vertical distance from apex of the tumor to the deepest invasive front is measured. Width invasion measures the greatest width of submucosal of invasion [68] (Fig. 2.18).

Numerical data regarding the extent of submucosal invasion aids in identifying tumors with very little risk for nodal involvement in patients with an absence of unfavorable parameters [68]. Width of submucosal invasion less than 4 mm, and depth of submucosal invasion less than 2 mm, in the absence of unfavorable parameters, identify tumors with very little risk for nodal involvement, and metastatic capability close to zero per cent [68]. Moreover, by estimating the extent of submucosal invasion, it is possible to identify within the low-risk early colorectal cancer, a subgroup of lesions with virtually no risk of nodal metastasis: depth of submucosal invasion less than0.3 mm (sm1), or depth of submucosal invasion less than 2 mm joined at a width of submucosal invasion less than 4 mm with negative budding [54,68].

These pathological parameters define two groups of early colorectal cancer with different risk of nodal and/or local recurrence: low- and high-risk early colorectal cancer. A low-risk early colorectal cancer is defined as being a completely excised Haggitt level 1 to 3 or Kikuchi Sm1 and possibly Sm2 depth of invasion, with no evidence of poorly differentiated adenocarcinoma or lymphatic or vascular invasion [62] and low-grade tumor budding [54,80]. It is now generally accepted that local excision, by either endoscopic polypectomy or transanal surgery, is adequate treatment for a low-risk early colorectal cancer [62].

A high-risk early colorectal cancer is defined as one that has one or more of the following characteristics: a positive resection margin, a high tumor grade, an Sm3 or possibly an Sm2 depth of invasion, presence of lymphatic or vascular invasion, or a high grade of tumor budding [54,62,80].

Low- and high-risk early colorectal cancers differ, not only with regard to lymph node metastases, but also to distant metastasis and mortality rates [71]. Such adverse clinical outcomes occur despite the majority of high-risk patients undergoing surgical resection. This observation strengthens the usefulness of this classification not only for addressing the therapeutic choice, but also as a staging procedure.

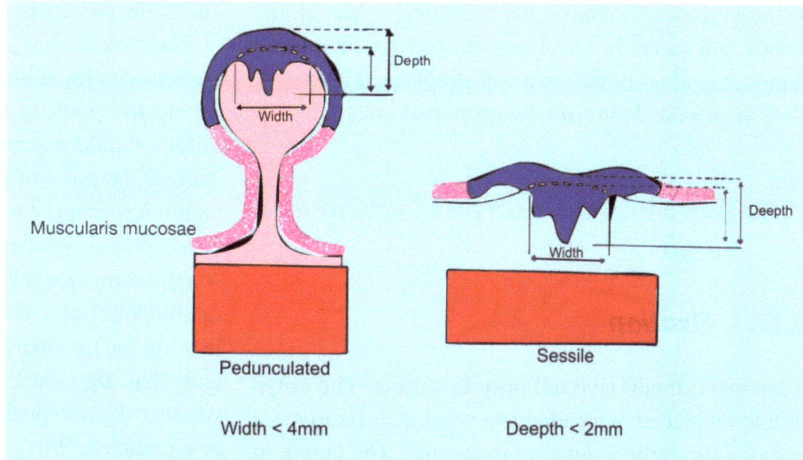

Fig. 2.18 Width and depth of submucosal invasion. Modified from [68] with permission from Elsevier

2.15 Reporting Malignant Polyps

Gross features:

- Macroscopic growth pattern: pedunculated/sessile/flat polyp.
- Greatest dimension.

Histological features:

- Grade of adenocarcinoma: low-grade/high-grade dysplasia.
- Status of the resection margin.
- Presence or absence of lymphatic or venous invasion.
- Tumoral budding.
- Microstaging:
  - volume of adenoma replaced by the carcinoma;
  - levels of invasion of the submucosa:
    - Haggitt's levels (pedunculated polyp)
    - Kikuchi's classification (sessile polyp)
    - depth and width of infiltration of the submucosa.

2.16 Pseudocarcinomatous Entrapment (Pseudoinvasion)

A recognized histological pitfall in diagnosing adenoma is the presence of "entrapped" (pseudoinvasive) dysplastic glands in the submucosa mimicking invasive adenocarcinoma.

Pseudocarcinomatous entrapment, variously termed colitis cystica profunda, submucosal cysts, pseudocarcinomatous invasion, or epithelial misplacement, affects a small proportion of pedunculated adenomas usually located in the sigmoid colon (64% to 85%) [8]. Repeated episodes of torsion lead to hemorrhage, inflammation, and ulceration of the adenoma. As a result, the adenomatous glands herniate through the muscularis mucosae into the underlying submucosa. Forceps biopsies may also cause epithelial displacement: the adenomatous tissue may be pulled further into the stalk by contraction of fibrous tissue as the biopsy site heals [86].

Histologically, areas of pseudoinvasion can be recognized by the presence of adenomatous glands in a submucosa without cytological evidence of malignancy.

Pseudoinvasion is characterized by the presence of entrapped adenomatous glands in a submucosa surrounded by normal lamina propria with hemosiderin deposits, as opposed to a desmoplastic response in invasive carcinoma (Fig. 2.19). The degree of dysplasia

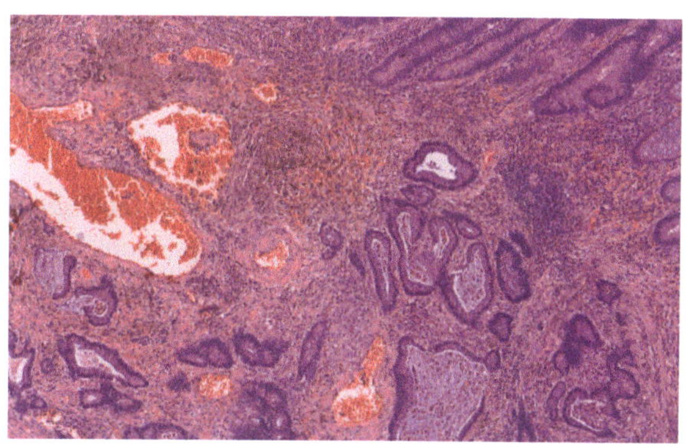

Fig. 2.19 Adenoma with pseudocarcinomatous entrapment: at high magnification the displaced pseudoinvasive glands demonstrated low-grade dysplasia and are surrounded by lamina propria. Siderogenous desmoplasia is present within the submucosa (brownish color comes from the presence of hemosiderin-laden macrophages)

in the displaced glands often resembles that of the glands immediately overlying it, and the displaced glands may also coexist with non-neoplastic glands that were displaced along with the neoplastic ones.

2.17 Gross Examination and Cutting of Polyps

2.17.1 Fixation

Adenomas should be fixed prior to cutting. The polyp should be placed in an adequate volume of fixative (at least ten times the volume of the tissue). The length of time needed for adequate fixation varies with the size of the polyp (i.e. larger polyps need longer fixation). The pathologist can often appreciate when the tissue is adequately fixed and firm enough for subsequent sectioning, by careful palpation of the polyp [9]. Ideally, the endoscopist should indicate the stalk of larger adenomas by placing a needle at its base when the polyp is removed from the endoscope. Realistically, this almost never happens. Occasionally, the pathologist and the endoscopist disagree as to whether a stalk is present or how long the stalk is, since the stalk often retracts into the head of the adenoma. Occasionally, the precise orientation of the polyp cannot be identified clearly; sectioning at several levels may then be needed to recognize the exact anatomical relationships.

However, some specimens defy accurate orientation so that the assessment of margins may be impossible. In this case, the margins are reported as not evaluable.

2.17.2 Sampling

Once fixed, the entire lesion should be examined histologically. When one receives polyp biopsies or polypectomy specimens, it is important to record all of the pathological features, including the number of tissue fragments received, their size, their gross morphology (i.e. pedunculated or sessile), and their locations. The stalk of a pedunculated polyp or the point of transection of sessile or semipedunculated polyps should be identified. In sessile and semipedunculated polyps, the point of transection can often be identified as an ashen white area of discoloration. It is possible to identify the excision edge of the specimen due to the presence of a prominent cautery effect [62].

The endoscopist should identify the point of transection with India ink; a pin is another method of identifying the point of transection in sessile polyps [9]. Polyps should be cut in the sagittal plane through the stalk or the point of transection, such that all the relevant microscopic landmarks will be easily assessable. If piecemeal polypectomy is unavoidable, the endoscopist can place the true transected margin in a separately identified container, or use a pin or India ink to identify the true margin of transection. It is also important that the endoscopist informs the pathologist whether the polypectomy was believed to be complete or incomplete [9].

Tissue fixation ensures retention of the ball shape, making identification of the resection site difficult. This artifact can be avoided by having the endoscopist place sessile polyps on a firm matrix, such as a piece of paper or Gelfoam, before placing the specimen in the fixative.

If the lesion is pedunculated and received in a fresh state, it can be fixed in such a way that the stalk is pinned to a piece of cork.

The histological classification of fractional biopsies of smaller adenomas (<1.7 cm) are in 88.9% agreement with the final diagnosis in the polypectomy specimen, whereas the reliability of the biopsies in accurately diagnosing adenomas >1.7 cm is only 27.68%. Invasive carcinomas are frequently missed in biopsies taken of larger lesions [8].

This diagnosis is made on either a polypectomy specimen or a biopsy of sessile lesions. Diagnosing areas of invasive carcinoma on a midsagittal section of a pedunculated adenoma is often easier than making a diagnosis of invasion on a small forceps biopsy of a larger lesion.

Biopsy fragments in which the neoplastic cells mingle with the fat, medium-sized blood vessels, nerve trunks, ganglia, or large lymphatics can be diagnosed as invasive lesions.

2.18 Pathology of Post-Polypectomy Resection Specimens

The reports from pathologists with regard to the post-polypectomy resection specimens have to focus on two significant statements: (1) the presence of residual neoplastic cells at the site of previous polypectomy, and (2) the presence of lymph-nodal metastases.

On gross examination of the resection specimens, it is important to identify the actual polypectomy site. If the resection is performed within approximately 10 days post-polypectomy, the polypectomy site will usually be apparent as an area of erosion, ulcer, or induration. When resections are performed more than 10 days post-polypectomy, it is often difficult to identify the polypectomy site, which has probably healed and re-epithelialized.

In a fresh unfixed specimen, if the polypectomy site is not grossly obvious, the pathologist, by careful palpation, can often find an area of induration that corresponds to the polypectomy site. The polypectomy site should be confirmed microscopically.

In instances of delayed resection with re-epithelialization of the polypectomy site, one should look for focal fibrosis, thrombosed submucosal blood vessels, occasional giant cells, disruption of the muscularis mucosae, etc, to confirm the polypectomy site. If, after taking the routine number of sections, one is unsuccessful in finding the polypectomy site, more random sections should be taken. No specific number is recommended, but the sampling should be extensive. If, after extensive sampling, the site is not found, the pathology report should clearly indicate that the polypectomy site was not found. This should indicate to the surgeon that there is a possibility that the correct area of bowel may not have been removed.

To facilitate finding the polypectomy site, the endoscopist might tattoo the area with India ink. This tattoo remains for several months [9].

With respect to lymph nodes, it should be remembered that all nodes present must be sampled. It has been shown that a minimum of 12 to 18 lymph nodes must be examined to accurately predict regional node negativity in colorectal cancer [87–90]; moreover it has been suggested that 12 lymph nodes be considered the minimum number that is acceptable [72,88].

References

1. Parkin DM, Bray F, Ferlay J, et al (2005) Global cancer statistics, 2002. CA Cancer J Clin 55:74–108.
2. Leslie A, Carey FA, Pratt NR, et al (2002) The colorectal adenoma–carcinoma sequence. Br J Surg 89:845–860.
3. Muto T, Bussey HJ, Morson BC (1975) The evolution of cancer of the colon and rectum. Cancer 36:2251–2270.
4. Vogelstein B, Fearon ER, Hamilton SR, et al (1988) Genetic alterations during colorectal-tumor development. N Engl J Med 319:525–532.
5. Fearon ER, Vogelstein B (1990) A genetic model for colorectal tumorigenesis. Cell 61:759–767.
6. Leppert M, Burt R, Hughes JP, et al (1990) Genetic analysis of an inherited predisposition to colon cancer in a family with a variable number of adenomatous polyps. N Engl J Med 322:904–908.
7. Day WD, Jass JR, Price AB, et al (2003) Epithelial tumours of the large intestine. In: Morson BC, Dawson IMP, Day DW (eds) Morson and Dawson's gastrointestinal pathology. Blackwell Science, Malden, pp 551–609.
8. Fenoglio CM (2008) Epithelial neoplasms of the colon. In: Fenoglio-Preiser CM (ed) Gastrointestinal pathology. Lippincott Williams & Wilkins, Philadelphia.
9. Cooper HS, Deppisch LM, Kahn EI, et al (1998) Pathology of the malignant colorectal polyp. Hum Pathol 29:15–26.
10. Moss SF, Liu TC, Petrotos A, et al (1996) Inward growth of colonic adenomatous polyps. Gastroenterology 111:1425–1432.
11. Cole JW, McKalen A (1963) Studies on the morphogenesis of adenomatous polyps in the human colon. Cancer 16:998–1002.
12. Hoff G (1987) Colorectal polyps. Clinical implications: screening and cancer prevention. Scand J Gastroenterol 22:769–775.
13. O'Brien MJ, O'Keane JC, Zauber A, et al (1992) Precursors of colorectal carcinoma. Biopsy and biologic markers. Cancer 70:1317–1327.
14. Riddell RH, Petras RE, Williams GT, et al (2003) Epithelial neoplasia of the intestines. In: Rosai J (ed) Tumors of the intestines. Armed Forces Institute of Pathology, Washington.
15. Clark JC, Collan Y, Eide TJ, et al (1985) Prevalence of polyps in an autopsy series from areas with varying incidence of large-bowel cancer. Int J Cancer 36:179–186.
16. Johannsen LG, Momsen O, Jacobsen NO (1989) Polyps of the large intestine in Aarhus, Denmark. An autopsy study. Scand J Gastroenterol 24:799–806.
17. Neugut AI, Jacobson JS, Ahsan H, et al (1995) Incidence and recurrence rates of colorectal adenomas: a prospective study. Gastroenterology 108:402–408.
18. Giacosa A, Frascio F, Munizzi F (2004) Epidemiology of colorectal polyps. Tech Coloproctol 8(suppl 2):s243–247.
19. Winawer SJ, Zauber A, Diaz B, et al (1988) The National Polyp Study: overview of program and preliminary report of patient and polyp characteristics. Prog Clin Biol Res 279:35–49.
20. Vatn MH, Stalsberg H (1982) The prevalence of polyps of the large intestine in Oslo: an autopsy study. Cancer 49:819–825.
21. Chapman I (1963) Adenomatous polypi of large intestine: incidence and distribution. Ann Surg 157:223–226.
22. Arminski TC, McLean DW (1964) Incidence and distribution of adenomatous polyps of the colon and rectum based on 1000 autopsy examinations. Dis Colon Rectum 7:249–261.
23. Winawer SJ, O'Brien MJ, Waye JD, et al (1990) Risk and surveillance of individuals with colorectal polyps. WHO Collaborating Centre for the Prevention of Colorectal Cancer. Bull World Health Organ 68:789–795.
24. Bernstein MA, Feczko PJ, Halpert RD, et al (1985) Distribution of colonic polyps: increased incidence of proximal

lesions in older patients. Radiology 155:35–38.
25. Eide TJ, Stalsberg H (1978) Polyps of the large intestine in Northern Norway. Cancer 42:2839–2848.
26. Winawer SJ, St John J, Bond J, et al (1990) Screening of average-risk individuals for colorectal cancer. WHO Collaborating Centre for the Prevention of Colorectal Cancer. Bull World Health Organ 68:505–513.
27. Tripp MR, Morgan TR, Sampliner RE, et al (1987) Synchronous neoplasms in patients with diminutive colorectal adenomas. Cancer 60:1599–1603.
28. Winawer SJ, Schottenfeld D, Flehinger BJ (1991) Colorectal cancer screening. J Natl Cancer Inst 83:243–253.
29. Atkin WS, Morson BC, Cuzick J (1992) Long-term risk of colorectal cancer after excision of rectosigmoid adenomas. N Engl J Med 326:658–662.
30. Stryker SJ, Wolff BG, Culp CE, et al (1987) Natural history of untreated colonic polyps. Gastroenterology 93:1009–1013.
31. Neugut AI, Johnsen CM, Forde KA, et al (1985) Recurrence rates for colorectal polyps. Cancer 55:1586–1589.
32. Rex DK, Cutler CS, Lemmel GT, et al (1997) Colonoscopic miss rates of adenomas determined by back-to-back colonoscopies. Gastroenterology 112:24–28.
33. Rex DK, Lehman GA, Ulbright TM, et al (1994) The yield of a second screening flexible sigmoidoscopy in average-risk persons after one negative examination. Gastroenterology 106:593–595.
34. Sobin LH (1985) The histopathology of bleeding from polyps and carcinomas of the large intestine. Cancer 55:577–581.
35. Stulc JP, Petrelli NJ, Herrera L, et al (1988) Colorectal villous and tubulovillous adenomas equal to or greater than four centimeters. Ann Surg 207:65–71.
36. Soetikno R, Friedland S, Kaltenbach T, et al (2006) Nonpolypoid (flat and depressed) colorectal neoplasms. Gastroenterology 130:566–576; quiz 588–569.
37. Fujii T, Rembacken BJ, Dixon MF, et al (1998) Flat adenomas in the United Kingdom: are treatable cancers being missed? Endoscopy 30:437–443.
38. Wolber RA, Owen DA (1991) Flat adenomas of the colon. Hum Pathol 22:70–74.
39. Riddell RH (1992) Flat adenomas and carcinomas: seeking the invisible? Gastrointest Endosc 38:721–723.
40. Kuramoto S, Ihara O, Sakai S, et al (1990) Depressed adenoma in the large intestine. Endoscopic features. Dis Colon Rectum 33:108–112.
41. Iishi H, Tatsuta M, Tsutsui S, et al (1992) Early depressed adenocarcinomas of the large intestine. Cancer 69:2406–2410.
42. Hunt DR, Cherian M (1990) Endoscopic diagnosis of small flat carcinoma of the colon. Report of three cases. Dis Colon Rectum 33:143–147.
43. Kuramoto S, Oohara T (1989) Flat early cancers of the large intestine. Cancer 64:950–955.
44. Hamilton SR (1993) Flat adenomas: what you can't see can hurt you. Radiology 187:309–310.
45. Kubota O, Kino I, Nakamura S (1994) A morphometrical analysis of minute depressed adenomas in familial polyposis coli. Pathol Int 44:200–204.
46. Konishi F, Muto T, Kamiya J, et al (1984) Histopathologic comparison of colorectal adenomas in English and Japanese patients. Dis Colon Rectum 27:515–518.
47. Konishi F, Morson BC (1982) Pathology of colorectal adenomas: a colonoscopic survey. J Clin Pathol 35:830–841.
48. Fung CH, Goldman H (1970) The incidence and significance of villous change in adenomatous polyps. Am J Clin Pathol 53:21–25.
49. Richter H, Slezak P, Walch A, et al (2003) Distinct chromosomal imbalances in nonpolypoid and polypoid colorectal adenomas indicate different genetic pathways in the development of colorectal neoplasms. Am J Pathol 163:287–294.
50. Jorgensen OD, Kronborg O, Fenger C (1993) The Funen Adenoma Follow-Up Study. Characteristics of patients and initial adenomas in relation to severe dysplasia. Scand J Gastroenterol 28:239–243.
51. O'Brien MJ, Winawer SJ, Zauber AG, et al (1990) The National Polyp Study. Patient and polyp characteristics associated with high-grade dysplasia in colorectal adenomas. Gastroenterology 98:371–379.
52. Pascal RR (1994) Dysplasia and early carcinoma in inflammatory bowel disease and colorectal adenomas. Hum Pathol 25:1160–1171.
53. Haggitt RC, Glotzbach RE, Soffer EE, et al (1985) Prognostic factors in colorectal carcinomas arising in adenomas: implications for lesions removed by endoscopic polypectomy. Gastroenterology 89:328–336.
54. Risio M, Baccarini P, Casson P, et al (2006) [Histopathologic diagnosis in colorectal cancer screening: guidelines.] Pathologica 98:171–174.
55. Groden J, Thliveris A, Samowitz W, et al (1991) Identification and characterization of the familial adenomatous polyposis coli gene. Cell 66:589–600.
56. Bodmer WF, Bailey CJ, Bodmer J, et al (1987) Localization of the gene for familial adenomatous polyposis on chromosome 5. Nature 328:614–616.
57. Muto T, Kamiya J, Sawada T, et al (1985) Small "flat adenoma" of the large bowel with special reference to its clinicopathologic features. Dis Colon Rectum 28:847–851.
58. Spratt JS Jr, Ackerman LV (1962) Small primary adenocarcinomas of the colon and rectum. JAMA 179:337–346.
59. Enterline HT, Evans GW, Mercudo-Lugo R, et al (1962) Malignant potential of adenomas of colon and rectum. JAMA 179:322–330.
60. Eide TJ (1986) The age-, sex-, and site-specific occurrence of adenomas and carcinomas of the large intestine within a defined population. Scand J Gastroenterol 21:1083–1088.
61. Hamilton SR, Vogelstein B, Kudo S, et al (2000) Carcinoma of the colon and rectum. In: Hamilton SR, Aaltonen LA (eds) Tumours of the digestive system. IARC, Lyon.
62. Tytherleigh MG, Warren BF, Mortensen NJ (2008) Management of early rectal cancer. Br J Surg 95:409–423.
63. Kikuchi R, Takano M, Takagi K, et al (1995) Management of early invasive colorectal cancer. Risk of recurrence and clinical guidelines. Dis Colon Rectum 38:1286–1295.
64. Colacchio TA, Forde KA, Scantlebury VP (1981) Endoscopic polypectomy: inadequate treatment for invasive colorectal carcinoma. Ann Surg 194:704–707.
65. Lipper S, Kahn LB, Ackerman LV (1983) The significance of microscopic invasive cancer in endoscopically removed polyps of the large bowel. A clinicopathologic study of 51 cases. Cancer 52:1691–1699.
66. Compton CC (1999) Pathology report in colon cancer: what is prognostically important? Dig Dis 17:67–79.

67. Cooper HS (1983) Surgical pathology of endoscopically removed malignant polyps of the colon and rectum. Am J Surg Pathol 7:613–623.
68. Ueno H, Mochizuki H, Hashiguchi Y, et al (2004) Risk factors for an adverse outcome in early invasive colorectal carcinoma. Gastroenterology 127:385–394.
69. Volk EE, Goldblum JR, Petras RE, et al (1995) Management and outcome of patients with invasive carcinoma arising in colorectal polyps. Gastroenterology 109:1801–1807.
70. Mitchell PJ, Haboubi NY (2008) The malignant adenoma: when to operate and when to watch. Surg Endosc 22:1563–1569.
71. Risio M, Fiocca R (2004) Malignant adenoma: diagnosis, staging, risk factors, lymph node involvement and problems of sampling. Tech Coloproctol 8(suppl 2):s253–256.
72. Compton CC, Fielding LP, Burgart LJ, et al (2000) Prognostic factors in colorectal cancer. College of American Pathologists Consensus Statement 1999. Arch Pathol Lab Med 124:979–994.
73. Jass JR (1995) Malignant colorectal polyps. Gastroenterology 109:2034–2035.
74. Turner RR, Li C, Compton CC (2007) Newer pathologic assessment techniques for colorectal carcinoma. Clin Cancer Res 13:6871s–6876s.
75. Morodomi T, Isomoto H, Shirouzu K, et al (1989) An index for estimating the probability of lymph node metastasis in rectal cancers. Lymph node metastasis and the histopathology of actively invasive regions of cancer. Cancer 63:539–543.
76. Ueno H, Mochizuki H, Shinto E, et al (2002) Histologic indices in biopsy specimens for estimating the probability of extended local spread in patients with rectal carcinoma. Cancer 94:2882–2891.
77. Okuyama T, Nakamura T, Yamaguchi M (2003) Budding is useful to select high-risk patients in stage II well-differentiated or moderately differentiated colon adenocarcinoma. Dis Colon Rectum 46:1400–1406.
78. Tanaka M, Hashiguchi Y, Ueno H, et al (2003) Tumor budding at the invasive margin can predict patients at high risk of recurrence after curative surgery for stage II, T3 colon cancer. Dis Colon Rectum 46:1054–1059.
79. Ueno H, Murphy J, Jass JR, et al (2002) Tumour 'budding' as an index to estimate the potential of aggressiveness in rectal cancer. Histopathology 40:127–132.
80. Park KJ, Choi HJ, Roh MS, et al (2005) Intensity of tumor budding and its prognostic implications in invasive colon carcinoma. Dis Colon Rectum 48:1597–1602.
81. Sohn DK, Chang HJ, Park JW, et al (2007) Histopathological risk factors for lymph node metastasis in submucosal invasive colorectal carcinoma of pedunculated or semi-pedunculated type. J Clin Pathol 60:912–915.
82. Zamboni G, Lanza G, Risio M (1999) [Colorectal adenoma-carcinoma. Guidelines and minimal diagnostic criteria. Italian Group for Pathology of the Digestive System.] Pathologica 91:286–294.
83. Hassan C, Zullo A, Risio M, et al (2005) Histologic risk factors and clinical outcome in colorectal malignant polyp: a pooled-data analysis. Dis Colon Rectum 48:1588–1596.
84. Coverlizza S, Risio M, Ferrari A, et al (1989) Colorectal adenomas containing invasive carcinoma. Pathologic assessment of lymph node metastatic potential. Cancer 64:1937–1947.
85. Williams B, Saunders BP (1993) The rationale for current practice in the management of malignant colonic polyps. Endoscopy 25:469–474.
86. Dirschmid K, Kiesler J, Mathis G, et al (1993) Epithelial misplacement after biopsy of colorectal adenomas. Am J Surg Pathol 17:1262–1265.
87. Brown HG, Luckasevic TM, Medich DS, et al (2004) Efficacy of manual dissection of lymph nodes in colon cancer resections. Mod Pathol 17:402–406.
88. Compton CC, Greene FL (2004) The staging of colorectal cancer: 2004 and beyond. CA Cancer J Clin 54:295–308.
89. Goldstein NS (2002) Lymph node recoveries from 2427 pT3 colorectal resection specimens spanning 45 years: recommendations for a minimum number of recovered lymph nodes based on predictive probabilities. Am J Surg Pathol 26:179–189.
90. Goldstein NS, Turner JR (2000) Pericolonic tumor deposits in patients with T3N+MO colon adenocarcinomas: markers of reduced disease free survival and intra-abdominal metastases and their implications for TNM classification. Cancer 88:2228–2238.

Serrated Neoplasia Pathway

3

Mauro Risio

Abstract Epithelial serration, namely the saw-toothed outline derived from infolded epithelial tufts in the Lieberkühn's crypt, typically features hyperplastic polyps of the large bowel as a result of inhibition of the apoptotic homeostatic control. Dysplasia can affect the serrated epithelium of hyperplastic polyps, featuring right-sided colonic neoplastic polyps, the serrated adenomas. Whereas the diagnosis of sessile serrated adenomas is based mainly on architectural dysplasia, cytologic and nuclear dysplasia define traditional (polypoid) serrated adenomas. In mixed serrated polyps, both tubular or tubulo-villous and serrated adenomatous tissue can be seen, indicating the fusion of the classical adenoma–carcinoma sequence with the serrated pathway. Activating mutation of the *BRAF* gene is the triggering event, and the CpG islands methylator phenotype is the molecular genetic mechanism driving serrated tumorigenesis. Serrated adenomas are potentially evolving neoplastic lesions, maximally in hyperplastic polyposis, and need to be histologically classified and appropriately managed.

Keywords Colorectal polyps • Colorectal tumorigenesis • Hyperplastic polyps • Serrated neoplasia

3.1 Serration and Colonic Epithelium Homeostasis

Lieberkühn's crypt of intestinal mucosa is a finely tuned homeostatic unit whose morphological features are strictly dependent on the balance between cell proliferation and cell death. Under normal conditions, epithelial cells are generated in the proliferative compartment at the lower portion of the crypt, migrate, while differentiating, toward the mucosal surface, and are progressively deleted at different check-points by a scheduled programme of cell death. A single apoptotic programme could be hypothesized, which is under different temporal, microenvironmental, and genetic controls in the various sectors of the crypt. Apoptotic activation in the basal regions is elicited by exposure to genotoxic or cytotoxic agents, whereas in the surface epithelium extrusive apoptosis (anoikis) is damage independent, linked to cell senescence, triggered by modifications occurring in colonocyte–cell matrix interactions, and barely influenced by homeostatic fluctuations. Furthermore, there is mathematical evidence that apoptosis in the middle regions of the crypt (regulated and compensatory cell death) is the actual regulator of homeostasis in the gut [1,2]. As a result of such a plastic and modulable homeostatic balance, the entire colonic mucosa is composed of

M. Risio (✉)
Unit of Pathology, Institute for Cancer Research and Treatment (IRCC), Turin, Italy

G.G. Delaini, T. Skřička, G. Colucci (eds.), *Intestinal Polyps and Polyposis*,

uniform, straight, perpendicular, parallel crypts, which do not exhibit any branching and are lined by simple columnar epithelium.

Epithelial serration, namely the saw-toothed outline derived from infolded epithelial tufts in the crypt and in the luminal surface, occurs as a consequence of the undue accumulation of cells, following loss of homeostatic trophism. Transient serration associated with hyperproliferation can be found in reparative and regenerating colonic epithelium (Fig. 3.1): in this case, the reversible cellular heaping up is aimed at restoring the epithelial barrier, and serration disappears in a short time [3]. Conversely, steady serration in the absence of hyperproliferation is displayed in the common hyperplastic polyps of the large bowel, as a result of the inhibition of regulated cell death and anoikis [4]. To adapt to the cell accumulation and the slower migration rate, the epithelium protrudes into the lumen in the form of micropapillary infoldings, giving the crypt a serrated profile.

The migration of epithelial cells along the axis of the crypt is finely synchronized with both maturation and differentiation, so that changes in the former influence the latter, particularly in the mucus-producing process. Goblet cell, microvescicular, and mucus-poor cell components can therefore be found in serrated epithelium, allowing the identification of different subtypes of serrated polyps [5].

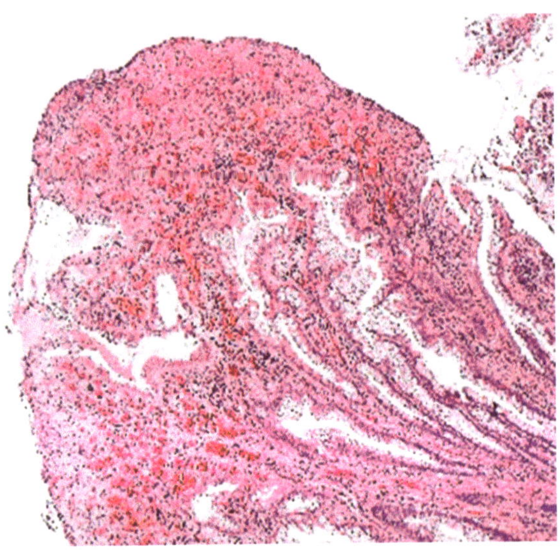

Fig. 3.1 Serrated profile of the actively regenerating epithelium in the crypts of an infiammatory polyp, myoglandular type

3.2 Serration in the Absence of Dysplasia: The Hyperplastic Polyp

Epithelial serration together with minor architectural changes are typical features of hyperplastic polyps of the large bowel. They are small-sized (0.2–0.5 cm) mucosal bumps of the sigmoid colon and rectum, consisting of straight, parallel, slightly elongated crypts lined by serrated epithelium in the intermediate and upper third, and by undifferentiated cells in the lower third (Fig. 3.2). The nuclei are round or oval, and located at the base with little or no stratification. Kinetic studies have shown that, in contrast to adenomas, cell renewal in hyperplastic polyps resembles that in the normal mucosa: there is no hyperproliferation, nor any abnormal shift of the proliferative compartment along the axis of the hyperplastic crypt

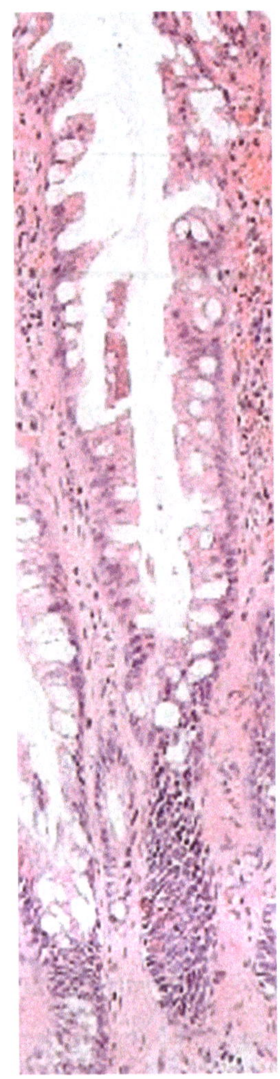

Fig. 3.2 A typical crypt of hyperplastic polyp. Epithelial serration is limited to the intermediate and upper regions

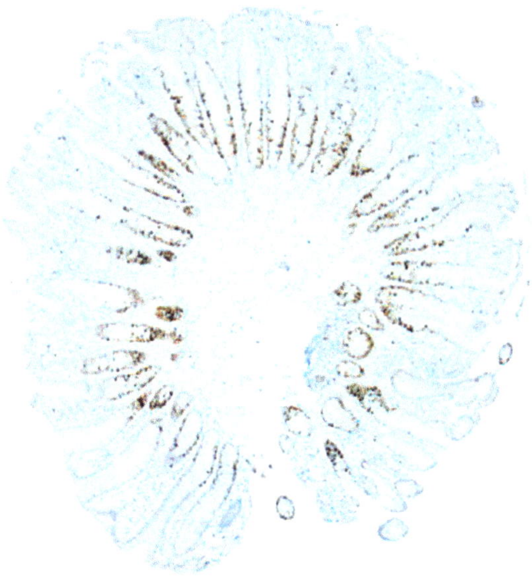

Fig. 3.3 Cell proliferation in hyperplastic polyps. The proliferative compartment (*brown nuclei*) is strictly confined to the lower third of the hyperplastic crypts, as in normal mucosa

(Fig. 3.3) [6]. Upward migration from the base to the surface of the crypt, on the other hand, is slower, the turnover time is longer, and the cell population is hypermature [7]. From the histological standpoint, therefore, the serrated epithelium of hyperplastic polyps is characterized by exaggerated cell maturation and differentiation without dysplasia.

Hyperplastic polyps are regarded as non-neoplastic lesions and, although there have been sporadic reports of their malignant transformation [8], no firm evidence has yet been presented of a greater likelihood of these polyps becoming cancerous, in comparison with the normal mucosa. On the other hand, hyperplastic polyps are more frequent in colonic segments with carcinoma, share some phenotypic features with colorectal adenocarcinoma, and are associated with lifestyle risk factors linked with neoplastic polyps [9]. Hyperplastic polyps are likely to be markers of the action of an environmental factor interacting in the initiation stage of colorectal tumorigenesis but not influencing promotion: they are paraneoplastic rather than neoplastic lesions, even though the specific mechanisms that orientate the mucosa towards hyperplastic differentiation remain to be elucidated [10].

3.3 Serrated Intraepithelial Neoplasia: Serrated Adenomas

It is well known that most adenocarcinomas of the large bowel are preceded by a pre-invasive stage of intraepithelial neoplasia that lasts for years. The neoplastic process is therefore a single, indivisible continuum that begins and is confined within the epithelium (hence the name "intraepithelial neoplasia") until invasion across the muscularis mucosa occurs, at which time the term carcinoma applies. The severity of intraepithelial neoplasia is estimated from the extent of the lesion as well as the degree of deviation from normal cellular morphology and differentiation pattern: the neoplastic process is thought to progress toward carcinoma through dysplasia of increasing severity. From the morphological standpoint, the term dysplasia is conventionally applied to the collection of changes in tissue architecture (branching and budding of the crypts, cribriform growth, back-to-back glands) and cellular and nuclear morphology (altered nucleocytoplasmic ratio, hyperchromatic nuclei, prominent nucleolus, crowded/stratified nuclei) that define intraepithelial neoplasia. Although the two categories have turned out to be differently associated with genomic changes in colorectal tumor progression [11], they have not been separately evaluated in pathological reports since they are deeply intermingled within a single neoplasia in most cases. Dysplasia in serrated epithelium has been recently demonstrated to represent a unique precancerous morphogenetic pathway, in that architectural features of dysplasia precede and are often uncoupled from nuclear and cytological dysplastic changes [12].

Sessile serrated adenomas (SSA) (serrated adenoma, superficial type; serrated adenoma type 2; serrated polyp with abnormal proliferation) [12–14] are right-sided, large-sized (>1 cm) sessile serrated lesions displaying patchy or diffuse distortions of tissue organization consistent with architectural dysplasia (i.e. branching of crypts, serration or foveolar cell phenotype at the base of the crypt, dilatation at the base of the crypt, horizontal crypt growth) (Fig. 3.4). Most findings are localized in the deeper sectors of the mucosa and require well-oriented samples to be identified. Subtle nuclear changes, when present, are focal and include small prominent nucleoli, open chromatin, and irregular contours [12].

On the other hand, in traditional serrated adenomas

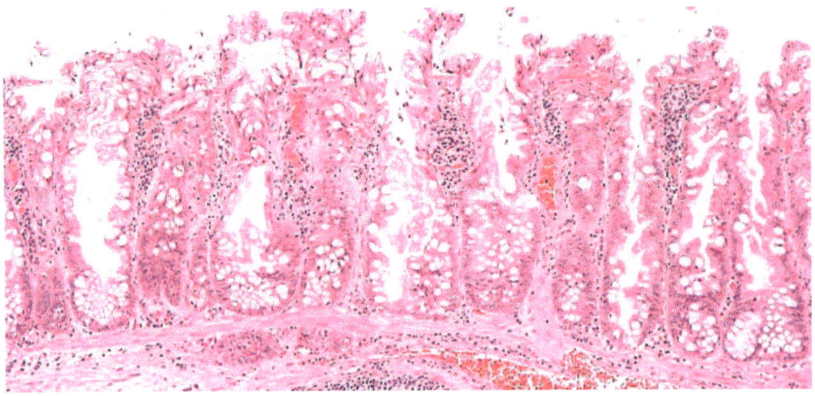

Fig. 3.4 Sessile serrated adenoma (SSA). Serrated epithelium lining the entire length of the crypts, which show dilatation at the base

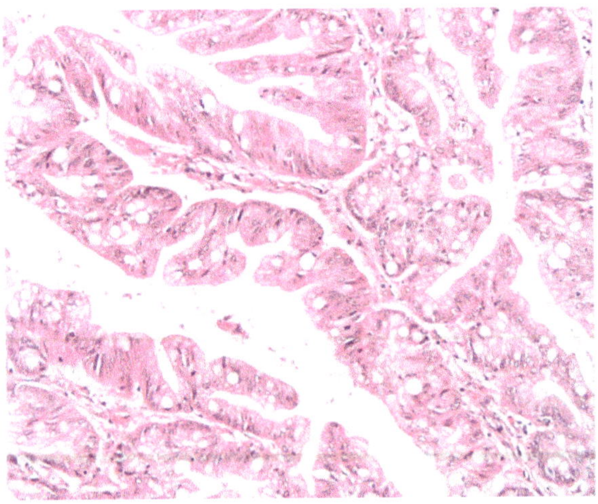

Fig. 3.5 Traditional serrated adenoma. Irregularly shaped, serrated crypts lined by frankly dysplastic epithelium

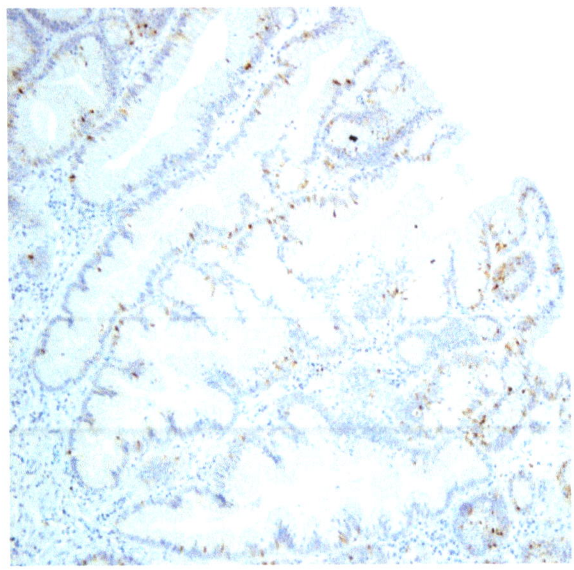

Fig. 3.6 Cell proliferation in traditional serrated adenoma. Proliferating cells (*brown nuclei*) are scattered throughout the entire length of the crypts, and focally clustered at the upper third ("upward shift")

(TSA) (serrated adenoma, polypoid type; serrated adenoma, type 1) [12–14], nuclear (elongated, hyperchromatic, stratified nuclei) and cytological frankly dysplastic features are seen within the serrated epithelium, besides the architectural ones (Fig. 3.5). Rapid and protuberant growth parallels nuclear dysplasia and, grossly, TSA can be indistinguishable from villous or tubulo-villous colorectal adenomas.

Although dysregulation of apoptotis is the basic event triggering and sustaining serrated tumorigenesis ("bottom-up morphogenesis") [15], cell proliferation changes can be detected in serrated adenomas. The asymmetric and irregular expansion of the proliferative compartment through the entire length of the crypt is seen in both SSA and TSA, and the upward shift of the major zone of proliferative activity ("top-down morphogenesis") [16] can also occur in TSA (Fig. 3.6), but to a less remarkable extent with respect to classical tubular colorectal adenomas [17,18].

3.4 Serrated Polyps: Diagnostic Accuracy and Reproducibility

Studies evaluating concordance in the morphological diagnosis of serrated polyps among pathologists have demonstrated an excellent interobserver agreement for TSA ($\kappa = 0.78–0.83$) and that the main source of discordance is in distinguishing SSA from hyperplastic polyps ($\kappa = 0.32–0.47$) [19, 20]. Distinction between the two serrated lesions mainly relies on the architectural features, some of which cannot be effectively recognized in superficial and tangentially cut biopsy, or in polyps fragmented by endoscopic removal [21]. Intermingling and/or intermediate features of SSA and hyperplastic polyps can often be seen in the same histological section, particularly in small polyps, impairing diagnostic reproducibility [20–22]. The terms "sessile serrated polyp" [14] and "serrated mucosal lesion" [23] have therefore been suggested for lesions with equivocal features that cannot be definitively categorized. Improvement in the concordance is suggested to derive from standardization of nomenclature, training of pathologists, and identification of molecular markers [19]. In support of this, Owens et al [24] have recently demonstrated that the immunohistochemical expression of gastrin mucin MUC6 displays 100% specificity in distinguishing SSA from hyperplastic polyps, and the nuclear immunostaining for β-catenin has been suggested to have diagnostic usefulness for serrated lesions [25].

It has to be taken into account, however, that the prevalence of SSA in patients undergoing colonoscopy ranges from 1.9% to 9% [17,21,26], and that they represent 7–15% of serrated polyps [21,27]. It has been estimated that only 8.3% of the polyps previously diagnosed as hyperplastic polyps would now be reclassified as SSA [17], and such a level of diagnostic accuracy is felt acceptable to support clinical decision making [20].

3.5 Serrated Tumorigenesis: The Paradigm of Hyperplastic Polyposis

Hyperplastic polyposis is defined by the presence of multiple or large hyperplastic polyps of the large bowel, typically located proximally, and often associated with familial clustering, so that World Health Organization (WHO) diagnostic criteria rely on the number, size, location, and distribution of polyps [28]. The condition has been ultimately associated with an increased risk of colorectal cancer [29], and morphological re-evaluation has shown the neoplastic nature of the polyps: the term "hyperplastic polyposis" could therefore be a misnomer, with "serrated adenomatous polyposis" [30], "serrated polyposis" [31], or "mixed polyposis syndrome" [32] being more appropriate. In fact, even if most polyps are indistinguishable from common sporadic hyperplastic polyps, SSA, TSA, classical adenomas, and mixed serrated polyps (serrated + classical adenoma) have been detected in various proportions and with various grades and extensions of dysplasia [29,30,33]. Mixed serrated polyps (MSP) (Fig. 3.7), in particular, include within a single lesion a serrated (dysplastic or non-dysplastic) and a conventional dysplastic component (tubular, tubulovillous, villous adenoma) morphologically combining, in accordance with the recently proposed "fusion model" [35], the two major routes of colorectal tumorigenesis, the adenoma–carcinoma sequence, and the serrated pathway. Such polyps, occurring also in attenuated familial adenomatous polyposis and in *MUTYH* gene-associated polyposis [36–38], are thought to be relatively aggressive, exploiting both hyperproliferation of adenomas and apoptosis inhibition of serrated neoplasia [35]. The observation that most cancerous SSA display transition sectors of classical adenoma between SSA and invasive carcinoma [12] is also consistent with this interpretation.

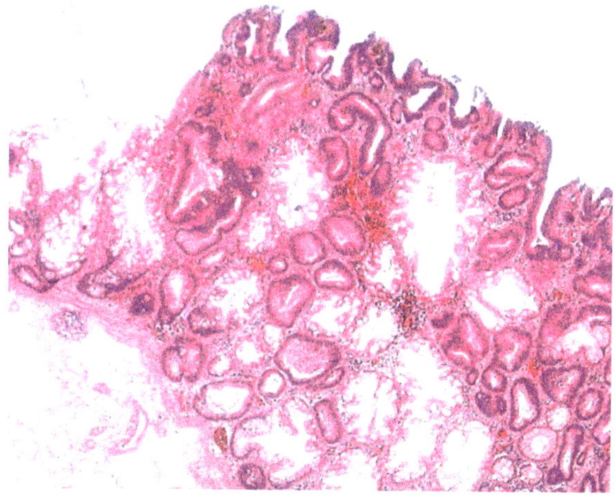

Fig. 3.7 Mixed serrated polyp (MSP). Serrated and conventional, tubular dysplastic crypts are intermingled. Reproduced from [34], with permission from GIED: Area Qualità Editor, Milan, Italy

Taken together, this evidence indicates that a multistep progression exists, leading from serrated adenomas toward colorectal carcinoma through the transition from architectural to nuclear and cytologic dysplasia or the conversion to classical adenomatous dysplasia, the latter being even more common than evolution to advanced dysplasia [12]. The following morphogenetic sequence might therefore be hypothesized:

SSA → TSA/MSP → invasive adenocarcinoma.

On the whole, 10–15% of colorectal cancers are expected to originate from serrated polyps [39]. The percentage of cancerization and speed of evolution are not yet completely known: roughly, the time elapsed from the diagnosis of SSA and the onset of advanced carcinoma is greater than 3 and 5 years in 90% and 55% of cases, respectively [40]. The overall association between serrated dysplasia and cancer (5.8%) [41] peaks at 55% in cancers showing microsatellite instability [42], and 54–70% in hyperplastic (serrated) polyposis [29,43].

Based on current knowledge, recommendations for treatment of serrated neoplasia should consider:

1. location and size of the lesion, as critical issues for endoscopic removal
2. evidence of nuclear and cytologic dysplasia, featuring high-risk serrated polyps (TSA or MSP) [12].

In consequence, the pathologic report should detail the type (architectural versus nuclear), grade (low versus high), and extension of dysplasia in order to orientate the follow-up and management [44]. Left-sided endoscopically removed SSA of any size are suitable to be followed-up as low-risk adenomatous polyps, and TSA and MSP as high-risk adenomatous polyps. It is advisable to follow up as low-risk adenomatous polyps the small (<1 cm), right-sided, endoscopically completely removed SSA; incompletely removed, large (>1 cm) SSA should be repeatedly biopsied until the onset of cytologic dysplasia: then surgical excision should be carried out and the patients followed up as high-risk for adenomatous polyps. Major surgery and follow-up as high-risk for adenomatous polyps is also recommended for right-sided TSA and MSP [12].

3.6 Molecular Genetics of Serrated Neoplasia

Somatic genetic instability in the form of chromosome instability (CIN) and microsatellite instability (MIN) has been demonstrated to sustain colorectal carcinogenesis [45] and orientate the morphogenesis of both premalignant and malignant lesions. The CpG islands methylator phenotype (CIMP) is a newly described mechanism for intestinal neoplasia operating extensive methylation of CpG sites (cytosine–guanine dinucleotide sequence), richly aggregated in the promoter regions of genes. Hypermethylation of CpG islands brings about promoter silencing, and this epigenetic change, in turn, results in functional suppression of the gene. Genes frequently methylated in colorectal cancer are *p14*, *p16*, estrogen receptor, DNA repair genes *MLH1* and O^6-methylguanine DNA methyltransferase (*MGMT*), *RASSF1*, *APC*, *COX2*, *CDH1*, and anonymous marker genes (*MNT1*, *MNT2*, *MINT31*) [15]. CIMP drives serrated tumorigenesis by silencing pro-apoptotic (e.g. *RASSF1*, *RASSF2*) and cell cycle inhibitory (*p16*, *p14*, *p19*, *Rb*) genes, while epigenetic suppression of DNA repair genes *hMLH1* and *MGMT*, inducing high-level MSI, would represent the rate-limiting step for tumor progression [15,46–48].

KRAS and *BRAF* genes participate in the mitogen-activated protein kinase signaling pathway, which mediates cellular responses to growth signals. Somatic mutations of these genes, leading to activation of the signaling pathway, confer a proliferative and invasive potential on the cells. In fact, KRAS plays a key role in the classical adenoma–carcinoma sequence, allowing the growth and progression of small adenomatous polyps [11], and *BRAF* mutations are likely to represent the initiating event, unlinked with the action of the CIMP machinery, in serrated tumorigenesis [12]. Acquisition of a *BRAF* mutation appears to be associated with the progression of hyperplastic polyps to SSA [15,48–50], whereas fusion pathways involving the sequential alterations of the genes *KRAS*, *APC*, *TP53*, and *MGMT* could lead from hyperplastic polyps to MSP and TSA [35].

Familial syndromes originating in the serrated pathway include an autosomal dominant condition ("serrated pathway syndrome") and a recessive one (hyperplastic polyposis), both of which are characterized by *BRAF* gene somatic activating mutations together with CIMP [51]. Although the underlying genetic alteration is at present unknown, a germline defect in epigenetic regulation is conceivable, causing diffuse stochastic methylation of CpG islands or targeting specific vulnerable loci among promoter regions.

References

1. Hall PA, Coates PJ, Ansari B, Hopwood D (1994) Regulation of cell number in the mammalian gastrointestinal tract: the importance of apoptosis. J Cell Sci 107:3569–3577.
2. Risio M, Sarotto I, Rossini FP, et al (2000) Programmed

cell death, proliferating cell nuclear antigen and p53 expression in mouse colon mucosa during diet/induced tumorigenesis. Anal Cell Pathol 2:87–94.
3. Nakamura S, Kino I, Akagi T (1992) Inflammatory myoglandular polyps of the colon and rectum. A clinicopathological study of 32 peduncolated polyps, distinct from other types of polyps. Am J Surg Pathol 16:772–779.
4. Tateyama H, Li W, Takahashi E, et al (2002) Apoptosis index and apoptosis-related antigen expression in serrated adenoma of the colorectum. The saw-toothed structure may be related to inhibition of apoptosis. Am J Surg Pathol 26:249–256.
5. O'Brien MJ, Yang S, Clebanoff JL, et al (2004) Hyperplastic (serrated) polyps of the colorectum. Relationship of CpG island methylator phenotype and k-ras mutation to location and histologic subtype. Am J Surg Pathol 28:423–434.
6. Risio M, Coverlizza S, Ferrari A, et al (1988) Immunohistochemical study of epithelial cell proliferation in hyperplastic polyps, adenomas and adenocarcinomas of the large bowel. Gastroenterology 94:899–906.
7. Hayashi T, Yatani R, Apostol J, Stemmermann GN (1974) Pathogenesis of hyperplastic polps. A hypothesis based on ultrastructural and in vitro kinetics. Gastroenterology 66:347–401.
8. Franzin G, Novelli P (1982) Adenocarcinoma occurring in a hyperplastic (metaplastic) polyp of the colon. Endoscopy 14:28–30.
9. Martinez ME, McPherson LS, Levin B, Glober GA (1997) A case-control study of dietary intake and other lifestyle risk factors for hyperplastic polyps. Gastroenterology 113:423–427.
10. Risio M, Arrigoni A, Pennazio M, et al (1995) Mucosal cell proliferation in patients with hyperplastic colorectal polyps. Scand J Gastroenterol 30:344–348.
11. Risio M, Malacarne D, Giaretti W (2005) KRAS transition and villous growth in colorectal adenomas. Cell Oncol 2:363–366.
12. Snover DC, Jass JR, Fenoglio-Preiser C, Batts KP (2005) Serrated polyps of the large intestine. A morphological and molecular review of an evolving concept. Am J Clin Pathol 124:380–391.
13. Oka S, Tanaka S, Hiyama T, et al (2004) Clinicopathologic and endoscopic features of colorectal serrated adenoma: differences between polypoid and superficial types. Gastrointest Endosc 59:213–219.
14. Torlakovic E, Skovlund E, Snover DC, et al (2003) Morphological reappraisal of serrated colorectal polyps. Am J Surg Pathol 27:65–81.
15. Jass JR, Whitehall VL, Young J, Leggett BA (2002) Emerging concepts in colorectal neoplasia. Gastroenterology 123:862–876.
16. Shih IM, Wang TL, Traverso G, et al (2001) Top-down morphogenesis of colorectal tumors. Proc Natl Sci USA 27:2640–2645.
17. Higuchi T, Sugihara K, Jass JR (2005) Demographic and pathological characteristics of serrated polyps of colorectum. Histopathology 47:32–40.
18. Torlakovic EE, Gomez JD, Driman DK, et al (2008) Sessile serrated adenoma (SSA) vs traditional serrated adenoma (TSA). Am J Surg Pathol 32:21–29.
19. Glatz K, Pritt B, Glatz D, et al (2007) A multinational, internet-based assessment of observer variability in the diagnosis of serrated colorectal polyps. Am J Clin Pathol 128:938–945.
20. Farris AB, Misdraji JM, Srivastava A, et al (2008) Sessile serrated adenoma. Challenging discrimination from other serrated colonic polyps. Am J Surg Pathol 32:30–35.
21. Sandmeier D, Seelentag W, Bouzourene H (2007) Serrated polyps of the colorectum: is sessile serrated adenoma distinguishable from hyperplastic polyp in a daily practice? Virchows Arch 450:613–618.
22. Chung SM, Chen Y-T, Panczykowski A, et al (2008) Serrated polyps with "intermediate features" of sessile serrated polyp and microvescicular hyperplastic polyp. A practical approach to the classification of nondysplastic serrated polyps. Am J Surg Pathol 32:407–412.
23. Cunningham KS, Riddell R (2006) Serrated mucosal lesions of the colorectum. Curr Opin Gastroenterol 22:48–53.
24. Owens SR, Chiosea SI, Kuan SF (2008) Selective expression of gastric mucin MUC6 in colonic sessile serrated adenoma but not in hyperplastic polyp aids in morphological diagnosis of serrated polyps. Mod Pathol 21:660–669.
25. Wu JM, Montgomery EA, Iacobuzio-Donahue CA (2008) Frequent beta-catenin nuclear labeling in sessile serrated polyps of the colorectum with neoplastic potential potential. Am J Surg Pathol 129:416–423.
26. Spring KJ, Zhao ZZ, Karamatic R, et al (2006) High prevalence of sessile serrated adenomas with BRAF mutations: a prospective study of patients undergoing colonoscopy. Gastroenterology 131:1400–1407.
27. Ruschoff J, Aust D, Hartmann A (2007) Colorectal serrated adenomas: diagnostic criteria and clinical implications. Verh Dtsch Ges Pathol 91:119–125.
28. Burt R, Jass JR (2000) Hyperplastic polyposis. In: Hamilton SR, Aaltonen AL (eds) Pathology and genetics, tumours of the digestive system. IARC Press, Lyon, pp 135–139.
29. Rubio CA, Stemme S, Jaramillo E, Lindblom A (2006) Hyperplastic polyposis coli syndrome and colorectal carcinoma. Endoscopy 38:266–270.
30. Torlakovic E, Snover CD (1996) Serrated adenomatous polyposis in humans. Gastroenterology 110:748–755.
31. Torlakovic E, Skovlund E, Snover DC, et al (2003) Morphologic reappraisal of serrated colorectal polyps. Am J Surg Pathol 27:65–81.
32. Rozen P, Samuel Z, Brazowski E (2003) A prospective study of the clinical, genetic, screening, and pathologic features of a family with hereditary mixed polyposis syndrome. Am J Gastroenterol 98:2317–2320.
33. Leggett BA, Devereaux B, Biden K, et al (2001) Hyperplastic polyposis. Association with colorectal cancer. Am J Surg Pathol 25:177–184.
34. Risio M (2008) Polipo iperplastico, polipo serrato ed adenoma serrato: classificazione morfologica. Giornale Italiano di Endoscopia Digestiva 31:16
35. Jass JR, Baker K, Zlobec I, et al (2006) Advanced colorectal polyps with the molecular and morphological features of serrated polyps and adenomas: concept of a "fusion" pathway to colorectal cancer. Histopathology 49:121–131.
36. Matsumoto T, Iida M, Kobori Y, et al (2002) Serrated adenomas in familial adenomatous polyposis: relation to germline *APC* gene mutation. Gut 50:402–404.
37. Kambara T, Whitehall VL, Spring KJ, et al (2004) Role of inherited defects of MYH in the development of sporadic

colorectal cancer. Genes Chromosomes Cancer 40:1–9.
38. Venesio T, Molatore S, Cattaneo F, et al (2004) High frequency of *MYH* gene mutations in a subset of patients with familial adenomatous polyposis. Gastroenterology 126:1681–1686.
39. Makinen MJ (2007) Colorectal serrated adenocarcinoma. Histopathology 50:131–150.
40. Goldstein NS, Bhanot P, Odish E, Hunter S (2003) Hyperplastic-like colon polyps that preceded microsatellite-unstable adenocarcinomas. Am J Clin Pathol 119:778–796.
41. Makinen MJ, George SM, Jernvall P, et al (2001) Colorectal carcinoma associated with serrated adenoma. Prevalence, histological features, and prognosis. J Pathol 193:286–294.
42. Hawkins NJ, Ward RL (2001) Sporadic colorectal cancer with microsatellite instability and their possible origin in hyperplastic polyps and serrated adenomas. J Natl Cancer Inst 93:1307–1313.
43. Hyman NH, Anderson P, Blasyk H (2004) Hyperplastic polyposis and the risk of colorectal cancer. Dis Colon Rectum 47:2101–2104.
44. Risio M, Baccarini P, Cassoni P, et al (2006) Histopathologic diagnosis in colorectal cancer screening: guidelines. Pathologica 98:171–174.
45. Lengauer C, Kinzler KW, Vogelstein B (1999) Genetic instabilities in human cancers. Nature 396:643–649.
46. Park SJ, Rashid A, Lee JH, et al (2003) Frequent CpG island methylation in serrated adenomas of the colorectum. Am J Pathol 162:815–822.
47. Winter CV, Walsh MD, Higuchi T, et al (2004) Methylation patterns define two types of hyperplastic polyp associated with colorectal cancer. Gut 53:573–580.
48. Minoo P, Jass JR (2006) Senescence and serration: a new twist to an old tale. J Pathol 210:137–140.
49. Chan TL, Zhao W, Cancer Genome Project, Leung SY, Yuen ST (2003) *BRAF* and *KRAS* mutations in colorectal hyperplastic polyps and serrated adenomas. Cancer Res 63:4878–4881.
50. Kambara T, Simms LA, Whitehall VLJ, et al (2004) BRAF mutation is associated with DNA methylation in serrated polyps and cancers of the colorectum. Gut 53:1137–1144.
51. Young J, Jass JR (2006) The case for a genetic predisposition to serrated neoplasia in the colorectum: hypothesis and review of the literature. Cancer Epidemiol Biomarkers Prev 15:1778–1784.

Genetic and Clinical Features of Familial Adenomatous Polyposis (FAP) and Attenuated FAP

4

Mauro Risio and Tiziana Venesio

Abstract Familial adenomatous polyposis (FAP) is a heterogeneous genetic syndrome characterized by the presence of hundreds to thousands of adenomas, leading to a malignant degeneration at young age. Affected individuals can differ in the number of polyps, the presence of carcinoma, and extraintestinal manifestations. In the last decade, a distinctive phenotype, with fewer than 100 polyps and a later onset of adenomas, termed attenuated familial adenomatous polyposis (AFAP), has been identified, but its prevalence remains unknown. The *APC* and *MUTYH* genes are associated with this syndrome in 70–80% of FAP cases, but in no more than 30–40% of AFAP patients. The structure and function of these two genes explain much of the clinical heterogeneity of this syndrome. *APC,* triggering both inherited and sporadic colorectal tumorigenesis by controlling the Wnt signaling pathway activation, leads to a severe phenotype with a dominant pattern of inheritance. On the other hand, *MUTYH*, controlling a DNA repair system for oxidative DNA damage, leads to a more attenuated expression of the disease by a recessive inheritance.

Keywords AFAP • *APC* • Dominant inheritance • FAP • Genotype–phenotype correlation • Germline mutations • *MUTYH* • Recessive inheritance

4.1 Introduction

Information and insights into the molecular pathogenesis of colorectal cancer have illuminated our general knowledge of human cancer genetics. Much of this has been catalyzed by the identification and characterization of probands and families affected by hereditary forms of colorectal cancer.

Among these, familial adenomatous polyposis (FAP) syndrome has served as a model to elucidate the molecular steps involved in the adenoma–carcinoma sequence, a paradigmatic colorectal initiation and progression process which involves most of the sporadic colorectal tumors and is characterized by mutations on three tumor-suppressor genes, *APC*, *SMAD4*, and *p53*, gain of function of one oncogene, *KRAS*, and acquisition of an aneuploid/polyploid karyotype [1].

FAP (OMIM 175100) accounts for about 1% of all colorectal cancer cases. It is a distinctive syndrome, both clinically and genetically, characterized by the presence of hundreds to thousands of adenomas in the colon and rectum. This generally leads to a malignant degeneration by the age of 40 to 50 years. In addition to colorectal polyps, which are prevalently adenomas, extraintestinal manifestations have also been reported: gastric fundic and duodenal polyps, and increased risk

T. Venesio (✉)
Unit of Pathology, Institute for Cancer Research and Treatment (IRCC), Turin, Italy

of malignancy of the brain (glioblastoma), thyroid gland (papillary carcinoma), and liver (hepatoblastoma). Other diagnostically important extraintestinal features of FAP include retinal lesions, known as congenital hypertrophy of the retinal pigment epithelium (CHRPE), found in 60–90% of FAP patients, epidermoid cysts, osteomas, and dental anomalies in about one-third of patients. Overall, individuals affected by this syndrome can be heterogeneous in relation to the number of adenomas and the presence of carcinomas.

In the last decade, studies performed to investigate the genotype–phenotype correlation have evidenced a subset of familial polyposis patients with a distinctive phenotype characterized by a lower number of colorectal adenomas (generally fewer than 100), a later onset of colorectal adenomatosis and carcinoma, and an extremely limited expression of extracolonic manifestations. This phenotype has been termed attenuated familial adenomatous polyposis (AFAP). After some revisions, a kind of consensus has been reached on a definition for AFAP, but its incidence and frequency are still unknown and vary between the different studies [2].

According to data in the literature, the genetic cause of this syndrome can be found in about 70–80% of FAP cases, but in no more than 30–40% of AFAP patients. Both the classical and attenuated form are mainly associated with the *APC* and *MUTYH* genes, but *APC* germline alterations are highly prevalent in FAP (up to 70% of the cases), whereas *MUTYH* constitutional mutations are more frequent in AFAP (20–30% of *MUTYH* versus 10% of *APC*). These percentages are only indicative, and depend on the selection of patients and on the type of molecular approach used in different studies to screen for germline mutations.

Although both these genes are associated with the transmission of familial polyposis, the pattern of inheritance and the role exerted in sustaining the initiation and progression of colorectal epithelial cells are considerably different. *APC* is a well-known tumor-suppressor with a key and specific role in both inherited and sporadic colorectal tumorigenesis, acting by controlling the activation of the Wnt signaling pathway. *MUTYH* belongs to a DNA repair system, BER, important in repairing oxidative DNA damage, which is particularly active, but not specific for colorectal epithelial cells. *APC* is linked to familial polyposis by an autosomal dominant pattern of inheritance, while *MUTYH* segregates in these families according to an autosomal recessive model.

4.2 The Adenomatous Polyposis Associated with *APC*: Dominant Inheritance

In 1991 the identification of a constitutional deletion on chromosome 5q in a patient affected by polyposis, and the subsequent familial linkage analysis, led to the cloning of the Adenomatous Polyposis Coli (*APC*) gene [3,4]. Since then, germline alterations of this gene have been found to be associated with 80% of cases affected by the classical FAP and with 10% of patients with the attenuated form of the syndrome (AFAP).

APC contains 15 exons (ORF of 8538 nucleotides), with exon 15 forming 75% of the coding region (6.5 Kb), and encodes a protein of 2843 amino acids. Its protein product occurs in several isoforms as a result of alternative splicings, mainly in the first 14 exons [5]. Since its identification, *APC* has been demonstrated to be a tumor-suppressor gene with a “gatekeeper” function in the initiation of colorectal tumorigenesis [6]. *APC* is a tumor-suppressor gene playing a key role in the Wnt signaling pathway where it binds and controls the degradation of β-catenin within the cell cytoplasm. Germline or somatic *APC* mutations render cytoplasmic β-catenin stable, resulting in its nuclear translocation where certain target genes, involved mainly in the control of cell proliferation and differentiation, such as *C-MYC* and *cyclin-D1*, are activated transcriptionally. This gene contains several domains to allow oligodimerization and interaction with several other molecules. In addition to β-catenin, GSK3-β, axin-conductin, tubulin, and EB-1 have been reported as major partners. APC protein takes part, directly or indirectly, in several cell functions: proliferation and differentiation via the Wnt pathway, cell–cell contact by modulating the complex β-catenin/e-cadherin, movement along the intestinal crypts by controlling the accumulation of β-catenin along the cell membranes, chromosomal segregation by binding with the checkpoint proteins Bub1 and Bub3, and also transcription-independent apoptosis by caspase cleavage of *APC* itself.

To date, 800 different *APC* mutations have been

reported [7]. Most (about 90%) of the identified germline alterations are truncations due to nonsense mutations or insertions/deletions of a few nucleotides along the *APC* coding sequence, including the acceptor/donor splice sites at the intron/exon boundaries. Genetic testing has exploited the notion of the highly preponderant presence of truncating germline mutations by using an *in vitro* transcription/translational assay (protein truncation test or PTT) to detect truncated APC protein [8]. Although these mutations can be scattered throughout the entire gene, they generally tend to cluster in exon 15, mainly in the central region of the gene, also known as the mutation cluster region (MCR), where the amino acid motifs, involved in the binding with β-catenin, are located between codons 1290 and 1400 (Fig. 4.1a). The

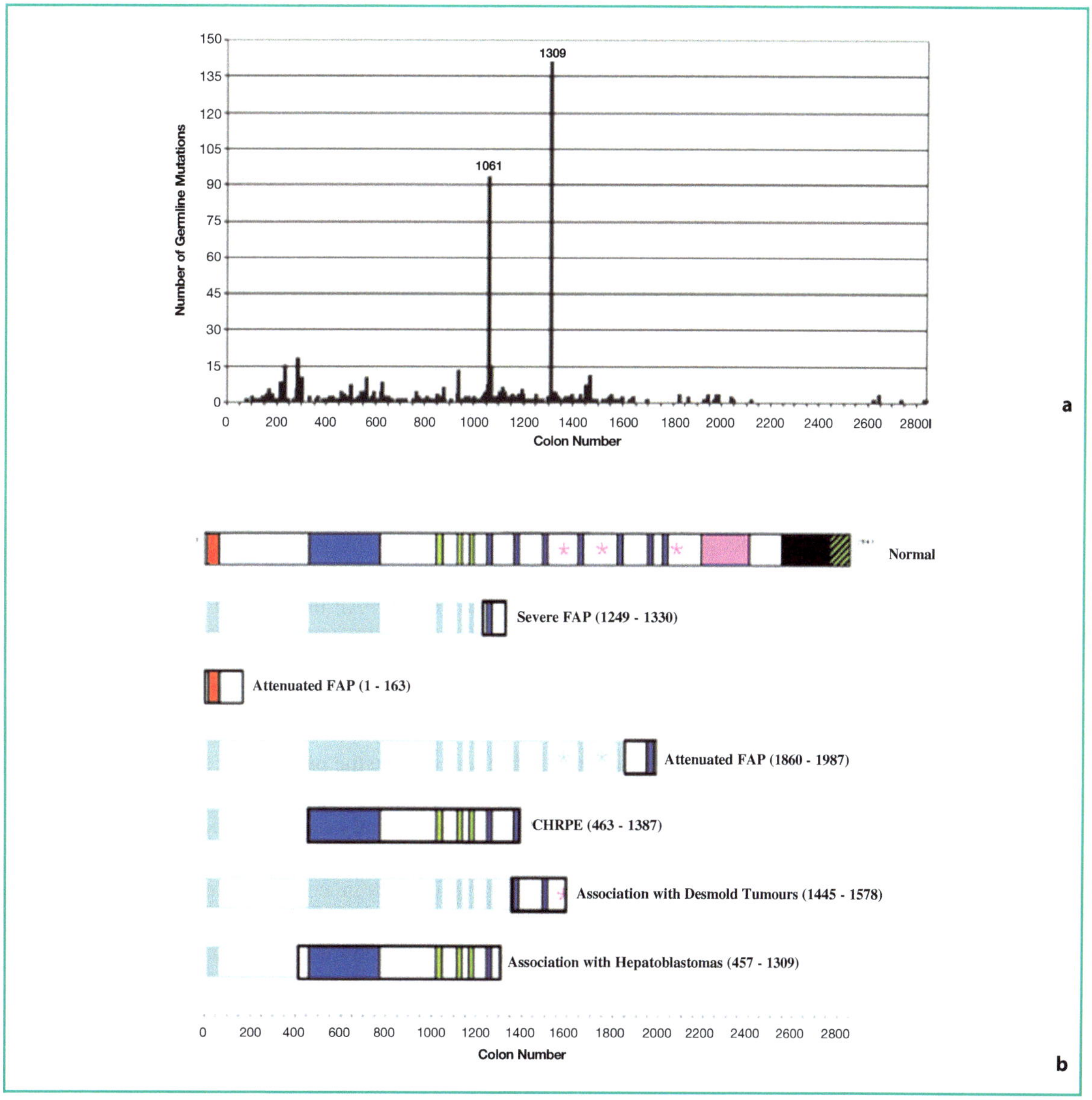

Fig. 4.1a,b The *APC* germline mutations. **a** The frequency and the distribution of *APC* germline mutations as reported in the *APC* Mutation Database (hptt://perso.curie.fr) by T. Soussi. Truncating mutations are distributed along the gene with a cluster in the central part of *APC*, with two "hot spots" on codons 1061 and 1309. **b** The *APC* genotype–phenotype correlation. Reproduced from [13], with permission from Oxford University Press

remaining subset of *APC* constitutional alterations identified so far (about 10%) is composed chiefly of intragenic or whole-gene deletions, and very few missense variants [9,10]. The preponderance of these types of alterations could be slightly underestimated since most of the molecular screenings performed on the *APC* gene have been focused on the search for truncating mutations. Recently, the detection of these kinds of alteration has been improved by the introduction in molecular diagnostics of multiplex ligation-dependent probe amplification (MLPA) to detect the partial and/or the complete loss of the gene, and denaturating high-performance liquid chromotography (dHPLC), specifically aiming to detect missense variants [11,12].

According to the autosomal dominant model, susceptibility to polyposis is inherited from one affected parent through the passage of his/her mutant allele to the offspring. Onset of the disease (appearance of adenomas) starts when the second wild-type *APC* allele undergoes inactivation in the epithelial cells of the colon. However, about 30% of APC polyposis seems to be sporadic since affected individuals are carriers of *de novo* mutations, acquired in parental germinal cells (especially male) or during embryo development. The penetrance of the *APC* mutations is 100%, meaning that all individuals carrying a germline mutation will develop the syndrome, although with a phenotype (age, number of adenomas, presence of carcinoma and/or extracolonic manifestations etc) that could vary in accordance with the type of constitutional mutation.

In fact, genotype–phenotype correlations have been reported with regard to the site of the truncating germline mutations within the *APC* gene (Fig. 4.1b). Generally, mutations located in the central part of the gene, between codons 1000 and 1400, are associated with the classical form of polyposis (FAP), whereas constitutional mutations occurring in the 5' (codons 78–167, exons 3,4) region, between codons 311 and 411 (alternatively spliced region of exon 9), and in the 3' (after codon 1595) region give an attenuated polyposis phenotype (Fig. 4.2a) [2]. Some of the alterations are related to the presence of extracolonic manifestations: mutations between codons 457 and 1444 can be associated with CHRPE, while osteomas and desmoid tumors are associated with mutations after codon 1400 [13].

Several models of *APC* function and corresponding phenotype have been developed, but no single model explains all current data. The "dominant negative model" assumes that the mutated *APC* alleles act in a dominant way because the truncated protein forms a homodimer with the wild-type allele product. This interaction lowers or abolishes tumor-suppressor activity, leaving only a few homodimers with normal function. According to this model, mutations located at the 5' end prevent interaction between mutated and wild-type *APC*, allowing residual normal gene activity as in the attenuated phenotype; at the same time, patients with mutations at the 3' end of the gene give an AFAP phenotype because the mutant allele generates an unstable product that does not interact with the wild-type protein [14,15].

In general, large deletions and missense variants have been found to be more frequently associated with the classical phenotype, whereas mutations at splice junctions and the abnormal expression of mRNA isoforms are more related to an attenuated form of the disease [16–18]. However, patients affected by all these type of alteration have been reported in both classical and attenuated groups, confirming that the phenotype is due more to the specific gene dosage effect of each *APC* alteration than to the different classes of mutation.

Phenotype variations can also occur as a result of a somatic mosaicism, which is defined as the co-presence in the same tissue of two genetically different cell populations. In FAP this type of somatic alteration is due to *de novo* mutations of the *APC* gene, arising during embryo development, and could probably account for 20% of sporadic polyposis cases. In some of these patients, an AFAP phenotype has been observed despite the presence of *APC* mutations associated with classical forms of FAP [19,20]. In any case, independently of the position or the type of the *APC* alteration, FAP/AFAP families are sometimes characterized by a considerable heterogeneity, and affected offspring, carriers of the same constitutional mutations, can show a variable disease phenotype with respect to the age of onset, the number of adenomas, and the presence of carcinoma. Mouse models have shown that these variations can be attributed to environmental and genetic modifier factors. However, at present, case-control studies have not provided conclusive results even if some good candidates have been identified such as the corresponding mouse *Mom1* and *Mom2*, on 1p35–36 (phospholipase or *PLA2G2*) and

Mutation location	Authors	Year
Mutations at the 5° end		
exon 3	Spirio et al.	1993
codon 208	Ficari et al.	2000
exons 3 + 4	Samowitz et al.	1995
codon 175	Iwama et al.	1999
exon 4	Kuwada et al.	2001
codon 157–175	Enomoto et al.	2000
codon 233	Smith-Ravin et al.	1994
prox. of codon 158	Giardiello et al.	1997
codon 99	Dobbie et al.	1994
codon 151	Dobbie et al.	1994
intron 3	Spirio et al.	1999
codon 163	Soravia et al.	1998
Mutations in exon 9		
exon 9	van der Luijt et al.	1995
codon 398	Young et al.	1998
APC-as9	Su et al.	2000
exon 9	Curia et al.	1998
codon 366	Bunyan et al.	1995
codon 363	Rozen et al.	1999
codon 332	Soravia et al.	1998
codon 332	Lamlum et al.	2000
codon 332	Andreutti-Zaugg et al.	1999
codon 394	Lamlum et al.	2000
intron 9	Varesco et al.	1994
Mutations at the 3° end		
codon 1597	Friedl et al.	1996
codon 1062	Scarano et al.	1999
codon 1979	Brensiger et al.	1998
codon 2644	Brensiger et al.	1998
codon 1581	Friedl et al.	1996
codon 2643 + 2644	Couture et al.	2000
codon 1860–1862	van der Luijt et al.	1997, 1995, 1996
codon 2662	Matsubara et al.	2000
codon 1924	Eccles et al.	1996
codon 1924	Lamlum et al.	2000
codon 1942	Lamlum et al.	2000

a

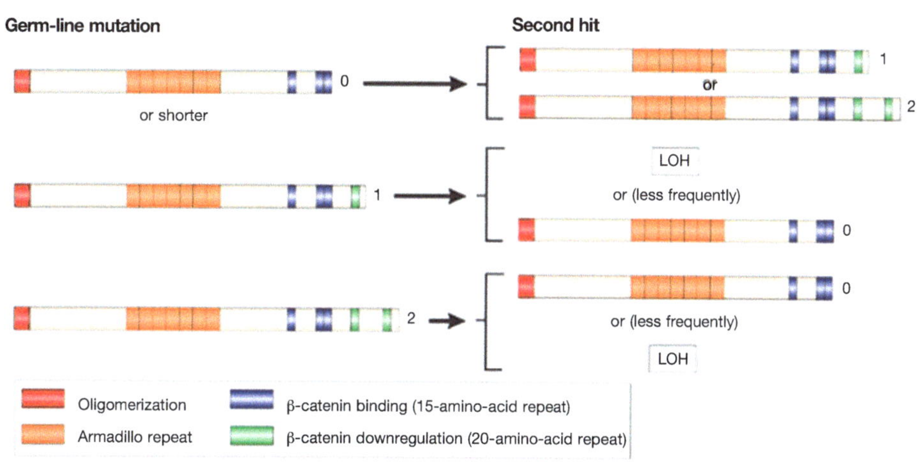

b

Fig. 4.2a,b The *APC* mutations in AFAP and the second hit; **a**, the list of the truncating mutations identified in AFAP patients. These alterations are located in exons 3, 4, 9 and in the 3' part of exons 15 [14–16]; **b**, the "just-right signaling model". The somatic alteration on the wild-type allele is selected according to the position of the first germline mutation to allow the upregulation of β-catenin for Wnt pathway activation. Figure 4.2a reproduced from [2], with permission from Kluwer Academic Publisher; Figure 4.2b reproduced from [6], with permission from Macmillan Publishers Ltd

18q21–23, respectively, and *N*-acetyl transferases, *Nat1* and *Nat2*, on 8p22.

According to the classical tumor-suppressor gene model, *APC* is expected to initiate colorectal progression when both alleles are inactivated. This mechanism has been widely investigated by using lesions derived from FAP/AFAP patients, carriers of characterized constitutional mutations. This analysis has evidenced that the somatic alteration affecting the wild-type allele is not independent of the first germline mutation but it occurs in such a way that one of the alterations truncates the protein in the β-catenin-binding domain. This type of truncating mutation would be selected since it allows the APC heterodimer to achieve the optimum level of β-catenin for tumor cell growth. Such a mechanism has been called "the just-right signaling model" (Fig 4.2b) [21]. Phenotype variability between FAP and AFAP and intra/inter AFAP has also been explained by the type of acquired somatic alterations in the adenomas. The optimum β-catenin level is easily reached in the adenomas of most FAP patients since they are frequently carriers of "suitable" germline mutations in the MCR; on the other hand, adenomas of AFAP patients can only acquire advantageous mutations on the β-catenin-binding domain later, by somatic alterations. In some cases, initiation of tumor growth can require the involvement of two somatic hits, which can affect both the wild-type and the mutant allele by loss or mutation. This mechanism, termed the "three hit model", can explain the heterogeneity of the phenotype on the different statistical chance that adenomas have to acquire the "right" *APC* mutation to upregulate the level of β-catenin and progress in the transformation [22].

4.3 The Adenomatous Polyposis Associated with *MUTYH*: Recessive Inheritance

In 2002 Al-Tassan and colleagues at the University of Wales in Cardiff showed, in a landmark study, that familial polyposis could be linked to a gene of the base excision repair (BER), *MUTYH*, via the autosomal recessive model [23].

This finding originated from the study of a British family (family N) with three siblings affected by multiple colorectal adenomas and carcinoma, but without evident vertical transmission of the disease. Analysis of the entire *APC* open reading frame (ORF) by using sequencing of constitutional DNA samples from two of the affected brothers, had excluded germline alteration of this tumor-suppressor gene segregating with the disease. However, a further examination of adenomas and colorectal tumors from family N evidenced a high proportion of somatic G:C →T:A mutations in *APC*. Since somatic G→T transversions are characteristically associated with a mutator phenotype, lacking proficient oxidative damage repair, the investigators examined the three genes of the human BER system at the constitutional level (*MUTYH*, *OGG1*, and *MTH1*). This analysis showed that the affected brothers were all germline compound heterozygotes for the missense variants Y165C and G382D in *MUTYH* (previously known as *hMYH* or *MYH*) (Fig 4.3a) [23]. BER is committed to repair oxidative damage originating as a byproduct of normal cellular metabolism or from an extrinsic source, including reactive oxygen species (ROS) and methylation, a type of damage that is particularly prevalent in the colon. This generally results in a DNA adduct, 8-oxo-G, that can mispair with adenine during subsequent DNA replication, and cause G:C→T:A transversion. The products of *MUTYH*, *OGG1*, and *MTH1* cooperate to prevent the fixation of these lesions. In this system, *MUTYH*, being an adenine-specific DNA glycosylase, acts by specifically removing adenines mispaired with 8-oxo-G. *MUTYH* is a 7.1 Kb gene, mapping between 1p32.1 and p34.3, in a region frequently affected by chromosomal rearrangements in colorectal tumorigenesis, and composed of 16 exons. *MUTYH* belongs to the group of genes involved in maintaining DNA integrity, thus it can be ascribed to the group of "caretaker" genes, which comprise the mismatch repair (MMR) genes, *BRCA1* and *BRCA2*. The ORF of full-length *MUTYH* translates into a 535-amino acid (aa) protein, characterized by key conserved domains for DNA binding, such as a helix–hairpin–helix domain (HhH-GPD) (aa114–273), an adenine recognition motif (aa 255–273), and the MutT-like (NUDIX) domain (aa 354–486). In addition, the product of *MUTYH* shows interaction sites for PCNA (aa 509–527), APE1 (aa 295–317), MSH6 (aa232–254), and RPA (aa 8–31), as well as the putative mitochondrial (MLS, 1–14) and nuclear targeting signals (NLS, 505–509) [24]. Recent structural studies have contributed to understanding of the glycosylase activity of MUTYH, but it still remains to clarify its involvement in other pathways of DNA repair by the interaction with MSH6, APE1, and PCNA.

Current information can allow assessment of a unique role for *MUTYH* among the BER enzymes in recognizing/removing a mismatch between a damaged 8-oxo-G and a normal adenine. In this regard the interaction with the MMR heterodimer MSH2/MSH6, via MSH6, has been shown to improve these activities

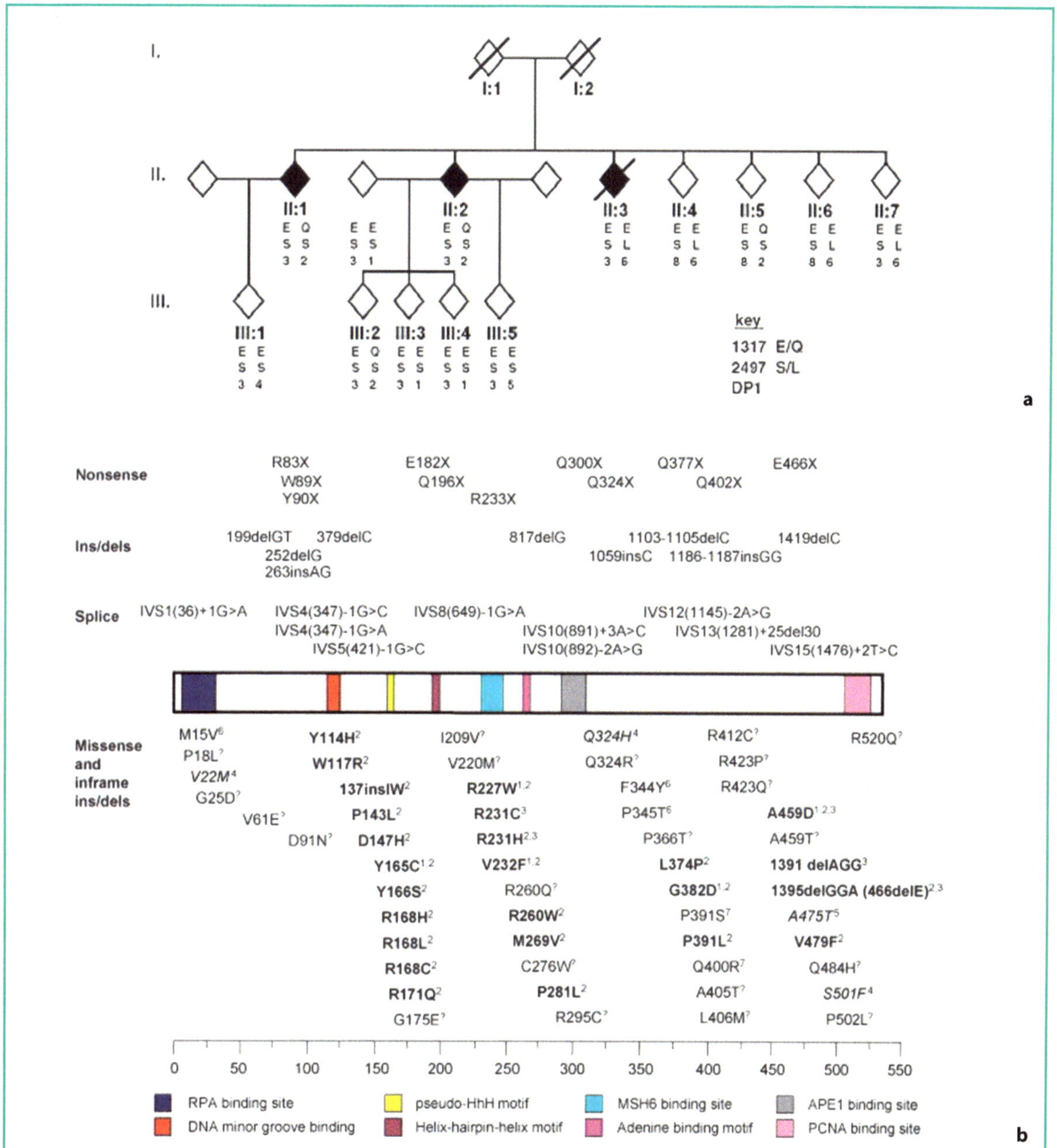

Fig. 4.3a,b The *MUTYH* germline mutations. **a** The pedigree of Family N. The three affected brothers were found to be carriers of the same *MUTYH* bi-allelic gemline mutations, Y165C and G382D, on exon 7 and 13, respectively. **b** Spectrum and distribution of all *MUTYH* variants. Missense variants proven or likely to be pathogenetic are in bold. Figure 4.3a reproduced from [23], with permission from Macmillan Publishers Ltd; Figure 4.3b reproduced from [29], with permission from Elsevier

[25]. On the other hand, the binding with AP endonuclease, PCNA, and RPA suggests a function also in long-patch BER and in replication-coupled repair [26]. *MUTYH* also exerts its repairing activity efficiently in mitochondria. These organelles need an effective means of repairing their DNA because of the abundant 8-oxo-G lesions caused by exposure to a high level of endogenous ROS [24].

To date, bi-allelic germline mutations of *MUTYH* have been reported in approximately 25% of polyposis patients, negative for inherited *APC* alterations, from different areas of Europe, North America, and Asia, with FAP-like and AFAP-like phenotypes. According to the model of autosomal recessive segregation, none of these patients' families showed vertical transmission of the syndrome, with many of the cases being apparently sporadic. As far as the colorectal phenotype, *MUTYH*-associated polyposis or MAP can resemble both AFAP (10–100 adenomas) and FAP (100–300 adenomas), but not severe FAP (>1000 adenomas) [27,28]. The data from molecular screenings, performed during the last 5 years, suggest that, at presentation, a common MAP patient is generally middle-aged (median age 50 years) with between 40 and 50 adenomas. In 50% of these cases, a colorectal cancer is already reported and, sometimes, the presence of this neoplasia is associated with few or no macroscopic adenomas. Apart from 10–15% of the patients with duodenal polyps, extracolonic manifestations are extremely rare [29]. An almost exclusive colorectal phenotype in individuals with an impaired constitutional *MUTYH* can be partly explained by the high level of oxidative damage occurring in the bowel. An additional proposed factor concerns the two bases immediately 3' to the mutated G that have been shown to be almost always AA. *APC*, the gatekeeper of colorectal transformation, seems to be particularly prone to undergoing G→T transversions since it has 216 GAA sequences. In contrast, key genes controlling the transformation of other tissues, subjected to oxidative damage, have considerably fewer target GAA sequences. This is the case for *TP53*, *RB1*, *NF1*, or *VHL* for lung, brain, retina, and kidney tumorigenesis [29].

To date, the mutations Y165C and G382D have been reported as homozygous or compound heterozygous in approximately 80% of MAP patients. However, a considerable spectrum of several other variants, scattered throughout the entire gene, has been identified: 30 mutations predicted to truncate the protein (nonsense substitutions, small insertions/deletions, and splice site variants), 52 missense variants, and three small inframe insertion/deletions (Fig. 4.3b) [30,31]. This means that Y165C and G382D can be frequently detected as compound heterozygous in association with other mutations, and 20% of the cases segregate the syndrome with two alternative bi-allelic mutations. Although most of the variants are rare, the recurrence of few specific mutations has been observed in different populations, such as 1395delGGA in Italian, 1186–1187insGG in Portugese, P391L in Dutch, or E466X in Gujarati families. The large spectrum of genomic variants identified in different geographical areas makes it necessary to adopt a mutational analysis screening covering the entire coding sequence as well as the flanking sequence of the splice site junctions. This is generally achieved by using dHPLC and direct sequencing. Overall, 70–80% of the patients carrying bi-allelic germline mutations in *MUTYH* show an attenuated polyposis, but no clear genotype–phenotype correlation has been evidenced. Differently from polyposis associated with *APC* mutations, in this case the phenotype results from the interaction of the products from two mutant alleles. Unfortunately, at present, functional studies are only available for few missense variants (including Y165C, R227W, V232F, G382D, and A459D), and crystal structures have been limited mainly to some domains of *MUTYH*.

Genetic testing of the *MUTYH* gene often shows the presence of constitutive variants of uncertain pathogenic significance or of mono-allelic mutations; in the latter case it is impossible to determine whether the mutation is responsible for the development of phenotypic manifestations, or whether another *MUTYH* mutation, not detectable with routine techniques, is present. Such results leave open questions about diagnostic and counseling management of the patients in whom they are detected.

The *MUTYH* gene produces three major classes of mRNA isoforms, each of which is also alternatively spliced, suggesting up to ten possible transcripts. These isoforms can differentially target the nucleus or the mitochondria, where an efficient BER is required for the high amount of 8-oxo-G lesions originating from exposure to ROS (Fig. 4.4a) [24]. Although these transcript variants are evident, the relative distribution of the isoforms, the immunological detection, and the

	Isoform	Amino acids	AUG 1, 2 or 3	Probable cellular location	Comments
Type 1	MutYα1	546 aa	1	mitochondira	33bp insert
	MutYα2	536 aa	1	mitochondira	3bp CAG insert
	MutYα3a	535 aa	1	mitochondira	same as Slupska et al. [17]
Type 2	MutYα4	429 aa	3	nucleus	similar to MutY*g*4 isoform
	MutYβ1	532 aa	2	nucleus	same 33-bp insert as MutY*a*1
	MutYβ3	521 aa	2	nucleus	similar to MutY*g*3 isoform
	MutYβ5	521 aa	2	nucleus	
	MutYγ2	522 aa	2	nucleus	
	MutYγ3	521 aa	2	nucleus	similar to MutY*b*3 isoform
	MutY*g*γ4	429 aa	3	nucleus	similar to MutY*a*4 isoform

a Reference Sample.
AUG 1, 2 or 3, from which translation starts; NLS, residues 505–509; MLS, residues 1–14.
cDNA not drawn to scale.

a

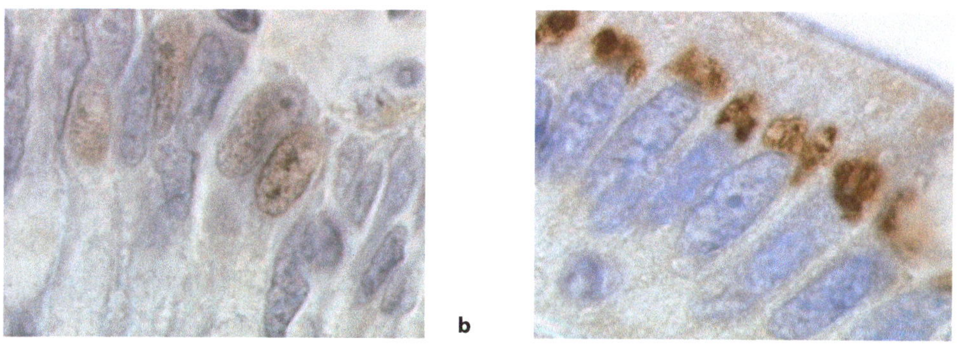

b **c**

Fig. 4.4a-c The *MUTYH* isoforms and protein expression in MAP adenomas. **a** The characteristics of the human *MUTYH* isoforms. AUG 1, 2, 3 transcription starts. NLS and MLS, nuclear and mitochondrial targeting signals. **b** Nuclear and cytoplasmic immunoreactivity of MUTYH protein in colonocytes of normal mucosa (magnification 40×). **c** Strong cytoplasmic immunoreactivity and absence of nuclear expression of MUTYH protein in normal mucosa of a patient with a biallelic *MUTYH* mutation (magnification 40×). Figure 4.4a reproduced from [24], with permission from Birkhäuser Verlag AG, Basel

function of each transcript are poorly known. However, *in vitro* studies have shown that the subcellular expression of different *MUTYH* transcripts can physiologically vary in rat neuronal cells according to the brain development, being mainly nuclear when the neurons are actively proliferating, and primarily cytoplasmatic in post-mitotic cells [32]. In line with other well-known genetic syndromes, an altered ratio of the different transcripts could affect in some way the function of the normal protein product, even in the absence of "canonical" mutations, and, in the long term, affect the phenotype of the disease. Although the characterization of *MUTYH* expression is still ongoing, in polyposis patients carrying different bi-allelic germline mutations, there is a disappearance of normal staining from the nucleus and segregation of the immunoreactivity in the cytoplasm in both neoplastic tissue and the surrounding normal mucosa (Fig. 4.4b,c) [33]. At present, this analysis is helpful for investigating genotype–phenotype correlation, but in the future, if validated on a considerable number of different types of mutations, could be adopted as diagnostic tool to screen for patient carriers of bi-allelic *MUTYH* germline mutations.

It has been proposed that the accumulation of G→T transversions in *APC* cause *MUTYH*-driven carcinogenesis to follow the adenoma–carcinoma sequence, but specific features of MAP have been evidenced [34,35]. As reported for other lesions subjected to a high oxidative damage, such as lung or ovary cancers, the higher frequency of G→T transversions has been found on codon 12 of *KRAS* (about 40% of the tested lesions). Moreover, no mutations have been reported in *p53* or *SMAD4*, which are frequently altered in the adenoma–carcinoma sequence, and aneuploidies, such as loss of 1p, 17, 19, and 22 or duplication of 7 and 13, are more common in MAP than in FAP (80% versus 60%). On the other hand, features associated with other pathways of colorectal tumorigenesis can be excluded, since neither microsatellite instability nor *BRAF* mutations were detected. Overall, the full genetic pathway of MAP tumorigenesis has not yet been elucidated, and knowledge of the effects of BER inactivation on neoplastic transformation is still limited.

4.4 Conclusion

APC and *MUTYH* are associated with most of the adenomatous polyposis cases identified so far, but with an important difference since *APC* germline alterations lead to a more severe phenotype, while *MUTYH* constitutional mutations lead to an attenuated expression of the disease. The structure and function of these two genes indicate that there are two alternative ways of polyposis tumorigenesis and explain much of the heterogeneity of this syndrome.

APC and *MUTYH* mutation-negative FAP is a numerical limited subset of cases (probably no more than 10%). It is conceivable that in the future an accurate mutation analysis, also including MLPA and allelic mRNA expression, could allow identification of these non-conventional alterations in linkage with the disease. Recently, germline mutation in *AXIN2,* a Wnt signaling regulator, has been found in 4% of selected Finnish FAP patients, proved to be negative for *APC* and *MUTYH* mutations [10]. To date, no other candidate has been identified.

In the future, adenomatous polyposis investigation, both genetically and clinically, will be more focused on germline mutation-negative AFAP. This concerns mainly *MUTYH.* Assessment of the genetic etiology of these patients will be improved by *in vitro* functional analysis and by investigation of non-conventional mutation of this gene. On the other hand, the characterization and identification of low-penetrant alleles will allow identification of other inherited attenuated polyposis from familial and sporadic cases, and lead to an understanding of the complex modulation of a variable phenotype.

References

1. Chung DC (2000) The genetic basis of colorectal cancer: insights into critical pathways of tumorigenesis. Gastroenterology 119:854–865.
2. Knudsen AL, Bisgaard ML, Bulow S (2003) Attenuated familial adenomatous polyposis (AFAP). A review of the literature. Fam Cancer 2:43–55.
3. Kinzler KW, Nilbert MC, Su LK, et al (1991) Identification of FAP locus genes from chromosome 5q21. Science 253:661–665.
4. Groden J, Thliveris A, Samowitz W, et al (1991) Identification and characterization of the familial adenomatous polyposis coli gene. Cell 66:589–600.
5. Sulekova Z, Reina-Sanchez J, Ballhausen WG (1995) Multiple APC messenger RNA isoforms encoding exon 15 short open reading frames are expressed in the context of a novel exon 10A-derived sequence. Int J Cancer 63:435–441.
6. Fodde R, Smits R, Clevers H (2001) APC, signal transduction and genetic instability in colorectal cancer. Nat Rev Cancer 1:55–67.
7. Laurent-Puig P, Béroud C, Soussi T (1998) *APC* gene: database of germline and somatic mutations in human tumors and cell lines. Nucleic Acids Res 26:269–270.
8. Powell SM, Petersen GM, Krush AJ, et al (1993) Molecular diagnosis of familial adenomatous polyposis. N Engl J Med 329:1982–1987.
9. Sieber OM, Lamlum H, Crabtree MD, et al (2002) Whole-gene *APC* deletions cause classical familial adenomatous polyposis, but not attenuated polyposis or "multiple" colorectal adenomas. Proc Natl Acad Sci U S A 99:2954–2958.
10. Montera M, Piaggio F, Marchese C, et al (2001) A silent mutation in exon 14 of the *APC* gene is associated with exon skipping in a FAP family. J Med Genet 38:863–867.
11. Renkonen ET, Nieminen P, Abdel-Rahman WM, et al (2005) Adenomatous polyposis families that screen *APC* mutation-negative by conventional methods are genetically heterogeneous J Clin Oncol 23:5651–5659.
12. Zhou XL, Eriksson U, Werelius B, et al (2004) Definition of candidate low risk *APC* alleles in a Swedish population. Int J Cancer 110:550–557.
13. Fearnhead NS, Britton MP, Bodmer WF (2001) The ABC of *APC*. Hum Mol Genet 10:721–733.
14. Spirio LN, Samowitz W, Robertson J, et al (1998) Alleles of *APC* modulate the frequency and classes of mutations that lead to colon polyps. Nat Genet 20:385–388.
15. van der Luijt RB, Meera Khan P, Vasen HF, et al (1996) Germline mutations in the 3‘ part of *APC* exon 15 do not result in truncated proteins and are associated with attenuated adenomatous polyposis coli. Hum Genet 98:727–734.
16. Varesco L, Gismondi V, Presciuttini S, et al (1994) Mutation in a splice-donor site of the *APC* gene in a family with polyposis and late age of colonic cancer death. Hum Genet 93:281–286.
17. Aretz S, Uhlhaas S, Sun Y, et al (2004) Familial adenoma-

tous polyposis: aberrant splicing due to missense or silent mutations in the *APC* gene. Hum Mutat 24:370–380.
18. Venesio T, Balsamo A, Sfiligoi C, et al (2007) Constitutional high expression of an *APC* mRNA isoform in a subset of attenuated familial adenomatous polyposis patients. J Mol Med 85:305–312.
19. Aretz S, Stienen D, Friedrichs N, et al (2007) Somatic *APC* mosaicism: a frequent cause of familial adenomatous polyposis (FAP). Hum Mutat 28:985–992.
20. Hes FJ, Nielsen M, Bik EC, et al (2008) Somatic APC mosaicism: an underestimated cause of polyposis coli. Gut 57:71–76.
21. Albuquerque C, Breukel C, van der Luijt R, et al (2002) The "just-right" signaling model: *APC* somatic mutations are selected based on a specific level of activation of the beta-catenin signaling cascade. Hum Mol Genet 11:1549–1560.
22. Sieber OM, Segditsas S, Knudsen AL, et al (2006) Disease severity and genetic pathways in attenuated familial adenomatous polyposis vary greatly but depend on the site of the germline mutation. Gut 55:1440–1448.
23. Al-Tassan N, Chmiel NH, Maynard J, et al (2002) Inherited variants of *MYH* associated with somatic G:C→T:A mutations in colorectal tumors. Nat Genet 30:227–232.
24. Parker AR, Eshleman JR (2003) Human MutY: gene structure, protein functions and interactions, and role in carcinogenesis. Cell Mol Life Sci 60:2064–2083.
25. Gu Y, Parker A, Wilson TM, et al (2002) Human MutY homolog, a DNA glycosylase involved in base excision repair, physically and functionally interacts with mismatch repair proteins human MutS homolog 2/human MutS homolog 6. J Biol Chem 277:11135–11142.
26. Hashimoto K, Tominaga Y, Nakabeppu Y, Moriya M (2004) Futile short-patch DNA base excision repair of adenine:8-oxoguanine mispair. Nucleic Acids Res 32:5928–5934.
27. Sieber OM, Lipton L, Crabtree M, et al (2003) Multiple colorectal adenomas, classic adenomatous polyposis, and germ-line mutations in *MYH*. N Engl J Med 348:791–799.
28. Sampson JR, Dolwani S, Jones S, et al (2003) Autosomal recessive colorectal adenomatous polyposis due to inherited mutations of *MUTYH*. Lancet 362:39–41.
29. Cheadle JP, Sampson JR (2007) MUTYH-associated polyposis – from defect in base excision repair to clinical genetic testing. DNA Repair (Amst) 6:274–279.
30. Gismondi V, Meta M, Bonelli L, et al (2004) Prevalence of the Y165C, G382D and 1395delGGA germline mutations of the *MYH* gene in Italian patients with adenomatous polyposis coli and colorectal adenomas. Int J Cancer 109:680–684.
31. Venesio T, Molatore S, Cattaneo F, et al (2004) High frequency of *MYH* gene mutations in a subset of patients with familial adenomatous polyposis. Gastroenterology 126:1681–1685.
32. Lee HM, Hu Z, Ma H, et al (2004) Developmental changes in expression and subcellular localization of the DNA repair glycosylase, *MYH*, in the rat brain. J Neurochem 88:394–400.
33. Di Gregorio C, Frattini M, Maffei S, et al (2006) Immunohistochemical expression of MYH protein can be used to identify patients with *MYH*-associated polyposis. Gastroenterology 131:439–444.
34. Lipton L, Halford SE, Johnson V, et al (2003) Carcinogenesis in *MYH*-associated polyposis follows a distinct genetic pathway. Cancer Res 63:7595–7599.
35. Cardoso J, Molenaar L, de Menezes RX, et al (2006) Chromosomal instability in *MYH*- and *APC*-mutant adenomatous polyps. Cancer Res 66:2514–2519.

Not Only FAP

Other Rare Polyposis Syndromes

5

Gabriele Riegler, Annalisa de Leone and Ilaria Esposito

Abstract The rare gastrointestinal polyposis syndromes are a group of disorders characterized by the presence of multiple polyps in the gastrointestinal tract, associated in some cases with extraintestinal manifestations. These syndromes can be inherited or non-inherited. The first group includes, on the basis of polyp histopathology, adenomatous polyposis (familial adenomatous polyposis (FAP) variants like Gardner's syndrome and Turcot's syndrome) and hamartomatous polyposis (Juvenile Polyposis syndrome, Peutz–Jeghers syndrome, Cowden's disease, and Bannayan–Ryley–Ruvalcaba syndrome), while the second group includes Cronkhite–Canada syndrome and Hyperplastic polyposis.

The molecular genetics have not been well defined for all the syndromes. However, these conditions present variable risk of gastrointestinal and extraintestinal invasive malignancy. Therefore management and surveillance programs are different for each syndrome.

Keywords Adenomatous polyposis • APC • Bannayan–Ryley–Ruvalcaba syndrome • Cowden's disease • Cronkhite–Canada syndrome • Desmoid tumor • Gardner's syndrome • Hamartomatous polyposis • Hyperplastic polyposis • Juvenile polyposis syndrome • *PTEN* • Peutz–Jeghers syndrome • Serrated adenoma • *SMAD-4* • *LKB1/STK11* • Trichilemmoma • Turcot's syndrome

5.1 Introduction

The rare gastrointestinal polyposis syndromes, inherited and non-inherited, are a group of disorders characterized by the presence of multiple polyps in the gastrointestinal tract, with or without extraintestinal manifestations, and can have devastating clinical effects, particularly if diagnosis and treatment are delayed.

G. Riegler (✉)
Unit of Gastroenterology, "Magrassi-Lanzara" Department of Clinical and Experimental Medicine, 2nd University of Naples, Naples, Italy

5.2 Inherited Polyposis Syndromes

These syndromes are classified as adenomatous or hamartomatous based on the main histopathologic features (Table 5.1).

5.2.1 Adenomatous Polyposis Syndromes

The vast majority of publications regarding gastrointestinal polyps are dedicated to adenomatous polyps. Adenomatous polyposis syndromes are characterized by benign epithelial neoplasms arising from or forming glandular-type elements [1]. The increased risk of malignant degeneration associated with these

G.G. Delaini, T. Skřička, G. Colucci (eds.), *Intestinal Polyps and Polyposis*,

Table 5.1 Categories of rare inherited polyposis syndromes and their genetic mutations

Syndromes	Mutations
Adenomatous polyposis	
Gardner's syndrome	*APC*
Turcot's syndrome	*APC, MMR*
Hamartomatous polyposis	
Juvenile Polyposis syndrome	*PTEN, SMAD-4*
Peutz–Jeghers syndrome	*LKB1/STK11*
Cowden's disease	*PTEN*
Bannayan–Ryley–Ruvalcaba syndrome	*PTEN*

syndromes is thought to be secondary to progression through the "adenoma-carcinoma" sequence initially outlined by Vogelstein and Kinzler [2].

The better-known adenomatous polyposis are the familial adenomatous polyposes (classical and attenuated FAP) and its rare variants [Gardner's syndrome (GS) and Turcot's syndrome (TS)].

5.2.1.1 Gardner's syndrome

GS is characterized by autosomally dominant inherited adenomatous polyposis of the colon, and colon carcinoma, and is associated with extracolonic lesions. Extracolonic polyps are observed in 5–7% of cases and involve mainly the stomach and duodenum. GS is caused by the same mutation of the *APC* gene as FAP, with a 100% penetrance. The clinical and radiological features of GS are identical to those of FAP. The extracolonic lesions observed in GS include: multiple osteomas, dental abnormalities, multiple epidermoid cysts, and soft fibromas of the skin, desmoid tumors, and mesenteric fibromatosis. These extracolonic lesions present in various ways among different affected families, but the genetic basis for this variable expression is still unknown. The osteomas normally cause no clinical symptoms and have no malignant potential. They may precede the appearance of polyposis of the gastrointestinal tract.

Main clinical features of Gardner's syndrome:

- Adenomatous polyposis of the gastrointestinal tract
- Multiple osteomas
- Dental abnormalities
- Multiple epidermoid cysts
- Skin fibromas
- Desmoid tumors
- Mesenteric fibromatosis

The osteomas are dense cortical lesions, varying in number, seen most commonly in the angle of the mandible, the sinuses, and the outer part of the skull. The size of the osteomas may vary from pinpoint to several centimeters in diameter. Another feature may be diffuse cortical thickening of the long bones [3].

Desmoids are benign, non-inflammatory fibroblastic tumors with a tendency to local invasion and recurrence but without metastasis; in GS they are mostly seen postoperatively, when they develop in the surgical scar. Frequent locations are the abdominal wall, the root of the mesentery, and the retroperitoneum. Desmoid tumors typically arise from fascial and musculoaponeurotic structures, they have a band-like or tendon-like consistency, and are mostly well circumscribed but not encapsulated. Histologically, the lesions consist of well-differentiated and richly vascularized collagen and fibrous tissue.

In many patients the tumors will not cause clinical symptoms, but it is not uncommon for desmoids located in the mesentery to provoke intestinal obstruction, due to extension into the bowel wall.

On ultrasound, desmoids present as homogeneous anechoic or hypoechoic masses. On computed tomography (CT) most abdominal desmoids present as well-circumscribed masses, but in a minority of patients they are ill-defined with fuzzy borders. On precontrast scan, most are relatively homogeneously or focally hyperattenuating, compared to soft tissue. This hyperattenuation may reflect in part the high physical density of collagen, since no calcium has been demonstrated in these tumors. Most desmoids will clearly enhance following intravenous contrast medium, which can be explained by their rich capillary vascularization.

On magnetic resonance imaging (MRI), desmoids will typically show a low-intensity signal on both T_1-weighted and T_2-weighted scans. The multiplanar capabilities of MRI are helpful in defining the exact extent and origin of the lesion.

When any suggestive signs or symptoms are present, patients with GS should usually undergo radiologic follow-up, which can show even small mesenteric desmoid tumors.

In a minority of families a mutation cannot be identified, and so annual flexible sigmoidoscopy should be offered to at-risk family members from the age of 13–15 years until 30 years, and at 3–5-year intervals until the age of 60 years.

Table 5.2 Main clinical features of Turcot's syndrome

True Turcot's syndrome (autosomal recessive)	FAP-associated type (autosomal dominant)
• Adenomatous polyposis of the gastrointestinal tract (<100) • Brain tumor (gliobastoma or astrocytoma)	• Adenomatous polyposis of the gastrointestinal tract (<100) • Brain tumor (medulloblastoma) • Extraintestinal manifestations of Gardner's syndrome

Surveillance might also be offered as a temporary measure for people who have documented *APC* gene mutations but wish to defer prophylactic surgery for personal reasons. Such individuals should be offered six-monthly flexible sigmoidoscopy and annual colonoscopy, but surgery before 25 years of age should be strongly recommended. After colectomy and ileorectal anastomosis, the rectum must be kept under review at least annually for life because the risk of cancer in the retained rectum is 12–29%. After restorative proctocolectomy, the anorectal cuff should also be kept under annual review for life.

Periampullary and duodenal carcinoma, as well as papillary carcinoma of the thyroid, occur more frequently in patients with GS. Three-yearly upper-gastrointestinal endoscopy is recommended from the age of 30 years, with the aim of detecting early curable cancers [4].

5.2.1.2 Turcot's Syndrome

TS is a rare hereditary disorder, in which central nervous system tumors (usually gliomas and medulloblastomas) are associated with colonic adenomatous polyposis. Café-au-lait spots, cutaneous port wine stain, as well as focal nodular hyperplasia, have been reported as associated anomalies.

Recent molecular evidence suggests that TS could be divided into the following two entities based on the distinct genetic backgrounds:

1. true Turcot's syndrome (autosomal recessive): intestinal polyps are fewer in number (<100), large in size, and apt to transform to the malignant tumor. Brain tumor is mainly diagnosed as glioblastoma or astrocytoma, and mismatch repair genes might be involved;
2. FAP-associated type (autosomal dominant): predisposing to medulloblastoma.

Extraintestinal manifestations of GS have been reported in patients belonging to the FAP-associated type (Table 5.2) [5].

The mode of inheritance of TS is controversial; some authors support autosomal recessive inheritance, and others an autosomal dominant pattern. TS can be associated with two different types of germline genetic defects: mutation of the *APC* gene that is usually found in FAP, or mutation of a mismatch-repair gene that is usually found in hereditary non-polyposis colorectal cancer [6–12].

5.2.2 Hamartomatous Polyposis Syndromes

Conversely, hamartomatous polyposis syndromes are characterized by an overgrowth of cells or tissues native to the area in which they normally occur [1]. The classic hamartoma syndromes and related conditions show varying degrees of phenotypic and genetic overlap.

The syndromes associated with hamartomatous polyposis include the most-frequent Juvenile Polyposis syndrome (JPS) and Peutz–Jeghers syndrome (PJS), and the rare Cowden's disease (CD) and Bannayan–Riley–Ruvalcaba syndrome (BRRS) that are clinically overlapping autosomal dominant genodermatoses characterized by different hamartomatous lesions involving tissues of ectodermal, mesodermal, and endodermal origin.

5.2.2.1 Juvenile Polyposis Syndrome

JPS is about ten-fold less common than FAP, with a frequency of only 1 per 100,000 newborns.

JPS was first described by McColl in 1964 [13]. It is the most common of the hamartomatous syndromes, and is inherited in an autosomal dominant manner (variable penetrance), with approximately 20–50% of cases having a family history of juvenile polyposis [14]. The average age of onset is approximately 18 years.

Jass and colleagues [15] classified juvenile polyps into two main categories: (1) isolated juvenile polyps of childhood, and (2) juvenile polyposis of the colon or

Table 5.3 Main clinical features of Juvenile Polyposis syndrome

Isolated juvenile polyps of childhood	Juvenile Polyposis of entire gastrointestinal tract
• Hamartomatous polyposis (≤ 5 polyps)	• Hamartomatous polyposis (50–200 polyps) • Pancreatic carcinoma

entire gastrointestinal tract. The latter category has a subgroup, juvenile polyposis of infancy (Table 5.3).

The isolated juvenile polyp of childhood is considered to be a hamartoma with a low risk of malignancy [15,16].

The multiple polyps of juvenile polyposis are also considered hamartomas but have some histologic features that differ from those of isolated juvenile polyps. The most important of these features is the presence of foci of dysplasia.

Isolated juvenile polyps refer to one or more polyps, not more than five, evenly distributed throughout the rectum and colon.

The polyps may be either sessile or pedunculated, they vary in size from 2 mm to greater than 5 cm in diameter, and are often superficially ulcerated [17].

In infancy, patients present with gastrointestinal bleeding, either acute or chronic, intussusception, rectal prolapse, or a protein-losing enteropathy [18].

Congenital anomalies such as hydrocephalus and pulmonary arteriovenous malformations have a higher prevalence in the non-familial cases [19–23].

In adulthood, these patients will more commonly present with gastrointestinal blood loss, either acute or chronic. Most of these patients will be shown to have between 50 and 200 polyps, most commonly in the rectosigmoid region.

Histopatologically, there is a gross infiltration of the lamina propria by chronic inflammatory cells (lymphocytes and plasma cells), leading to attenuation of the underlying smooth muscle layer. Cystic dilation of glandular-type structures lined, at least initially, by a normal-appearing columnar epithelium is pathognomonic.

JPS was initially thought to be associated with mutations in the *PTEN* (phosphatase with tensin homology) gene (10q22–23) [24–25]. More recently, it has been shown that germline mutations in the *SMAD4* gene (18q21) account for approximately 50% of the reported familial cases of the syndrome [26]. In particular, a well-described four-base pair (bp) deletion, can be used to confirm a clinical diagnosis of JPS. This gene encodes a cytoplasmic mediator involved in the transforming growth factor beta (TGF-β) signal transduction pathway. Mutations in the *SMAD4* gene presumably lead to a loss of heteromeric complex formation and resultant growth inhibition and neoplastic progression.

The cumulative cancer risk in JPS has been estimated at up to 30% or 50% in the colorectum, and 10% in the upper gastrointestinal tract [27,28]. Patients with JPS are at increased risk for developing malignant gastrointestinal tumors. Most cancers are diagnosed in the second or third decade [15,29,30].

Carcinoma of the stomach, duodenum, and pancreas has also been reported in patients with juvenile polyposis of the entire gastrointestinal tract [31–33].

The management and surveillance of these individuals is predicated on their increased risk of upper- and lower-gastrointestinal malignancies.

Large-bowel surveillance for at-risk individuals is recommended at intervals of one to two years from the age of 15–18 years, or earlier if the patient presents with symptoms, and upper-gastrointestinal surveillance from the age of 25 years. Screening intervals could be extended at the age of 35 years in at-risk individuals. However, documented gene carriers or affected individuals should be kept under surveillance until the age of 70 years, and prophylactic surgery discussed.

Development of invasive colorectal adenocarcinoma mandates total abdominal colectomy with ileorectal anastomosis or restorative proctocolectomy, depending on the extent of rectal polyposis [4].

Like hereditary non-polyposis colorectal cancer (HNPCC), JPS presents the phenomenon of anticipation [20]; this relates to the increasing severity of disease with each successive generation, either by virtue of an earlier age of onset or a more-severe phenotype. Although the age of diagnosis is earlier with successive generations, this can be explained, at least in part, by a heightened awareness of this condition among family members.

5.2.2.2 Peutz–Jeghers Syndrome

PJS is the second most common hamartomatous syndrome, occurring as an autosomal dominant condition with variable penetrance. The original description of this rare hereditary syndrome is credited first to Peutz in 1921 [34], and then, in 1949, to Jeghers' description of cutaneous melanin deposition associated with gastrointestinal polyposis, neoplasms outside the alimentary tract, and the risk of invasive carcinoma [35].

The most common location of PJS polyps is the upper gastrointestinal tract. The jejunum and ileum are most frequently involved, followed by the duodenum, colon, and stomach [3]. It is more common for these polyps to occur in clusters rather than to spread evenly throughout the bowel. Individual polyps vary in size and may be either sessile or pedunculated. Larger lesions characteristically have a lobulated surface [3].

A polyp is a hamartoma, with a smooth-muscle core arising from the muscularis mucosae and extending into the polyp, and the mucosa covering the polyp is similar to the mucosa normally found in that portion of the gut [2,3].

The initial presentation of a patient with PJS is most commonly abdominal pain secondary to obstruction or impending obstruction with polyp intussusception or gastrointestinal blood loss [36].

Mucocutaneous pigmentation is one of the most characteristic features and usually appears after the first or second year of life. Brown or bluish-black macules occur most commonly on the lips and buccal mucosa, and less commonly on the eyelids and dorsal surfaces of the fingers and soles of the feet.

Main clinical features of Peutz–Jeghers syndrome:

- Hamartomatous polyposis of gastrointestinal tract
- Mucocutaneous pigmentation
- Pancreatic carcinoma
- Breast carcinoma
- Ovarian carcinoma
- Testicular carcinoma
- Uterine cervix carcinoma

Genetic alterations in the *LKB1/STK11* (19p13) gene are responsible for approximately 50% of the PJS cases [37]. This gene encodes for a multifunctional serine-threonine kinase, important in second messenger signal transduction. The function of this protein product is likely to be important in growth inhibition as has been shown with SMAD4 but, unlike SMAD4, the genetic alterations are confined to the epithelial component [38].

There is an increased risk of gastrointestinal cancer (stomach, duodenum, and colon), reported to range from 2% to 20% [3–10], and extraintestinal malignancy (pancreatic carcinoma, breast carcinoma, ovarian carcinoma, testicular carcinoma in prepubescent males, and adenoma malignum, a well-differentiated multicystic adenocarcinoma of the uterine cervix) with prevalence from 10% to 30%, associated with PJS [39,40].

Pancreatic cancer has a tendency to develop at an unusually early age, and the risk is estimated to be 100 times that expected in the general population [41,42]. Breast carcinoma is found with increased frequency, and is usually bilateral and ductal in origin [43]. Screening mammography for women with PJS should be encouraged, to detect breast carcinoma at an early stage.

Appropriate surveillance of the proband and first-degree relatives has not been extensively investigated or validated because the syndrome is rare and so experience is limited.

Large-bowel surveillance is recommended at three-year intervals from the age of 18 years, and upper-gastrointestinal surveillance is recommended at three-year intervals from the age of 25 years [4].

Particular attention to the pancreas and reproductive tract is necessary whenever abdominal or pelvic imaging is performed in these patients [3].

5.2.2.3 Cowden's Disease

CD, also named multiple hamartoma syndrome, is a rare autosomal dominant condition with variable expression that results from a mutation in the tumor suppressor gene *PTEN* on chromosome arm 10q [44]. The syndrome was first described in 1963 by Lloyd and Dennis who named it according to the patient's surname [45]. In a population-based Dutch clinical epidemiological study, the incidence of CD before gene identification was estimated to be one in a million [46,47]. However, after gene identification, this figure was revised to 1 in 200,000, which is almost certainly an underestimate [48]. It is slightly more common in females than in males [49].

The syndrome causes hamartomatous neoplasms of the skin and mucosa of the gastrointestinal tract, bones, eyes, and genitourinary tract. Skin lesions, most frequent in the head and neck area, are facial papules, oral mucosal papillomatoses, acral keratoses, and multiple sclerotic fibromas. Eighty per cent of patients present with dermatologic manifestations, the most common being a benign tumor of the hair shaft: a trichilemmoma (benign tumor of the hair follicle infundibulum).

Other hamartomas include breast fibroadenomas in 70% of affected females, which usually become bilateral, and ductal carcinoma in 30–50% of cases; the age of onset is 38–40 years, which is earlier than in the general population [50]. The second most common area of involvement is the central nervous system. CD in conjunction with cerebellar gangliocytomatosis is referred to as Lhermitte–Duclos syndrome.

Thyroid adenomas and multinodular goiter occur in 65% of all patients, while thyroid follicular adenocarcinoma has been reported in 3–12% of patients. All these cases have been in women.

Genitourinary lesions are frequent and include uterine and cervical carcinomas and transitional cell carcinoma of the renal pelvis and urinary bladder [47].

Main clinical features of Cowden's disease:

- Polyposis of the gastrointestinal tract (hamartoma, adenoma, ganglioneurofibroma)
- Mucocutaneous lesions (trichilemmomas, mucosal papillomatosis, acral keratoses)
- Breast diseases (fibroadenomas, carcinoma)
- Thyroid diseases (multinodular goiter, follicular adenocarcinoma)
- Macrocephaly
- Lhermitte–Duclos disease
- Endometrial carcinoma
- Genitourinary tract malformations or tumors

Gastrointestinal polyps occur in 35–40% of affected individuals; the polyps are usually sessile, smaller, and have a less exophytic and arborizing proliferation of the muscularis mucosae. They are most commonly located in the rectosigmoid colon, followed, in decreasing frequency, by the stomach (Fig. 5.1), duodenum, small bowel (Fig. 5.2), and esophagus, and show a variety of histopathologic appearances like hamartomas, adenomas, and ganglioneurofibromas, particularly in the colon. Often the esophagus presents small protrusions diagnosed as glycogenic acanthosis, and in the small bowel lymphangiectasia and lymphoid polyps can be found.

There has been no reported increased risk of invasive gastrointestinal malignancy to date, but gastrointestinal polyps should be assessed by endoscopic surveillance [51–53].

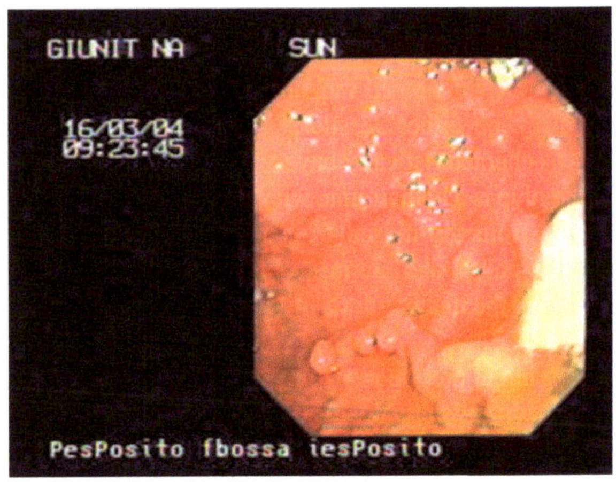

Fig. 5.1 Gastric polyps in a patient with Cowden's disease

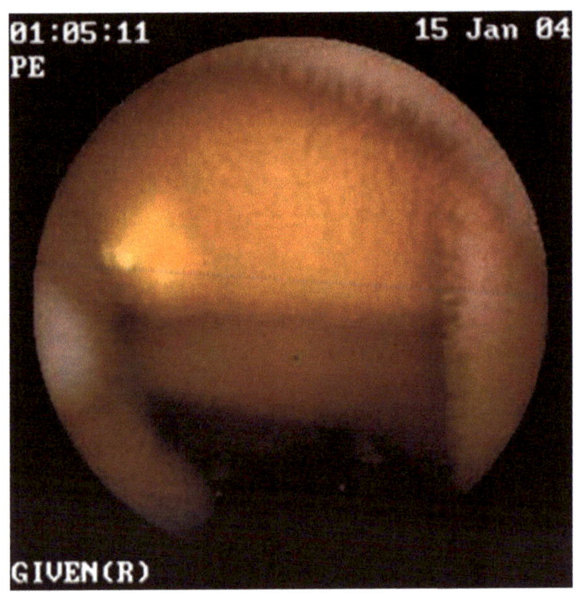

Fig. 5.2 Jejenum polyp in Cowden's disease diagnosed by video capsule endoscopy

Using the International Cowden Consortium operational criteria for the diagnosis of CD, germline region *PTEN/MMAC1/TEP1* mutations are identified in CD families with a frequency of 81% [54]. PTEN is a tumor suppressor; it is a dual-specificity phosphatase that dephosphorylates both protein and lipid substrates. PTEN is a negative regulator of the PI3K/AKT signaling pathway that is required for cell survival and proliferation. PTEN expression leads to downregulation of AKT activation and increased apoptosis [55].

Screening and surveillance for breast malignancies should include monthly self-examination of the breasts; mammography should be implemented at the age of 25 years as previously outlined for PJS syndrome. A thyroid ultrasound may be used in parallel every 1 to 2 years.

5.2.2.4 Bannayan–Ryley–Ruvalcaba Syndrome

BRRS is an autosomal dominant condition characterized by the classic triad of macrocephaly, lipid storage myopathy, and hyperpigmentation of the skin of the genitalia [56,57]. CD and BRRS share several mucocutaneous features, including tricholemmomas, oral papillomas, and acral keratoses, while lipomas and vascular malformations typify BRRS as well as acrochordons, acanthosis nigricans, and café-au-lait macules [54,58]. Other BRRS features are mental retardation, delayed psychomotor development, and Hashimoto's thyroiditis; pseudopapilledema and amblyopia have also been described in individuals with BRRS.

Main clinical features of Bannayan–Ryley–Ruvalcaba syndrome:

- Hamartomatous polyposis of the gastrointestinal tract
- Mucocutaneous lesions (hyperpigmentation of the genitalia, trichilemmomas, acral keratoses, lipomas)
- Macrocephalia
- Lipid storage myopathy
- Vascular malformations
- Mental retardation
- Hashimoto thyroiditis
- Pseudopapilledema and amblyopia

BRRS presents with gastrointestinal polyps in 45% of patients; these lesions are hamartomas limited to the distal part of the ileum and colon.

This syndrome has a *PTEN* mutation frequency of 50–60%. Furthermore there are CD/BRRS overlap families; these are associated with a single germline *PTEN* mutation, suggesting that they are allelic. Although patients with BRRS were not originally considered to have increased risk for cancer, the identification of germline *PTEN* mutations in over 50% of these patients suggests a risk for cancers related to Cowden syndrome [58–60].

All the gastrointestinal polyposis syndromes described include small-bowel polyps as part of their clinical presentation. Video capsule endoscopy (VCE) is a simple, safe, and non-invasive procedure for the detection of small-bowel polyps. VCE has been found to have higher accuracy compared to barium studies, and similar accuracy compared to MRI for the detection of large polyps (>15 mm); however, the detection rate for small (5–15 mm diameter) and diminutive (<5 mm) polyps is much higher with VCE [61,62].

5.3 Non-Inherited Polyposis Syndromes

In this section we describe unusual entities of non-inherited, non-adenomatous polyposis syndromes; little is known about the aetiology or pathogenic mechanisms of these diseases.

5.3.1 Cronkhite–Canada Syndrome

The syndrome, first described in 1955, is not familial and occurs in older adults. Patients of European or Asian descent are most frequently affected, with an average age of onset of 60 years, and an age distribution of 31–86 years [63,64].

The aetiology of this syndrome remains unknown; some investigators consider that mast cell dysfunction may play a role in the pathogenesis of this unusual disease; other proposed theories have included an infectious cause, nutritional deficiency, and altered intestinal mucin production [65].

These patients commonly have abdominal pain, a severe, protein-losing diarrhea, weight loss, anorexia, nausea, vomiting, and hypogeusia. The physical examination usually identifies unique ectodermal abnormalities associated with the disease, such as nail dystrophy, with thinning, splitting, and separation from the nailbeds, and alopecia of the scalp and body hair. Diffuse hyperpigmentation of the skin, manifested by light to dark brown macular lesions, is seen most frequently on the extremities, face, palms, soles, and neck [64,66].

The syndrome is distinguished by the diffuse distribution of polyps throughout the entire gastrointestinal tract, apart from a characteristic sparing of the esophagus. Polyps are usually small and are most commonly sessile rather than pedunculated. In the stomach, small-to moderate-sized polyps carpet the mucosal surface and are usually superimposed on thickened rugal folds [67].

Histopathologically, polyps are most frequently described as being of the hamartomatous or juvenile type, and it is difficult to differentiate between the two on the basis of histologic criteria alone [64,65,68].

Microscopic examination of the intestinal mucosa also reveals edema of the lamina propria, cystic dilation of glands, and inflammatory cell infiltration with mononuclear cells, eosinophils and, in some specimens, mast cells [69].

Complications of the syndrome include potentially fatal gastrointestinal bleeding, intussusception, prolapse, electrolyte abnormalities, dehydration, protein-losing enteropathy, and other nutritional deficiencies due to malabsorption.

The question of whether polyps in Cronkhite–Canada syndrome possess malignant potential is controversial; cancers in the stomach and colon have been reported [70,71].

Treatment includes aggressive nutritional support with electrolyte replacement and parenteral nutrition. Most patients, however, do not respond to nutritional therapy alone, and other treatments are required. Various antibiotics (ampicillin, tetracycline, metronidazole, clindamycin, and trimethoprim-sulfamethoxazole) are used, with remission in up to 40% of cases when combined with other therapies. Other treatments are prednisone and hydrocortisone, with complete response in 33% of cases; partial response is seen in an additional 22%. Surgical treatment for the complications of Cronkhite–Canada syndrome has been advocated, as resection of the specific sections of the gastrointestinal tract that seem to be responsible for particular complications appears to be beneficial in some cases [64].

5.3.2 Hyperplastic Polyposis

Hyperplastic polyposis (HPP) was first described in 1980 [72]. Some authors have also referred to it as metaplastic polyposis [73].

The youngest reported patient with HPP was 11 years of age, but the condition most commonly occurs in adults during the sixth or seventh decade of life [74,75].

This condition, which often exhibits familial clustering, is characterized by multiple and/or large hyperplastic polyps distributed throughout the colon, and smaller numbers of coexisting serrated adenomas (Fig. 5.3), traditional adenomas, and mixed polyps [76].

While these polyps are confined to the large bowel, they may be found scattered throughout the colon or localized to the left or right side. Although most solitary hyperplastic polyps are usually only 3–5 mm in diameter, it is not uncommon for "giant" polyps, more than 3 cm in diameter, to be encountered in patients with HPP. Patients with HPP may be asymptomatic or may present with gastrointestinal bleeding with resulting anemia, diarrhea, abdominal pain, and weight loss. Other patients may exhibit symptoms of intestinal obstruction, which in some cases occurs with polyp-associated intussusception [72,77]. Polyp numbers range from 5 to well over 100, with most patients having between 40 and 100 lesions [78].

This polyposis does not seem to have any consistent extraintestinal manifestations.

The diagnostic criteria for HPP generally include the presence of: (1) at least five histologically diag-

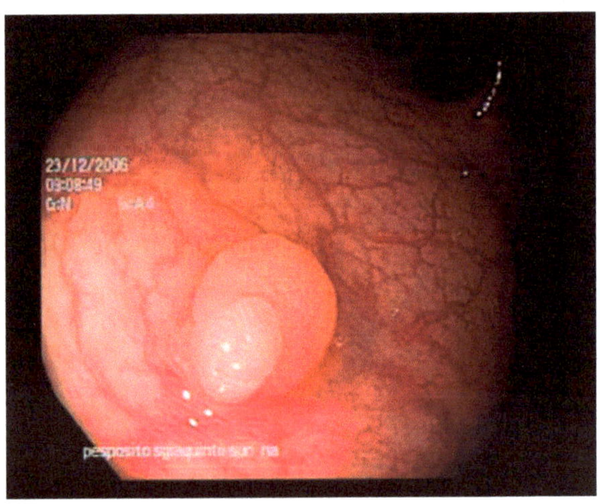

Fig. 5.3 Serrated adenoma of the colon in a patient with hyperplastic polyposis

nosed hyperplastic polyps proximal to the sigmoid colon, of which two are greater than 10 mm in diameter, or (2) any number of hyperplastic polyps proximal to the sigmoid colon in an individual who has a first-degree relative with HPP, or (3) more than 30 hyperplastic polyps of any size that are distributed throughout the colon [79].

Initially, HPP was assumed to have no malignant potential. Subsequently, cases of HPP with synchronous adenocarcinoma have been increasingly reported and, in the late 1990s, HPP was acknowledged as a condition carrying an increased risk for malignant transformation. Individuals with HPP present with synchronous cancers of the colorectum in approximately half of cases [80–83]. However, patients with large, atypical, and dysplastic polyps appear to be at the highest risk for presenting with a synchronous colorectal cancer [76].

It was initially thought that this polyposis was not associated with a family history of HPP or colorectal cancer in first-degree relatives. However, HPP may sometimes occur in several members of the same family, whereas colorectal cancer occurs in 27% of relatives of individuals with HPP [84–86].

Although the underlying genetic cause of HPP is currently unknown, a reasonable explanation would implicate a genetic predisposition (either directly or indirectly) to hypermethylate multiple gene promoters.

The somatic mutation of BRAF (protein involved in cellular processes such as proliferation, differentiation, survival, and apoptosis) and CIMP (CpG island methylator phenotype) are probably the most important changes for subsequent development of cancer [87–89].

Methylation events, occurring in vulnerable tumor-suppressor genes, will synergize with somatic oncogenic activation of *BRAF* and result in the development of serrated premalignant lesions.

Importantly, the discovery of the germline alteration underlying HPP will identify the individuals and their families who are most at risk for the development of serrated neoplasia, and then it will be possible to propose guidelines for management of this polyposis.

References

1. Haggitt RC, Reid BJ (1986) Hereditary gastrointestinal polyposis syndromes. Am J Surg Pathol 10:871–887.
2. Vogelstein B, Kinzler KW (1993) The multistep nature of cancer. Trends Genet 9:138–141.
3. Foulkes WD (1995) A tale of four syndromes: familial adenomatous polyposis, Gardner syndrome, attenuated APC and Turcot syndrome. QJM 88:853–863.
4. Dunlop MG (2002) Guidance on gastrointestinal surveillance for hereditary non-polyposis colorectal cancer, familial adenomatous polyposis, juvenile polyposis, and Peutz–Jeghers syndrome. Gut 51(suppl V):v21–v27.
5. Sunahara M, Nakagawara A (2000) Turcot syndrome. Nippon Rinsho 58:1484–1489.
6. Itoh H, Hirata K, Ohsato K (1993) Turcot's syndrome and familial adenomatous polyposis associated with brain tumor: review of related literature. Int J Colorectal Dis 8:87–94.
7. Jarvis L, Bathurst N, Mohan D, Beckly D (1988) Turcot's syndrome: a review. Dis Colon Rectum 31:907–914.
8. McKusick VA (1962) Genetic factors in intestinal polyposis. JAMA 182:271–277.
9. Baughman FA Jr, List CF, Williams JR, Muldoon JP, Segarra JM, Volkel JS (1969) The glioma-polyposis syndrome. N Engl J Med 281:1345–1346.
10. Rustgi AK (1994) Hereditary gastrointestinal polyposis and nonpolyposis syndromes. N Engl J Med 331:1694–1702.
11. Lewis JH, Ginsberg AL, Toomey KE (1983) Turcot's syndrome: evidence for autosomal dominant inheritance. Cancer 51:524–528.
12. Costa OL, Silva DM, Colnago FA, Vieira MS, Musso C (1987) Turcot syndrome: autosomal dominant or recessive transmission? Dis Colon Rectum 30:391–394.
13. McColl I, Bussey HJR, Veale AMO, Morrison BC (1964) Juvenile polyposis coli. Proc R Soc Med 57:896–897.
14. Desai DC, Neale KF, Talbot IC, Hodgson SV, Phillips RKS (1995) Juvenile polyposis. Br J Surg 82:14–17.
15. Jass JR, Williams CB, Bussoy HA, Morson BC (1988) Juvenile polyposis: a precancerous condition. Histopathology 13:619–630.
16. Jass JR, Sobin LH (1989) Histological typing of intestinal tumors, 2nd ed. (WHO international histological classification of tumors) Springer-Verlag, Berlin.
17. Mestre JR (1986) The changing pattern of juvenile polyps. Am J Gastroenterol 81:312–314.
18. Sachatello CR, Hahn IS, Carrington CB (1974) Juvenile gastrointestinal polyposis in a female infant: report of a case and review of the literature of a recently recognized syndrome. Surgery 75:107–114.
19. Bussey HJ, Veale AM, Morson BC (1978) Genetics of gastrointestinal polyposis. Gastroenterology 74:1325–1330.
20. Smilow PC, Pryor CK, Swinton NW (1966) Juvenile polyposis coli: a report of three patients in three generations of one family. Dis Colon Rectum 9:248–254.
21. Sachatello CR, Pickren JW, Grace JT (1970) Generalized juvenile gastrointestinal polyposis. Gastroenterology 58:699–708.
22. Cox KU, Frates AC, Wong A (1980) Hereditary generalized juvenile polyposis associated with pulmonary artoniovonous hypertension. Gastroenterology 78:1566–1570.
23. Prioto G, Polanco I, Sarria J, Larruri J, Lassaletta L (1990) Association of juvenile and adonomatous polyposis with pulmonary arteriovenous malformation and hypertrophic osteoarthropathy. J Pediatr Gastroenterol Nutr 11:133–137.

24. Jacoby RF, Schlack S, Cole CE, et al (1997) A juvenile polyposis tumor suppressor locus at 10q22 is deleted from nonepithelial cells in the lamina propria. Gastroenterology 112:1398–1403.
25. Olschwang S, Serova-Sinilnikova OM, Lenoir GM, Gilles T (1998) PTEN germ-line mutations in juvenile polyposis coli. Nat Genet 18:12–14.
26. Howe JR, Roth S, Ringold JC, et al (1998) Mutations in the *SMAD4/DPC4* gene in juvenile polyposis. Science 280:1086–1088.
27. Howe JR, Mitros FA, Summers RW (1998) The risk of gastrointestinal carcinoma in familial juvenile polyposis. Ann Surg Oncol 5:751–757.
28. Longo WE, Touloukian RJ, West AB, Ballantyne GH (1990) Malignant potential of juvenile polyposis coli: report of a case and review of the literature. Dis Colon Rectum 135:980–984.
29. Giardiello FM, Hamilton SA, Kern SE, et al (1991) Colorectal neoplasia in juvenile polyposis or juvenile polyps. Arch Dis Child 66:971–975.
30. Stemper TJ, Kent THE, Summers AW (1975) Juvenile polyposis and gastrointestinal carcinoma: a study of a kindred. Ann Intern Med 83:639–646.
31. Yoshida T, Haraguchi A, Tanaka A, et al (1988) A case of generalized juvenile gastrointestinal polyposis associated with gastric carcinoma. Endoscopy 20:33–35.
32. Walpole IA, Cullity G (1989) Juvenile polyposis: a case with early presentation and death attributable to adenocancinoma of the pancreas. Am J Med Genet 32:1–8.
33. Sassatelli A, Bortoni G, Serra U, et al (1993) Generalized juvenile polyposis with mixed pattern and gastric cancer. Gastroenterology 104:910–915.
34. Peutz JLA (1921) Very remarkable case of familial polyposis of mucous membrane of intestinal tract and nasopharynx accompanied by peculiar pigmentations of skin and mucous membrane. Nederl maandschr v geneesk 10:134–146.
35. Jeghers H, McKusick VA, Katz KH (1949) Generalized intestinal polyposis and melanin spots of the oral mucosa, lips and digits. N Engl J Med 241:993–1005.
36. Guillem JG, Smith AJ, Puig-La Calle J, Ruo L (1999) Hamartomatous polyposis. In: Wells SA, Creswell LL (eds) Gastrointestinal polyposis syndromes. Mosby Inc, St Louis, pp 286–299.
37. Hemminki A, Tomlinson I, Markie D, et al (1997) Localization of a susceptibility locus for Peutz–Jeghers syndrome to 19p using comparative genomic hybridization and targeted linkage analysis. Nat Genet 15:87–90.
38. Wang ZJ, Ellis I, Zauber P (1999) Allelic imbalance at the LKB1 (STK11) locus on 19p13.3 in hamartomas, adenomas and carcinomas from patients with Peutz–Jeghers syndrome provides evidence for a hamartoma-(adenoma)-carcinoma sequence. J Pathol 188:613–617.
39. Spigelman AD, Murday V, Phillips RKS (1989) Cancer and the Peutz–Jeghers syndrome. Gut 30:1588–1590.
40. Hizawa K, Iida M, Matsumoto T, et al (1993) Cancer in Peutz–Jeghers syndrome. Cancer 72:2777–2781.
41. Konishi F, Wyse NE, Mutot TV, et al (1987) Peutz–Jeghers polyposis associated with carcinoma of the digestive organs: report of three cases and review of the literature. Dis Colon Rectum 30:790–799.
42. Bowlby US (1986) Pancreatic adenocarcinoma in an adolescent male with Peutz–Jeghers syndrome. Hum Pathol 17:97–99.
43. Lehun PA, Madarnas P, Devroede G (1984) Peutz–Jeghers syndrome: association of duodenal and bilateral breast cancers in the same patient. Dig Disc Sci 29:178–182.
44. Scala S, Bruni P, Lo Muzio L, et al (1998) Novel mutation of the PTEN gene in an Italian Cowden's disease kindred. Int J Oncol 13:665–668.
45. Lloyd KM 2nd, Dennis M (1963) Cowden's disease. A possible new symptom complex with multiple system involvement. Ann Intern Med 58:136–142.
46. Nelen MR, Kremer H, Konings IB, et al (1999) Novel PTEN mutations in patients with Cowden disease: absence of clear genotype-phenotype correlations. Eur J Hum Genet 7:267–273.
47. Starink TM, van der Veen JP, Arwert F, et al (1986) The Cowden syndrome: a clinical and genetic study in 21 patients. Clin Genet 29:222–233.
48. Fistarol SK, Anliker MD, Itin PH (2002) Cowden disease or multiple hamartoma syndrome-cutaneous clue to internal malignancy. Eur J Dermatol 12:411–421.
49. Williard W, Borgen P, Bol R, Tiwari R, Osborne M (1992) Cowden's disease. A case report with analyses at the molecular level. Cancer 69:2969–2674.
50. Longy M, Lacombe D (1996) Cowden disease. Report of a family and review. Ann Genet 39:35–42.
51. Chen YM, Ott DJ, Wu WC, Gelfand DW (1987) Cowden's disease: a case report and literature review. Gastrointest Radiol 12:325–329.
52. Carlson GJ, Nivatrongs S. Snovor DC (1984) Colorectal polyps in Cowden's disease (multiple hamartoma syndrome). Am J Pathol 8:703–770.
53. Riegler G, Esposito I, Esposito P, et al (2006) Wireless capsule enteroscopy (Given) in a case of Cowden syndrome. Dig Liver Dis 38:151–152.
54. Marsh DJ, Coulon V, Lunetta KL, et al (1998) Mutation spectrum and genotype-phenotype analyses in Cowden disease and Bannayan-Zonana syndrome, two hamartoma syndromes with germline PTEN mutation. Hum Mol Genet 7:507–515.
55. Waite KA, Eng C (2002) Protean PTEN: form and function. Am J Hum Genet 70:829–844.
56. Ruvalcaba RH, Myhre S, Smith DW (1980) Sotos syndrome with intestinal polyposis and pigmentary changes of the genitalia. Clin Genet 18:413–416.
57. Gorlin RJ, Cohen MM Jr, Condon LM, Burke BA (1992) Bannayan–Riley–Ruvalcaba syndrome. Am J Med Genet 44:307–314.
58. Marsh DJ, Kum JB, Lunetta KL, et al (1999) PTEN mutation spectrum and genotype-phenotype correlations in Bannayan-Riley-Ruvalcaba syndrome suggest a single entity with Cowden syndrome. Hum Mol Genet 8:1461–1472.
59. Pilarski R, Eng C (2004) Will the real Cowden syndrome please stand up (again)? Expanding mutational and clinical spectra of the PTEN hamartoma tumor syndrome. J Med Genet 41:323–326.
60. Celebi JT, Tsou HC, Chen FF, et al (1999) Phenotypic findings of Cowden syndrome and Bannayan–Zonana syndrome in a family associated with a single germline mutation in PTEN. J Med Genet 36:360–364.
61. Mata A, Llah J, Castells A, et al (2005) A prospective trial comparing wireless capsule endoscopy and barium contrast

series for small bowell surveillance in hereditary GI polyposis syndromes. Gastrointest Endosc 61:721–725.
62. Gheorghe C, Iacob R, Bancila I (2007) Olympus capsule endoscopy for small bowel examination. J Gastrointestin Liver Dis 16(3):309–313.
63. Cronkhite LW, Canada WJ (1955) Generalized gastrointestinal polyposis: an unusual syndrome of polyposis, pigmentation, alopecia and onychontrophia. N Engl J Med 252:1011–1015.
64. Daniel ES, Ludwig SL, Lewin KJ, et al (1982) The Cronkhite–Canada syndrome: an analysis of clinical and pathologic features and therapy in 55 patients. Medicine (Baltimore) 61:293–309.
65. Daniel ES (1993) The Cronkhite–Canada syndrome. Probl Gen Surg 10:699–706.
66. Nyam DC, Ho MS, Goh HS (1996) Progressive ectodermal changes in the Cronkhite–Canada syndrome. Aust N Z J Surg 66:780–781.
67. Dachman AH, Buck JU, Burke AP, Sobin UH (1989) Cronkhite–Canada syndrome: radiologic features. Gastrointest Radiol 14:285–290.
68. Goto A, Mimoto H, Shibuya C, et al (1988) Cronkhite–Canada syndrome: an analysis of clinical features and follow-up studies of 80 cases reported in Japan. Nippon Geka Hokan 57:506–526.
69. Johnson GK, Soergel KH, Hensley GT, et al (1972) Cronkite-Canada syndrome: gastrointestinal pathophysiology and morphology. Gastroenterology 63:140–152.
70. Watanabe T, Kudo M, Shirane H, et al (1990) Cronkhite–Canada syndrome associated with triple gastric cancers: a case report. Gastrointest Endosc 50:688–691.
71. Malhotra R, Sheffield A (1988) Cronkhite–Canada syndrome associated with colon carcinoma and adenomatous changes in C-C polyps. Am J Gastroenterol 83:772–776.
72. Williams GT, Arthur JF, Bussey HJ, et al (1980) Metaplastic polyps and polyposis of the colorectum. Histopathology 4:155–170.
73. Orii S, Nakamura S, Sugai T, et al (1997) Hyperplastic (metaplastic) polyposis of the colorectum associated with adenomas and an adenocarcinoma. J Clin Gastroenterol 25:369–372.
74. Keljo DJ, Weinberg AG, Winick N, et al (1999) Rectal cancer in an 11-year-old girl with hyperplastic polyposis. J Pediatr Gastroenterol Nutr 28:327–332.
75. Jorgensen H, Mogensen AM, Svendsen LB (1996) Hyperplastic polyposis of the large bowel. Three cases and a review of the literature. Scand J Gastroenterol 31:825–830.
76. Leggett BA, Devereaux B, Biden K, et al (2001) Hyperplastic polyposis: association with colorectal cancer. Am J Surg Pathol 25:177–184.
77. Sumner HW, Wasserman NF, McClain CJ (1981) Giant hyperplastic polyposis of the colon. Dig Dis Sci 26:85–89.
78. Lage P, Cravo M, Sousa R, et al (2004) Management of Portuguese patients with hyperplastic polyposis and screening of at-risk first-degree relatives: a contribution for future guidelines based on a clinical study. Am J Gastroenterol 99:1779–1784.
79. Burt RW, Jass JR (2000) Hyperplastic polyposis. In: Hamilton SR, Aaltonen LA (eds) World Health Organization classification of tumours. Pathology and genetics. Tumours of the digestive system. Springer-Verlag, Berlin, pp 135–136.
80. Hawkins NJ, Bariol C, Ward RL (2002) The serrated neoplasia pathway. Pathology 34:548–555.
81. Jass JR, Young J, Leggett BA (2000) Hyperplastic polyps and DNA microsatellite unstable cancers of the colorectum. Histopathology 37:295–301.
82. Abeyasundara H, Hampshire P (2001) Hyperplastic polyposis associated with synchronous adenocarcinomas of the transverse colon. Aust N Z J Surg 71:686–687.
83. Whittaker MA, Carr NJ, Cripps N (2001) Hyperplastic polyposis coli associated with dysplasia. Postgrad Med J 77:532, 541–542.
84. Jeevaratnam P, Cottier DS, Browett PJ, et al (1996) Familial giant hyperplastic polyposis predisposing to colorectal cancer: a new hereditary bowel cancer syndrome. J Pathol 179:20–25.
85. Jass JR, Cottier DS, Pokos V, et al (1997) Mixed epithelial polyps in association with hereditary non-polyposis colorectal cancer providing an alternative pathway of cancer histogenesis. Pathology 29:28–33.
86. Rashid A, Houlihan PS, Booker S, et al (2000) Phenotypic and molecular characteristics of hyperplastic polyposis. Gastroenterology 119:323–332.
87. Weisenberger D, Siegmund K, Campan M, et al (2006) A Distinct CpG island methylator phenotype in human colorectal cancer is the underlying cause of sporadic mismatch repair deficiency and is tightly associated with BRAF mutation. Nat Genet 38:787–793.
88. Beach R, Chan AO, Wu TT, et al (2005) BRAF mutations in aberrant crypt foci and hyperplastic polyposis. Am J Pathol 166:1069–1075.
89. Chan AO, Issa JP, Morris JS, et al (2002) Concordant CpG island methylation in hyperplastic polyposis. Am J Pathol 160:529–536.

Diet, Polyps, and Cancer

Where is the Truth?

Luigi Benini, Anna Rostello, Chiara Scattolini,
Laura Peraro, Luca Frulloni and Italo Vantini

6

Abstract Colorectal cancer (CRC) is among the leading causes of cancer-related mortality in Western countries. As for most other cancers, it is likely that CRC represents an interaction between genetic and environmental factors. Since genetic factors are a cause in only a minority of cases, environmental factors, particularly diet, are probably prevalent. Many studies have shown that an increase in meat consumption produces a clear increase in CRC. However, the difference is mostly in the extreme classes of meat intake (i.e. between people who eat a large amount of meat every day and those who hardly ever eat red meat). Moreover, the difference in risk is only 12–17% for an increase of 100 g/day in meat intake, probably not enough in itself to warrant large-scale campaigns to reduce meat intake. Similar conclusions can be drawn from the relationship between an increase in fruit, fiber, or milk intake and a decrease in CRC. However, all the suggested changes move towards a healthier diet, which also has positive effects on other highly prevalent disorders (cardiovascular, degenerative, and neoplastic). Sound dietary advice should therefore be offered even in the absence of formal evidence-based proof.

Keywords Colorectal cancer • Diet • Fruit • Meat • Milk

6.1 Introduction

Colorectal cancer (CRC) is among the leading causes of cancer-related mortality in Western countries. The colon is easily accessible to endoscopic and radiological imaging technologies (for example, computerised axial tomography to obtain the so-called virtual colonoscopy). In the last decade the hope was that screening programs (with occult fecal blood) and early detection of neoplastic lesions in populations at risk (patients over 50 years, with positive family history of polyps or CRC, or with inflammatory bowel disease) may reduce the burden of this disease. The efficacy of such expensive programs in the reduction of mortality is still debatable, mainly because the compliance of most target populations is still far from optimal [1]. It must also be considered that screening programs may easily detect preneoplastic lesions or cancers with an inherent slow growth, which take long periods to become invasive, while anaplastic cancers, which are characterized by an "explosive" growth, may not benefit from screening programs. Moreover, because of the high incidence of polyps during screening colonoscopies or virtual colonoscopies, attempts are being made to reduce the recurrence of polyps in screened individuals.

These considerations explain the persistent interest

L. Benini (✉)
Gastroenterology Unit, Department of Biomedical and Surgical Sciences, University of Verona, Verona, Italy

in the causes of colonic polyps and their progression to CRC, in an attempt to limit its occurrence. The recently reported, small decline in CRC incidence might indirectly confirm that the attention to healthy lifestyles is effective in the long run [2].

As for most other cancers, it is likely that CRC represents an interaction between genetic and environmental factors.

The identification of mutations on genes regulating proliferation and homeostasis of cell replication in some well-characterized familial forms of cancer-prone disorders (such as familial multiple adenomatosis coli) has fostered studies on the role of a genetic component even in cases of sporadic cancers. These studies have clearly shown that only a limited number of sporadic cases have an underlying genetic component, suggesting that the environment has a vital role.

This environment is mainly related to the dietary components that may reach the colon in an unabsorbed form, and to the hormonal and metabolic responses to the ingested macronutrients.

There is a famous observation by Burkitt of a reduced incidence of colonic disorders in general, and of colonic cancer in particular, in people attending his mission infirmary in Africa, compared with Caucasian populations [3]. The hypothesis he formulated on the effect of diet content and unabsorbable fiber is well known and widely accepted, even in the general population. He suggested that the low incidence of non-infective colonic disorders in Black subjects on a "traditional" diet was related to its high content of local corns, rich in indigestible fiber. Fiber could possibly produce a fast colonic transit, with reduced production of carcinogens, and reduced contact with the colonic mucosa.

Many concordant epidemiological studies seem to confirm the predominance of the environment, and therefore of diet, in the development of CRC. On the other hand, firm proof of this relationship between cancer and particular constituents of the diet is weak and conflicting.

In the next sections, we will discuss some of these studies, trying to analyze the reasons for these discrepancies, and to provide a rational suggestion for everyday clinical practice.

The most frequently quoted studies on the epidemiology of CRC are those conducted on populations that changed their alimentary habits in short and well-characterized period. The differences in cancer incidence could easily be related to a particular environmental change (in our case, the diet), since a consistent genetic change could not take place in such a short interval.

The strongest body of evidence in this respect is the difference in the incidence of CRC in Black inhabitants of South Africa and Black individuals (both males and females) living in the United States. In Africans, the incidence of CRC is still very low, of the order of less than one per 100,000 for year. In Africa, CRC ranks only as the 12th cause of cancer death [4]. These numbers are lower by far than those reported in the United States, where they are of the order of 61 and 45 per 100,000 for White males and females respectively. What is even more impressive, is that in the United States, Black subjects have an incidence of CRC that is even higher than for Caucasians, for both males (73 versus 61/100,000/year) and females (56 versus 45/100,000/year) [5]. This difference cannot be fully explained by different medicalisation, with an easier diagnosis of premalignant conditions in White individuals. The diagnosis of this cancer, in fact, is not easily missed, at least in advanced stages. The change may rather be due to the fact that Black individuals in the United States changed their diet, which became richer in meat and fat [6]. Recent reports have shown that African-Americans are at present the greatest consumers of meat in the United States [7].

A similar change in diet has recently been witnessed in Japan. The traditional Japanese cooking used to consist mainly of fish, with small amounts of meat. Under these conditions, the incidence of CRC in Japan used to be much lower than in Western countries. In recent decades the Japanese diet has changed, becoming more similar to the Western diet, and red meat has been introduced in larger quantities. This was accompanied by a sharp increase in the incidence of CRC, to levels comparable to those of the Western Caucasian population. The same concomitant change in diet and in cancer was also seen in Japanese individuals migrating to the United States, which again showed, after only one generation, an incidence of CRC similar to that of White Americans [8,9].

Some studies have questioned the role of reduced fiber consumption in the pathogenesis of colonic cancer, since even Africans now consume fiber at levels well below the recommended daily intake [6]. It has been reported that Africans currently have a fiber intake that is only one-half of the average daily intake of all Americans, without any evident deleterious effect on the incidence of the disease. This means that the onset of CRC is probably not linked to the effects of a deficiency of some nutrients, but rather to an excessive presence of some other constituents (meat, animal fat, or their metabolic derivatives in the colon).

A major factor might be not the amount of animal products eaten, but rather the imbalance between energetic needs and foods eaten. Epidemiological evidence suggests, for example, some relationship between CRC and obesity in different populations [10]. The frequency of obesity is higher in those populations with greater incidence of CRC; for example, both obesity and CRC are more prevalent in northern Europe or the United States than in sub-Saharan nations. Clearly, this type of evidence is rather weak. In fact, it is open to debate whether the main difference between populations that are completely different is the simple incidence of obesity. Other studies have confirmed a marginal but significant association between body mass index (BMI) and CRC incidence even in the same population. For example, in a large prospective study of the onset of new cases of CRC, which followed 368,000 patients for 6 years, 984 and 586 new cases of colon and rectal cancer were diagnosed. The parameters that were suggestive of obesity (waist circumference, waist-to-hip ratio) carried a 50% increase of CRC between subjects in the highest and lowest quintiles, although, for unknown reasons, this was only true for males [11]. This may be due to higher amounts of fat and proteins, or of highly refined food, eaten by obese patients.

6.2 Meat

The role of red meat in the development of cancer, suggested by epidemiologic studies in populations, would be more convincingly confirmed by large-scale, longitudinal studies of selected populations, or, even more clearly, by interventional studies in selected populations. These studies should be carried out in homogeneous populations, to keep differences in other variables as small as possible.

Clear data are available on this aspect from large observational studies. For example, for 6 years, Willett and coworkers [12] followed a cohort of almost 90,000 female nurses, aged 34–59 years, who completed an alimentary questionnaire at the beginning of the study. At entry, they were free from colonic cancer or from diseases predisposing to CRC (inflammatory bowel diseases, polyps, familial polyposis). By the end of the study, 150 new cases of CRC were diagnosed. A clear relationship was found between newly developed CRC and the intake of animal (but not vegetable) fat. The relative risk of cancer was 2.49 in individuals with higher intake than those in the lowest quintile of meat intake. The increased risk was only for people eating pork, beef, or lamb, while chicken without skin and fish had a protective effect. This difference is relevant, but it must be underlined that it can only be demonstrated between, on the one hand, subjects eating red and processed meat as their main food every day, and on the other hand, subjects eating it less than once a month. The different risk in individuals with intermediate meat intake is not apparent even in such a large population. This study confirms also that the intake of fiber derived from fruit is associated with a decreased frequency of CRC, but this effect is strictly linked with meat intake (in that people eating only small amounts of meat more frequently eat large amounts of fruit).

The same results have been obtained by another large, multinational survey. Again, information was obtained from 478,000 individuals, who completed an accurate alimentary questionnaire over a 6-year period. During the subsequent 5-year period, CRC was diagnosed in 1230 of these subjects. Again these cancers were more frequently found in people with higher meat intake. The relative risk of cancer was 35% higher in people eating more than 120 g/day of meat than in people eating less than 20 g/day. A change in the opposite direction was found for fish (hazard ratio 0.69 for a fish intake >80 g/day than for an intake <10 g/day) [13]. We must underline the finding that the absolute risk of developing CRC for a study subject aged 50 years was 1.71% for the highest category of red meat intake and 1.28% for the lowest category of intake, while it was 1.86% and 1.28% for subjects in the lowest and highest quintiles of fish intake. Again, this means that relevant differences are only demonstrated for groups with extreme dietary habits.

The same study also provided important information on the relevance of cooking for CRC incidence [14]. A total of 30,000 people from 10 European countries were interviewed on the type of cooking and on the method of preservation of (salting, smoking, addition of nitrates) for meat they consumed. It demonstrated a more frequent use of risky methods of preservation (addition of nitrate/nitrite; smoking) and of hazardous cooking (frying, charcoal) in central and northern European centers, where a higher incidence of CRC is reported.

These data are important since they are similar to those for gastric cancer, for which a strict correlation has been reported with the use of salted meat. Moreover, cooking methods using high temperatures, such as frying or barbecue cooking may change the arrival of undigested constituents and mutagens at the colon [15]. It has been shown that higher consumption of mutagens from meats cooked at higher temperature and longer duration may be associated with higher risk of distal colon adenoma, independent of overall meat intake.

A recent systematic review of the studies on the relationship between meat consumption and CRC [16] has confirmed a 12–17% increase in CRC for an increase in red meat consumption of 100 g/day. This change in CRC incidence in itself is of marginal impact in individual patients (the individual risk is in any case rather low and hardly appreciable), and it is only important for a population as a whole. In any case, a reduction in meat intake should be advised if we accept that is produces other positive effects (cardiovascular, on pancreatic and gastric cancer, and so on).

Clearly, interventional studies on the effect of a "pure" change in meat intake on CRC incidence are not even conceivable. The incidence of this carcinoma peaks after the age of 50 years, and therefore exposure to the putative carcinogens requires very long periods. In the studies discussed above, less than 2% of subjects above 50 years of age developed CRC during the 5-year follow-up, so that huge number of patients need to be included in each group to obtain meaningful data. Finally, the change in daily meat intake that needs to be suggested to show a possible difference is so large that it is almost impossible to maintain it for any length of time.

6.3 Fiber and Fruit

As we have previously seen, analysis of the data from studies on the correlation between CRC and increased meat consumption have also found a correlation with a decrease in fruit consumption. A higher intake of meat is invariably associated with a lower intake of fruits and of vegetables, which may in itself be detrimental [17].

The reduction of the risk of developing recurrent adenomas was found to be 40% in women consuming large quantities of fruit, but it was only 18% for vegetable intake. The reasons for this difference are not clear, but suggest a role for other constituents of fruit, such as carotenoids, folate, vitamin C, flavonoids, organosulfides, isothiocyanates, and protease inhibitors, which may mitigate DNA damage and thus reduce mutations [18].

Some observational studies have addressed the relationship between fruit and vegetable consumption and colorectal adenomas. Seven of the ten case-control studies available [17,19–27] report significant inverse associations between fruit and vegetable consumption and the risk of adenomas.

On the other hand, the results from a large interventional study (the Polyp Prevention Trial) did not indicate any benefit for a diet high in fiber, fruits, and vegetables in preventing the recurrence of colorectal adenomas. However, this study was restricted to participants who had previous adenomas removed endoscopically, and their period of intervention and follow-up was 4 years, which may be insufficient to influence the occurrence or recurrence of adenomas [28].

We must underline that the lack of effect of fiber and fruit on polyp recurrence is not equivalent to an absence of effect on cancer development, which might require different factors.

Ultimately, fiber should probably be promoted by doctors as a protective factor against CRC, even in the absence of a firm proof of its efficacy [29,30]

6.4 Milk

The association between milk intake and CRC is rather complex. Many studies have reported that milk consumption (with the inherent increase in vitamin D and calcium intake) is associated with a reduced risk of CRC [31]. A recent meta-analysis of the literature on this subject has confirmed a 15% reduction of the risk in subjects with higher milk intake (>250 ml/day) [32]. Clearly, such doses carry a risk of cardiovascular disease, and cannot be recommended in the general population.

Another interesting study has been carried out on a cohort of subjects who had an accurate estimation of milk and dairy intake in childhood (70 years ago). Surprisingly, the lifelong risk of developing a CRC was clearly raised in people with high intake of milk. While the mechanisms involved are purely hypothetical at this stage, they may be linked with an increase in insulin growth factor-1 or with an interference with mucosal immunocompetence [33]. The possible role of vitamin D and of its receptors on cancer prevention has been challenged by a large interventional trial of 7 years' duration, which did not show any effect [34].

6.5 Conclusion

Suggestions about the importance of diet in the prevention of CRC have not been confirmed by large interventional trials. When present, this protective effect is rather weak, so that this prevention in itself cannot justify vigorous and long-term dietary changes in the population. However, the suggested changes are often in the direction of a healthier diet, with possible positive effects on other highly prevalent disorders (cardiovascular, degenerative, and neoplastic), so sound dietary suggestions may be offered even in the absence of formal evidence-based proof of their effect in CRC prevention.

References

1. Seeff LC, Shapiro JA, Nadel MR (2002) Are we doing enough to screen for colorectal cancer? Findings from the 1999 Behavioral Risk Factor Surveillance System. J Fam Pract 51:761–766.
2. Jemal A, Siegel R, Ward E, et al (2006) Cancer statistics 2007. CA Cancer J Clin 57:43–66.
3. Burkitt DP (1971) Epidemiology of cancer of the colon and rectum. Cancer 28:3–13.
4. Mqoqi N, Kellet P, Sitas F, Jula M (2004) The incidence of histologically diagnosed cancer in South Africa, 1998–1999. The National Cancer Registry, National Health Laboratory Service, Johannesburg.
5. O'Keefe SJD (2008) Nutrition and colonic health: the critical role of microbiota. Curr Opinion Gastroenterol 24:51–58.
6. O'Keefe SJD, Kidd M, Espitalier-Noel G, Owire P (1999) Rarity of colon cancer in Africans is associated with low animal product consumption, not fiber. Am J Gastroenterol 94:1373–1380.
7. Anonymous (2000) Continuing survey of food intakes by individuals: US meat consumption figures. USDA Agricultural Research Services, Washington.
8. Oba S, Shimizu N, Nagata C, et al (2006) The relationship between the consumption of meat, fat and coffee and the risk of colon cancer: a prospective study in Japan. Cancer Lett 244:260–267.
9. Haenszel W, Kurihara M (1968) Studies of Japanese migrants: mortality from cancer and other diseases among Japanese in the United States: J Natl Cancer Inst 40:43–68.
10. Parkin DM, Whelan SL, Ferlay J, Teppo L, Thomas DB (2002) Cancer incidence in five continents. Iarc Press, Lyon.
11. Pishon T, Lahmann PH, Boeing H, et al (2006) Body size and risk of colon and rectal cancer in the European prospective investigation into cancer and nutrition (EPIC). J Natl Cancer Inst 98:920–931.
12. Willett WC, Meir J, Stampfer MD, et al (1990) Relation of meat, fat, and fiber on the risk of colon cancer in a prospective study among women. N Eng J Med 3232:1664–1672.
13. Norat T, Bingham S, Ferrari P, et al (2005) Meat, fish, and colorectal cancer risk: the European Prospective Investigation into cancer and nutrition. J Natl Cancer Inst 97:906–916.
14. Linseisen J, Rohrmann S, Norat T, et al (2006) Dietary intake of different types and characteristics of processed meat which might be associated with cancer risk – results from the 24-hour diet recalls in the European Prospective Investigation into Cancer and Nutrition (EPIC). Public Health Nutr 9:449–464.
15. Wu K, Giovannucci E, Byrne C, et al (2005) Meat mutagens and risk of distal colon adenoma in a cohort of U.S. men. Cancer Epidemiol Biomarkers Prev 15:1120–1125.
16. Sandhu MS, White IR, McPherson K (2001) Systematic review on the prospective cohort studies on meat consumption and colorectal cancer risk: a meta-analytical approach. Cancer Epidemiol Biomarkers Prev 10:439–446.
17. Michels KB, Giovannucci E, Chan AT, et al (2006) Fruit and vegetable consumption and colorectal adenomas in the Nurses' Health Study. Cancer Res 66:3942–3953.
18. World Cancer Research Fund/American Institute for Cancer Research (1997) Food, nutrition and prevention of cancer: a global perspective, 1st ed. American Institute for Cancer Research, Washington (DC).
19. Witte JS, Longnecker MP, Bird CL, et al (1996) Relation of vegetable, fruit and grain consumption to colorectal adenomatous polyps. Am J Epidemiol 144:1015–1025.
20. Kato I, Tominaga S, Matsuura A, et al (1990) A comparative case-control study of colorectal cancer and adenoma. Jpn J Cancer Res 81:1101–1108.
21. Benito E, Cabeza E, Moreno V, et al (1993) Diet and colorectal adenomas: a case-control study in Majorca. Int J Cancer 55:213–219.
22. Sandler RS, Lyles CM, Peipins LA, et al (1993) Diet and risk of colorectal adenomas: macronutrients, cholesterol, and fiber. J Natl Cancer Inst 85:884–891.
23. Macquart-Moulin G, Riboli E, Cornee J, et al (1987) Colorectal polyps and diet: a case-control study in Marseilles. Int J Cancer 40:179–188.
24. Kune GA, Kune S, Read A, et al (1991) Colorectal polyps, diet, alcohol, and family history of colorectal cancer: a case-control study. Nutr Cancer 16:25–30.
25. Hoff G, Moen IE, Trygg K, et al (1986) Epidemiology of polyps in the rectum and sigmoid colon. Evaluation of nutritional factors. Scand J Gastroenterol 21:199–204.

26. Smith-Warner SA, Elmer PJ, Fosdick L, et al (2002) Fruits, vegetables, and adenomatous polyps: the Minnesota Cancer Prevention Research Unit case-control study. Am J Epidemiol 155:1104–1113.
27. Mathew A, Peters U, Chatterjee N, et al (2004) Fat, fiber, fruits, vegetables, and risk of colorectal adenomas. Int J Cancer 108:287–292.
28. Schatzkin A, Lanza E, Corle D, et al (2000) Lack of effect of a low-fat, high-fiber diet on the recurrence of colorectal adenomas. N Engl J Med 342:1149–1155.
29. Marlett JA, McBurney MI, Slavin JL (2002) Position of the American Dietetic Association: health implications of dietary fibers. J Am Diet Assoc 102:993–1000.
30. Kim YI (2000) AGA technical review: impact of dietary fiber on colon cancer occurrence. Gastroenterology 118:1235–1257.
31. Larsson S, Bergkvist L, Rutegard J, et al (2006) Calcium and dairy food intakes are inversely associated with colorectal cancer risk in the cohort of Swedish men. Am J Clin Nutr 83:667–673.
32. Cho E, Smith-Warner SA, Spiegelman D, et al (2004) Dairy foods, calcium, and colorectal cancer: a pooled analysis of 10 cohort studies. J Natl Cancer Inst 96:1015–1022.
33. Van Der Pols JC, Bain C, Gunnell D, et al (2007) Childhood dairy intake and adult cancer risk: 65-y follow-up of the Boyd-Orr cohort. Am J Clin Nutr 86:1722–1729.
34. Wactawski-Wende J, Kotchen JM, Anderson GL, et al (2006) Women's Health Initiative Investigators. Calcium plus vitamin D supplementations and the risk of colorectal cancer. N Engl J Med 354:684–696.

Chemoprevention of Colonic Cancer

Is There a Foreseeable Future?

7

Raffaele Palmirotta, Patrizia Ferroni, Mario Roselli and Fiorella Guadagni

Abstract The high incidence and low percentages of survival of colorectal cancer make effective prevention an important public health issue. Colorectal carcinogenesis is a multistep process involving dietary factors and multiple genetic alterations in signaling pathways that control cell proliferation, differentiation, and apoptosis.

Predisposing genetic syndromes, such as familial adenomatous polyposis (FAP) and hereditary non-polyposis colon cancer (HNPCC), have provided an excellent model for studying the genetic alterations involved in the etiology and progression of sporadic colon cancer. These developments in the cellular and molecular mechanisms of the neoplasm provide new insights for developing selective agents with potential chemopreventive properties against colon carcinogenesis. Clinical and experimental evidence have demonstrated that long-term intake of non-steroidal anti-inflammatory drugs (NSAIDs) has a reduced risk of developing colorectal polyps and cancer. Similarly, several synthetic or natural compounds, such as folate or selenium prevent the biological events leading to the development of cancer. However, the clinical use of these drugs as chemopreventive agents is limited by many open questions about the optimal drug, dose, duration of therapy and knowledge about the mechanism(s) by which these drugs act.

Keywords Calcium • Colorectal cancer prevention • Colorectal cancer therapy • DMFO • Folate • Nitric oxide • NSAIDs • Review • Selenium • Vitamin D • WNT

7.1 Introduction

Colorectal cancer (CRC), the most common neoplasm in the industrialized world, with 900,000 new cases annually worldwide [1], is still the second leading cause of cancer death in Western countries, after lung cancer in men and breast cancer in women [2]. The percentages of 5-year survival, in fact, remain very low, despite the improvements in surgical techniques and radio- and chemotherapy treatments that have occurred in recent years. The reason for this lack of improvement in the prognosis is attributable to the fact that, due to late presentation of clinical symptoms, most cases are diagnosed at an advanced stage.

Colorectal cancer provides an excellent model for genetic studies because of the availability of precursor adenoma lesions and the existence of several clear-cut

F. Guadagni (✉)
Department of Laboratory Medicine and Advanced Biotechnologies, IRCCS San Raffaele Pisana, Rome, Italy

G.G. Delaini, T. Skřička, G. Colucci (eds.), *Intestinal Polyps and Polyposis*,

familial inherited susceptibilities.

The peculiarity of the disease is that it is preceded by precancerous adenoma or adenomatous polyps followed by intermediate stages through which normal colorectal tissue becomes cancerous within an "adenoma–carcinoma" sequence. These phenotypic manifestations are closely related to a multistep process characterized by multiple genetic changes that correspond to the histological progression from normal colon epithelium to adenoma to carcinoma [3,4].

A further feature is the presence of several familial aggregation and clear-cut familial inherited susceptibilities [5], but only 5% of colon cancers are caused by a specific gene mutation. Such hereditary forms include familial adenomatous polyposis (FAP) and hereditary non-polyposis colorectal cancer (HNPCC). The former led to the identification of the *APC* gene, an important component of the wingless-type MMTV integration site family (WNT) pathway, while the latter identified the role of the mismatch repair genes in colorectal and other cancers [6]. These findings not only provided a critical basis to understand principles of cancer genetics, molecular diagnostics, risk stratification, and prevention, but served also to clarify our knowledge of the sporadic counterpart.

Secondary prevention, through the colonoscopic detection and eradication of adenomatous polyps, can lead to a drastic reduction of mortality due to colorectal cancer. However, given the high cost and inconvenience caused to patients, the idea of applying colonoscopy is not acceptable for the whole population, but only for groups of high-risk patients, such as *APC* mutation carriers.

More promising is the application of primary prevention that consists in the identification of genetic factors and possible modification of environmental and dietary factors.

The quantity of biological data, obtained through numerous biological and epidemiological studies, now offers interesting perspectives for improving tertiary prevention. Indeed, the molecular and genetic pathways involved in the multistep process of colon carcinogenesis, underlying the early preneoplastic and neoplastic phenotypic lesions, represent an important end-point for efficacy of chemopreventive interventions through several classes of synthetic pharmacological agents, including non-steroidal anti-inflammatory drugs (NSAIDs), receptor antagonists, small molecule inhibitors, vitamins, minerals, or synthetic substances [7], all of which have demonstrated a direct correlation between the modulation of pathogenesis and the relevant molecular target for efficacy [8,9].

The ideal chemopreventive agent is safe and not toxic for long-term administration, characterized by easy administration and certain effect. At present, the most promising agents seem to be NSAIDs, but no data are available yet on their safety profile, dose, and duration of treatment, or side-effects. Other substances such as calcium, folate, and selenium are currently being studied. This review will describe some of these traditional and innovative molecules and will discuss the anticancer molecular mechanisms of action and their efficacy as chemopreventive and therapeutic agents in colorectal cancer.

7.2 The WNT Pathway

The WNT ligands are secreted lipid-modified glycoproteins belonging to a family of proto-oncogenes expressed in several species and considered to be one of the major families of signaling molecules [10].

The WNT extracellular signaling pathway describes an evolutionarily conserved signal transduction network of molecules, and plays an essential and complex role in embryogenesis, cell proliferation, migration, and differentiation. Mutations charged to the WNT pathway, including numerous ligands, receptors, and transcriptional effectors, seriously undermine embryonic development and are involved as a major factor in oncogenesis in the human colon and other tissues [11].

In the presence of a WNT ligand (Fig. 7.1), if not inhibited by secreted antagonists, the WNT ligand binds a seven-transmembrane frizzled protein (Fz) and a coreceptor of the low-density lipoprotein receptor-related protein (LRP), forming a ternary complex (WNT/Fz/LRP), and activating the cytoplasmic protein dishevelled (Dsh/Dvl). Upon WNT stimulation, Dsh/Dvl, activated through a phosphorylation mechanism, inhibits the multiprotein destruction complex adenomatous polyposis coli (APC)–axin–casein kinase 1 (CK1)–glycogen synthase kinase (GSK-3β), leading to the stabilization of β-catenin [12,13]. The APC–axin–CK1–GSK-3β complex normally promotes the proteolytic degradation of the β-catenin intracellular signaling molecule. Therefore, complex inhibition, determined by WNT binding, results in an accu-

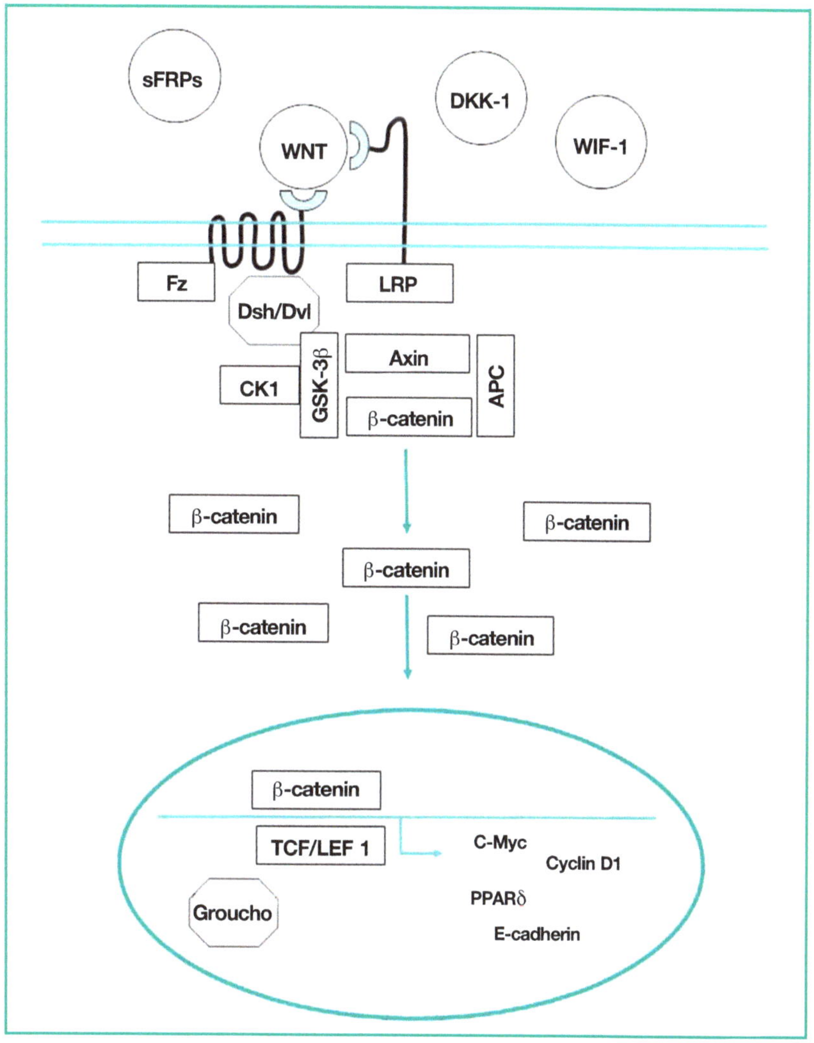

Fig. 7.1 Schematic representation of the WNT signaling pathway. In the presence of a WNT signal, LRP-Fz interaction and Dsh/Dvl activation block the APC–axin–CK1–GSK3 complex from phosphorylating β-catenin, resulting in an accumulation of stabilized cytosolic β-catenin. Stabilized β-catenin is able to enter the nucleus, where it interacts with TCF/LEF1, leading to the transcription of specific WNT target genes

mulation of stabilized cytosolic β-catenin. The latter forms a complex with E-cadherin and β-catenin, and participates in calcium-dependent cell adhesion [14]. Furthermore, stabilized β-catenin is able to enter the nucleus and interacts with the T-cell factor (TCF)/lymphoid enhancing factor 1 (LEF1) family of DNA-binding proteins, leading to the transcription of specific WNT target genes. The majority of WNT target genes are important mediators of development, cell proliferation, carcinogenesis, tumor cell migration, and invasion [15]. Indeed, several genes involved in human carcinogenesis, listed at The WNT Home Page [16], have been identified as targets for β-catenin/TCF complex transcriptional activation, such as cyclin Dl [17], the E-cadherin gene [18], peroxisome proliferator-activated receptors (PPARδ) [19], c-Myc [20], c-jun, and c-fra-1 [21].

In the absence of a WNT ligand, cytoplasmic β-catenin interacts with axin and APC and then is phosphorylated by kinases CK1 and GSK-3β. Phosphorylated β-catenin is then ubiquitinated and destroyed by the proteosome while TCF/LEF factors repress transcription binding of DNA at WNT-responsive genes, and interact with other factors such as groucho and histone deacetylase [13].

Alterations of several proteins of the WNT signaling pathway have been implicated in the development of sporadic and hereditary colorectal cancer and other neoplasms such as hepatocellular carcinomas (HCCs), melanomas, and uterine and ovarian carcinomas. For example, mutations of the tumor-suppressor gene APC, a key regulator of the WNT signaling pathway, are responsible for FAP and are also one of the earliest transforming events observed in sporadic colorectal cancer [11,22], while TCF4, one of the β-catenin-binding transcription factors, is mutated in an encoding polyA tract in almost half of the HNPCC and MSI+ (microsatellite instability) colorectal cancers [23–25].

Given its importance in colorectal cancer progression, the WNT signaling pathway is a logical attractive target for therapeutic intervention, and several experimental studies have demonstrated inhibition of colorectal cancer by targeting the TCF/β-catenin signaling [26,27]. For example the use of WNT antagonists or other inhibitors that interfere with cell surface interactions of WNT ligands and their receptors could provide a target for therapeutic intervention.

The secreted frizzled-related proteins (SFRPs), also known as secreted apoptosis-related proteins (SARPs), contain a cysteine-rich domain with similarity to the ligand-binding domain of the Fz transmembrane protein family, and compete with the Fz proteins for binding to secreted WNT ligands and antagonize the WNT function such as the WNT-inhibitory factor-1 (WIF-1) [28]. Recently, it has been demonstrated that the SFRP family gene *SFRP2* repressed WNT target genes and induced changes in the expression of numerous genes related to proliferation, growth and apoptosis in gastointestinal cells [29]. Moreover, SFRP function in colorectal cancer cells attenuates WNT signaling even in the presence of downstream β-catenin or APC mutations [30].

Another extracellular WNT inhibitor dickkopf 1 (DKK1) interacts with the WNT co-receptor LPR and prevents formation of an active WNT–Fz–LRP receptor complex [31]. Experimental evidence suggests that DKK1 acts as a tumor suppressor gene in this neoplasia, downregulating the WNT/β-catenin signaling and exerting anticancer activity [32,33].

The increasing knowledge of the WNT signaling pathway is leading to the identification of several new compounds that are useful for therapeutic purposes, such as WNT antagonists and synthetic molecules that act directly at the level of β-catenin or LEF1, and that are now being tested *in vivo* assays [34].

7.3 Non-Steroidal Anti-Inflammatory Drugs

NSAIDs are widely used as effective anti-inflammatory, antipyretic and analgesic drugs, and aspirin is also effective in both the primary and secondary prevention of cardiovascular diseases.

Since the first hypothesis in 1974 that NSAIDs might inhibit the occurrence or growth of CRC [35,36], and the epidemiological demonstration in 1988 that NSAIDs prevent human colon cancer [37], evidence has been accumulating suggesting that long-term use of NSAIDs can be regarded as an effective approach for cancer chemoprevention [38].

The accurate biological mechanisms by which NSAIDs exert their chemopreventive effects are not yet entirely clarified, but probably involve inhibition of cyclo-oxygenase (COX), the enzyme that converts arachidonic acid (AA) to prostaglandins.

AA is metabolized mainly through two major pathways (Fig. 7.2): the lipoxygenase pathway, leading to formation of leukotrienes, and the COX pathway, which consists essentially of two key enzymes: COX-1 and COX-2 [39]. Both isoforms convert AA to prostaglandin G_2 (PGG_2); the subsequent peroxidate reaction converts PGG_2 to PGH_2, leading to the formation of biologically important prostanoids [(PGI_2, PGE_2, PGD_2, thromboxane A_2 (TXA_2)], each of which exerts its unique functions coordinating signaling between the cell of origin (autocrine) and neighboring cells (paracrine) by binding to transmembrane G-protein-coupled receptors [40]. The production of these potent signaling molecules is strictly regulated, at the level of expression for COX-2, and at the level of catalysis for both COX-1 and COX-2 [41,42]. COX-1, in fact, is constitutively expressed in platelets and in the normal gastrointestinal mucosa, where it plays the physiological functions of platelet aggregation and gastrointestinal mucosa maintenance [43]. COX-2, in turn, is constitutively expressed in the human kidney and brain, but its expression is induced in many other tissues during inflammation, wound healing, and neoplasia [44]. Indeed, increased COX-2 expression has been shown in human CRC, in a close association with Duke's staging, indicating prognostic implications of COX-2 overexpression [45].

Non-selective NSAIDs inhibit both COX-1 and COX-2 [46], but exhibit different capabilities of inhibiting COX isoforms at different concentrations and in different tissues. For example, aspirin is a relatively selective inhibitor of COX-1 in platelets but inhibits COX-2 only at higher plasma concentrations [40,47]. Most other conventional NSAIDs, such as ibuprofen, sulindac, and indometacin, inhibit COX-1

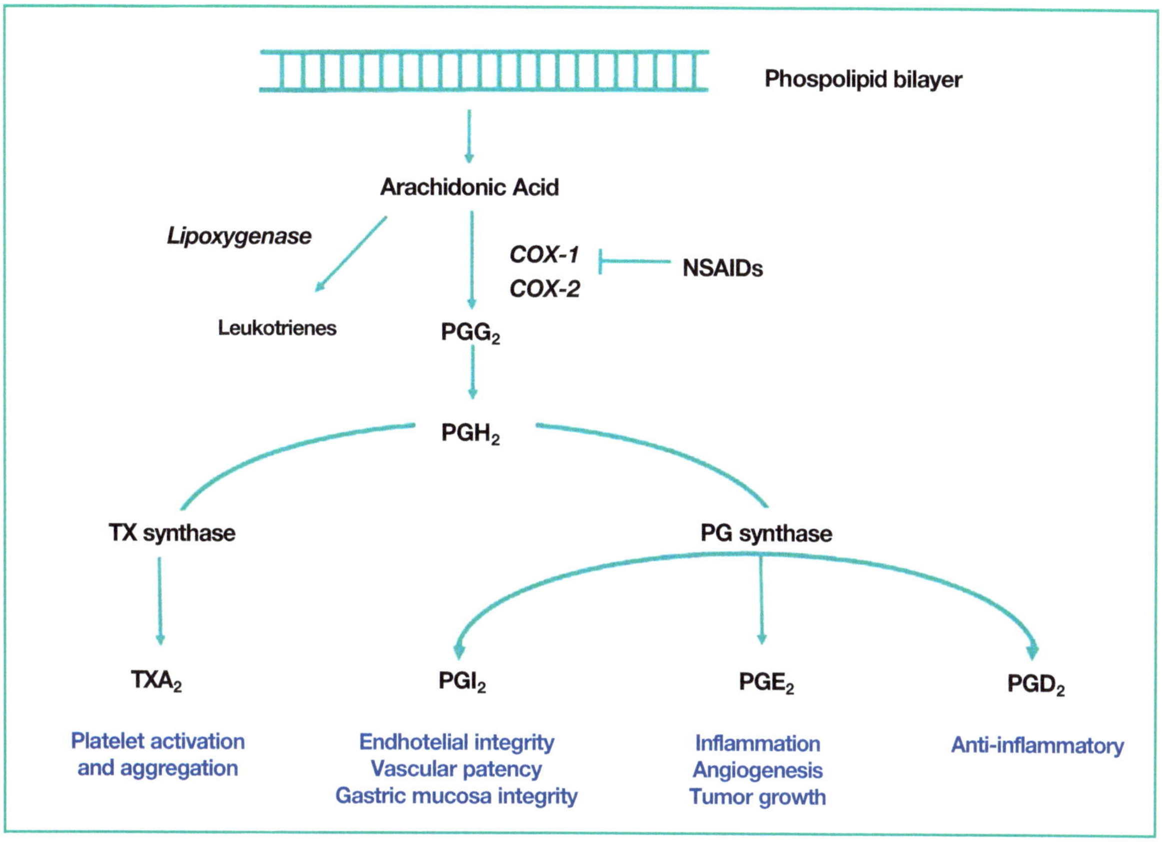

Fig. 7.2 Overview of prostaglandin (PG) synthesis pathways and their main functions. Arachidonic acid is metabolized mainly through two major pathways: the lipoxygenase pathway, leading to formation of leukotrienes, and the cyclo-oxygenase (COX) pathway, leading to formation of prostaglandins (PGI_2, PGE_2, PGD_2) or thromboxane A_2 (TXA_2), each of which exerts its unique functions. *NSAIDs* non-steroidal anti-inflammatory drugs

and COX-2 to the same extent, whereas a new class of NSAIDs, designated coxibs (such as celecoxib, rofecoxib, valdecoxib), selectively inhibits COX-2, while sparing COX-1, thus avoiding the most serious gastrointestinal toxic effects associated with chronic high-dose NSAID use [48,49].

COX-2 and PGE_2, a proinflammatory bioactive lipid produced in many human solid tumors including CRC [50], play an important role in tumorigenesis from the development to invasion and metastasis of carcinoma, through various mechanisms. In particular PGE_2 promotes tumor growth by stimulating PG receptor (EP) signaling pathways and downstream targets, such as epidermal growth factor receptor (EGFR), PPARδ, BCL2 and ERK2/JNK1, all of which are involved in promoting cellular proliferation, inhibiting apoptosis, and stimulating invasion and motility, as well as angiogenesis [51,52]. Overexpression of COX-2 in cancer tissues, in fact, may induce angiogenesis by increased production of PGE_2 and other COX-2-derived eicosanoids (i.e. TXA_2 and PGI_2), which stimulate endothelial cell migration and angiogenesis by modulating the expression of VEGF (vascular endothelial growth factor) and bFGF (basic fibroblast growth factor), and stimulating endothelial cell proliferation [53,54]. In addition, COX-2-independent mechanisms that contribute to neoangiogenesis in human cancer have also been described, acting through inhibition of EGR1 (early growth response protein) [55] or transcription factor Sp1 [56].

Several experimental results suggest that NSAIDs may prevent carcinogenesis through different pathways [51]. For example, NSAIDs can inhibit angiogenesis through increased endothelial cell apoptosis, inhibition of endothelial cell migration, recruitment of inflammatory cells and platelets, and/or TXA_2-mediated effects, all of which have been associated with growth inhibition and attenuation of the metastatic potential of cancer cells [53,57]. The expression of the anti-apoptotic proteins BCL2, BCL-XL, MCL1, and survivin decreases after treatment of cancer cells with celecoxib, whereas expression of the pro-apoptotic protein BAD increases [53,58], and rapid release of cytochrome c from mitochondria and activation of APAF1 and caspases 3, 8, and 9 are observed [53,59]. Furthermore, several studies indicate that NSAIDs induce the expression of proapoptotic molecules such as PAR4 (prostate apoptosis response) [60], ceramide [61], and 15-lipoxygenase-1 [62]. Aspirin has also been shown to decrease the expression of nuclear factor κB (NF-κB), a transcriptional factor that prevents apoptosis [63].

Several other mechanisms may be involved in the antineoplastic effects of NSAIDs, some of which may be independent of COX expression and prostaglandin biosynthesis [64]. For example, in animal models NSAIDs decrease β-catenin levels. As previously described, mutations in *APC* or β-catenin genes upregulate cytoplasmic β-catenin, which functions as a transcriptional activator of growth-promoting genes such as cyclin D1, c-myc, and PPARδ. NSAIDs decrease β-catenin levels, inducing a downregulation of cyclin D1 and PPARδ activity, and increasing *APC* mRNA expression [19,64,65].

The reduced expression of cyclins, which is only partially understood, resulting in a G1-phase arrest induced by NSAIDs seems to involve the inhibition of protein kinase B (PKB/AKT) or its upstream kinase phosphoinositide-dependent kinase 1 (PDK1) [66,67].

Since the first population-based case-control study was reported in 1988 [37], a growing body of evidence, from preclinical and clinical studies, has suggested a correlation between NSAID use and lower incidence of cancer. These results led to NSAIDs being considered as potential antineoplastic agents for cancer prevention [8,9,51]. However, the pharmacological effects of NSAIDs are complicated by the diverse functions of prostanoids in different tissues and by the variable effects of COX inhibition, depending on clinical context and drug dose.

At the present time, the available data concerning the use of non-specific NSAIDs for chemoprevention indicate that they would need to be ingested in doses greater than those used for cardiovascular prevention and for a duration of more than 10 years [9]. Consequently, the potential benefit of non-specific NSAIDs needs to be evaluated in relation to the gastrointestinal and renal toxicity caused by the inhibition of COX-1. These unwanted side-effects led to the development of coxibs. Indeed, highly selective COX-2 inhibitors retain the anti-inflammatory and antitumor effects of the NSAIDs while not interfering with COX-1, which is responsible for protection of the gastroduodenal mucosa from the effects of acid from the stomach [68].

For this reason, new expectations were raised for the use of selective COX-2 inhibitors in colorectal cancer prevention. Three international, multicenter studies were launched in 1999 and 2000: the Adenoma Prevention with Celecoxib (APC), the Adenomatous Polyp Prevention on Vioxx (APPROVe), and the Prevention of Colorectal Sporadic Adenomatous Polyps (PreSAP) trials [69–74]. The aims were to assess the effects of coxibs on the formation of adenomas and to evaluate the possibility of overcoming the gastrointestinal adverse events of non-selective NSAIDs. Although a reduced polyp recurrence and an improved risk profile for gastric toxicity was observed in all three trials, unexpectedly in the APPROVe and the APC studies this efficacy was associated with an increased risk of cardiovascular toxicity events [69]. On the advice of the Data Safety Monitoring Committee, the trials were stopped early due to a significant excess of adverse cardiovascular events (stroke, myocardial infarction, and heart failure) in patients receiving the drug for more than 18 months. These findings indicate that the potential benefit of COX-2 inhibitors for the prevention of colorectal adenomas needs to be carefully weighed against their potential cardiovascular effects, and cannot be routinely recommended for this indication. Similarly non-selective NSAIDs, despite ample evidence that their prolonged use substantially reduces the risk of developing colorectal cancer [75], show considerable unwanted side-effects such as damage to the gastric mucosa and gastrointestinal bleeding.

The available clinical evidence suggests that the therapeutic approach should take into account the optimal dose, starting age, and duration of NAISD use, in order to assess whether lower doses or other dose intervals may be associated with less cardiovascular or gastrointestinal risk. An innovative tool to predict NSAIDs' efficacy and/or toxicity is offered by the study of metabolic polymorphisms in NSAID targets or metabolizing enzymes. The use of pharmacogenetic approaches applied to epidemiological studies may allow the identification of reliable markers to predict drug-related adverse events, and indicates that the effectiveness of chemopreventive drugs can be modulated by the genotype of metabolizing enzymes. For example, the polymorphisms E158K and E308G in the flavin mono-oxygenase 3 gene (*FMO3*), a hepatic microsomal enzyme that inactivates sulindac, may reduce activity in catabolizing sulindac, and result in an increased efficacy for preventing polyposis in FAP [76].

The metabolic pathway of NSAIDs also involves glucuronidation and hydroxylation by the polymorphic enzymes UGT1A6 (uridine diphosphate-glucuronosyltransferase) and CYP2C9 (cytochrome P450). In a case-control study of adenoma recurrence and aspirin/NSAID use, the protective effect of aspirin was not seen in individuals who were homozygous wild type for *UGT1A6* and carrying at least one variant allele of *CYP2C9* [77]. The presence of a *CYP2C9*3* variant allele has been also associated with a significant high risk of gastroduodenal bleeding when treated with NSAIDs [78]. A similar correlation was reported between the NSAID-induced gastric ulcer and carrier of the *-1676T* allele in the COX-1 gene promoter [79], whereas other genetic variants in *COX-1* (*P17L* and *G230S*) significantly correlated to indometacin-mediated inhibition of COX-1 activity *in vitro* [80]. At present, the scientific evidence on these pharmacogenetic interactions is still very limited, and most studies were of limited sample size. Reliable detection of gene–NSAIDs interactions will require greater sample sizes, consistent definitions of NSAID use, and evaluation of clinical trial subjects of chemoprevention studies [81].

7.4 Nitric Oxide-Releasing NSAIDs

Nitric oxide-releasing non-steroidal anti-inflammatory drugs (NO-NSAIDs) are an innovative large family of pharmacologically active compounds, consisting of a classical NSAID to which an NO-releasing moiety is attached covalently, often via a spacer molecule [82]. NO-ASA (NO-aspirin) is the best-studied NO-NSAID to date, but the chemical linker spacer variation provides for a large number of variants of NO-NSAIDs, such as NO-sulindac, NO-ibuprofen, NO-indometacin, or NO-flurbiprofen [83].

Compared with their parent compounds, NO-NSAIDs inhibit the growth of cultured cancer cells several hundred times more potently, and prevent colon and pancreatic cancer in animal models. This enhanced chemopreventive effect is due to pleiotropic effects on cell signaling, inducing inhibition of proliferation, inhibition of cell-cycle-phase transitions, and induction of cell death involving redundant downstream pathways [84].

The early critical event in the action of NO-ASA, the best-studied NO-NSAID, is the induction of oxidative stress in target cells by a complex mechanism that is still unclear. The generation of a state of oxidative stress activates the caspase 9 of the intrinsic apoptosis pathway and other downstream reactive oxygen species (ROS)-responsive pathways critical to carcinogenesis, which inhibits cell proliferation and promotes cell death [83,84].

As previously described, the WNT pathway plays a crucial role in human colon carcinogenesis. In colon cancer cell lines the NO-ASA molecule inhibits β-catenin signaling, disrupting the association between TCF-4 and β-catenin in the nucleus, with reduction of transcriptional activity of genes modulated by this pathway, such as cyclin D1, c-Myc, and the antiapoptotic molecule PPARδ [85,86]. In a similar way, NO-ASA profoundly affects the nuclear interaction between the transcription factor NF-κB and DNA in cultured colon cancer cells, and inhibits the induction of NOS2, an enzyme whose activity is correlated with p53 mutations, and is overexpressed in cancer cells, and involved in the regulation of COX-2 [87].

An intriguing aspect of NO-ASA is the effect on COX-2 expression. *In vitro* experiments conducted on colon cancer and pancreatic cancer cell lines using NO-ASA for growth inhibition have demonstrated an increased expression of catalytically active COX-2, as demonstrated by an increased production of PGE_2 [87].

These results raise new questions about the biological role of COX-2 in colon carcinogenesis, suggesting that NSAIDs and NO-ASA prevent colorectal cancer by mechanisms that are independent of COX [84,88].

As evidence of this, another biological mechanism was recently demonstrated that might explain the antineoplastic effects of the molecule. It was recently shown that the administration of NO-ASA in human colon cancer animal models induces an inhibition of angiogenetic processes due to a deletion of the *VEGF* gene [89].

At present, *in vivo* studies using colon cancer animal models demonstrate an extraordinary chemopreventive effect, since treatment with NO-ASA for 3 weeks decreased the number of tumors by 55% [90] and significantly suppressed tumor incidence [91]. More-impressive results have been obtained on pancreatic cancer animal models, where NO-ASA reduced the incidence of pancreatic cancer by 88.9% [92]. These consistent findings in animal studies, and the reliable preclinical results, justify the need for clinical trials that may test the chemopreventive effectiveness of NO-NSAIDs [83].

7.5 Folate

Folic acid and its anionic form, folate, occur naturally in food and are indispensable nutrients in humans. They are essential for the production of nucleotides required for DNA synthesis and replication, and in the maintenance of intracellular normal methylation for the generation of *S*-adenosylmethionine, a cofactor required for cellular methylation reactions. This is especially important during periods of rapid cell division and growth such as infancy and pregnancy [93].

The pathway begins when folate is reduced to dihydrofolate (DHF) (Fig. 7.3), which is then reduced to tetrahydrofolate (THF) by dihydrofolate reductase. Methylenetetrahydrofolates are formed from THF by the addition of methylene groups from one-carbon donors. Methylenetetrahydrofolate reductase (MTHFR) is a key enzyme in folate metabolism, converting the intracellular 5,10-methylenetetrahydrofolate to 5-methyltetrahydrofolate, the predominant form of folate in plasma. 5-Methyltetrahydrofolate serves as the methyl group donor for the conversion of homocysteine to methionine, which then reacts with adenosine triphosphate to form *S*-adenosylmethionine (SAM), which is the methyl donor for DNA methylation. The tetrahydrofolate compounds are also substrates in a number of single-carbon-transfer reactions, involved in the production of dTMP (2'-deoxythymidine-5'-phosphate) from dUMP (2'-deoxyuridine-5'-phosphate), catalyzed by the enzyme thymidylate synthase (TS) for the synthesis of pyrimidinic nucleotides required for DNA synthesis [8].

The association between folate and cancer appears to be complex. Inadequacy of folate in the diet leads to alterations in genome methylation patterns, inducing DNA hypomethylation and consequent abnormal activation of several oncogenes involved in carcinogenesis. In addition, folate deficiency alters the balance of DNA precursors by dTTP insufficiency, and subsequent uracil accumulation in DNA instead of thymine, with genome instability [94].

Several pieces of epidemiological evidence support the association of folate intake with colon cancer risk [95]. In the Nurses' Health Study conducted on 88,756 women, Giovannucci et al found the greatest reduction in the risk of developing colon cancer among women taking 400 Ìg of folate for at least 15 years [96]. Other studies have confirmed that taking multivitamins containing folic acid is associated with a lower risk of CRC [97,98]. In individuals with a family history of CRC, the use of folate supplements for more than 5 years decreases the risk of neoplasia by almost 55% [99].

Conversely, recent randomized clinical trials found that folate supplements had no effect on adenoma recurrence [100] and did not reduce the risk of colorectal adenomas [101]. Furthermore, these studies suggested an effect of enhancing carcinogenesis progression in patients already suffering from cancer or from a precancerous condition. These results are in agreement with other recent reports which show that high concentrations of serum folate levels are associated with the risk of promoter methylation in tumor-suppressor-specific genes [102]. This is probably one of the molecular mechanisms by which folate supplementation might have a dual effect on carcinogenesis by protecting against initiation of adenoma formation but advantaging the cancer progression once cellular transformation has begun [9].

Within the folate pathway, candidate metabolic polymorphisms with a functional impact on protein function have been described in several of the enzymes. The *MTHFR* gene is polymorphic with a SNP within codon 677 (C to T, Ala to Val), and this variant encodes a thermolabile enzyme with reduced function that leads to a reduced plasma folate level. Individuals with homozygous mutation T/T are at decreased risk of developing colorectal cancer, compared with carriers with homozygous wild-type C/C or heterozygous C/T genotype. However, this protective effect is diminished by folate deficiency. The inverse associations with genetically reduced MTHFR activity is probably due to an impaired

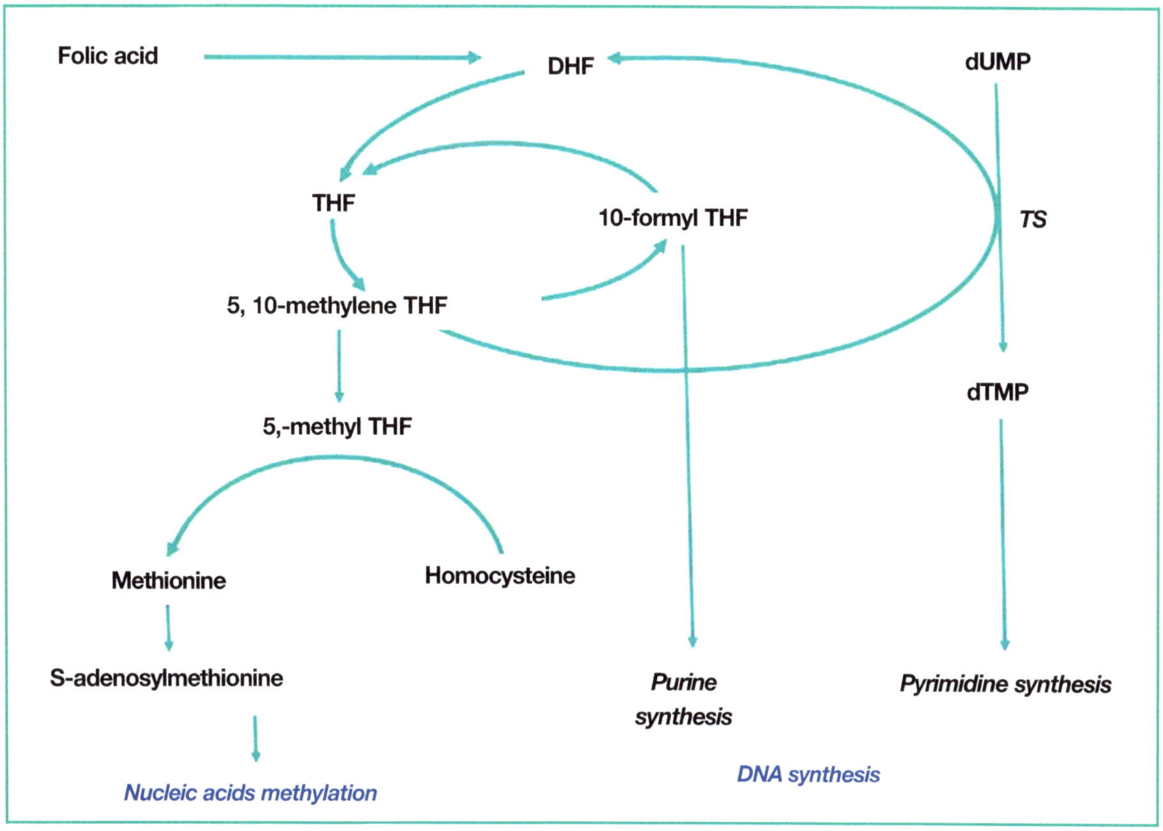

Fig. 7.3 Overview of folate metabolism. The folate is reduced to dihydrofolate (DHF), which is then reduced to tetrahydrofolate (THF) by dihydrofolate reductase. Methylenetetrahydrofolate reductase (MTHFR) converts 5,10-methylenetetrahydrofolate to 5-methyltetrahydrofolate. 5-Methyltetrahydrofolate serves as the methyl group donor for the conversion of homocysteine to methionine, which then reacts with adenosine triphosphate to form *S*-adenosylmethionine (SAM). The tetrahydrofolate compounds are also involved in the production of dTMP (2'-deoxythymidine-5'-phosphate) fromdUMP (2'-deoxyuridine-5'-phosphate) catalyzed by enzyme thymidylate synthase (TS) for the synthesis of pyrimidinic nucleotides requred for DNA synthesis

conversion and greater diversion of its substrate, 5,10-methylene THF, toward pyrimidine synthesis via the enzyme TS, and subsequent reduction of risk of uracil accumulation in DNA.

Studies performed in order to seek an interaction between 5'UTR of *TS* gene polymorphism and colorectal cancer risk showed that individuals with the TS 2R/2R genotype (low expression) are not at an increased risk of colorectal cancer, while surprisingly, individuals with the 3R/3R genotype (higher expression) appear more susceptible to colorectal adenoma in the presence of low folate or high alcohol intakes. These results suggest a further relevant mechanism linking folate metabolism to colorectal carcinogenesis, assuming that the increased availability of 5,10-methylene THF can also be directed towards the pathway of purinic synthesis [93].

7.6 Difluoromethylornithine

The polyamines putrescine, spermidine, and spermine are compounds that are necessary for normal cell proliferation, differentiation, and cell death, interacting with nucleic acids, membrane phospholipids, and several membrane-bound enzymes.

The biosynthesis of polyamines in eukaryotic cells begins from L-arginine (via L-ornithine) and L-methionine (Fig. 7.4). The key enzyme ODC (ornithine decarboxylase), via ornithine decarboxylation, produces putrescine, which forms spermidine and spermine, by the respective synthases, stable enzymes whose intracellular concentration is regulated by the availability of their substrates putrescine and spermidine.

The retroconversion pathway requires the action of SSAT enzyme (spermidine/spermine N^1-acetyltrans-

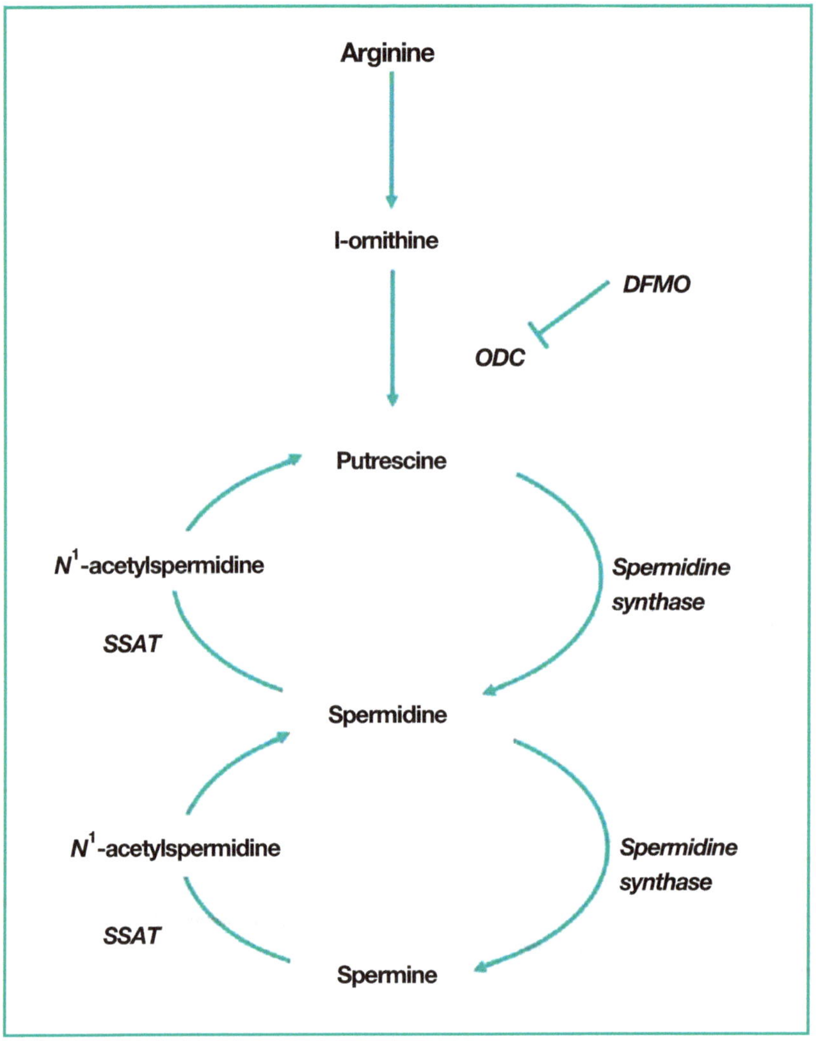

Fig. 7.4 Schematic representation of the polyamine metabolism pathway. Arginine is converted into l-ornithine, via the urea cycle enzyme arginase. The ornithine decarboxylase (ODC) produces putrescine, which forms spermidine and spermine, by the respective synthases. The retroconversion pathway requires the action of the enzyme spermidine/spermine N^1-acetyltransferase (SSAT) to recycle spermidine and putrescine from spermine and spermidine, respectively, forming intermediate products of polyamine catabolism, N^1-acetylspermidine and N^1-acetylspermine. Difluoromethylornithine (DFMO) is a selective and irreversible inhibitor of ODC

ferase) to recycle spermidine and putrescine from spermine and spermidine, respectively, forming intermediate products of polyamine catabolism, N^1-acetylspermidine and N^1-acetylspermine [103].

The presence of a link between alterations in polyamine metabolism and the processes of carcinogenesis has been highlighted by the presence of high levels of acetylpolyamines and ODC found in cancer tissues [103–105]. In particular, in much experimental evidence the activity of ODC was found to be elevated in sporadic colorectal carcinomas and adenomas and in the rectal mucosa of patients harboring *APC* gene mutations [8,106]. The molecular basis of this phenomenon relates to the fact that ODC is one of the transcriptional targets of the proto-oncogene c-myc [107], which in turn is repressed by the tumor suppressor gene *APC* pathway [65]. In other words, the inactivation of the gene *APC*, mutated/deleted in the majority of colorectal polyps, may increase the transcription of c-myc that can lead to induction and overexpression of ODC and consequent carcinogenesis [103].

This high activity in tumor cells has suggested ODC as a target of chemotherapy and chemoprevention. Difluoromethylornithine (DFMO) is a selective and irreversible inhibitor of ODC, inducing decreased production of putrescine and its derivate spermidine, and cell growth arrest.

Accordingly, in patients with a history of adenoma, the use of low-dose DFMO for 12 months suppressed the polyamine content of rectal mucosa with relatively few side-effects [108], mostly represented by a reversible hearing loss and thrombocytopenia [103].

The need to minimize the toxicity, the possibility of increasing the effectiveness of chemoprevention, and the observation that *in vitro* DFMO elevates COX-2 expression [8], has suggested the association of DFMO with NSAIDs for colorectal cancer prevention, and the combined administration of these compounds showed an increased efficacy compared to their use as single agents [109].

In this respect, the study of *ODC* gene metabolic polymorphisms has recently prompted further interest in combined therapy. The analysis of SNP +317 A/G in intron 1 of the *ODC* gene has, in fact, suggested a protective effect of aspirin on adenoma recurrence in patients harboring an A/A genotype, probably due to an increased catabolism of the polyamine pathway caused by an increased activity of the SST enzyme [110]. Large randomized clinical trials testing synergy between DFMO and several NSAIDs for colorectal cancer prevention are under way [9].

7.7 Ursodeoxycholic Acid

Cholic and chenodeoxycholic acids are the principal end-products of cholesterol metabolism and are secreted into the bile in the form of glycine and taurine conjugates. Under the action of anaerobic colonic microflora in the large bowel, they undergo enzymatic deconjugation and dehydroxylation and further biotransformation with the consequent production of the secondary lithocholic and deoxycholic acids (DCA) [111].

Recently, it has been reported that these secondary bile acids are cytotoxic to colonocytes, stimulate invasion and metastasis of colon carcinoma cells, and can have tumor-promoting capabilities via activation of multiple signaling pathways [112], mostly responsible for a suppression of the p53 protein levels in response to DNA-damaging agents (i.e. ionizing radiation), thus altering intracellular signaling [113]. Furthermore, DCA are capable of inducing mucin expression in human colon carcinoma cells by increasing MUC2 transcription through a process involving the protein kinase C (PKC) pathway [114].

Ursodeoxycholic acid (UDCA) is a synthetic bile acid, an epimer of tumor-promoting deoxycholic acid, and relatively more hydrophilic. The initially suggested mechanisms for the chemopreventive effect of ursodeoxycholic acid are induced reduction of the deoxycholic acid colonic concentration, and a cytoprotective action due to lower hydrophobicity compared to secondary bile acids [8].

Experimental animal studies have shown that UDCA significantly inhibits COX-2 protein and mRNA expression, through Ras/MAPK-dependent and -independent mechanisms [115]. More recently, a transcriptional pathway regulating COX-2 expression involving Ras, p38 and C/EBP, has been proposed as a target of UDCA's chemopreventive actions [116].

The chemopreventive effect of UDCA has been confirmed in preliminary clinical trials showing a long-term protective effect on relapse of colorectal cancer, a reduction in recurrence of adenomas with high-grade dysplasia [117,118], a short-term decrease of the proportion of DCA in fecal water and solids in relation to UDCA [119], and a lower gastrointestinal COX-2 expression [120].

7.8 Calcium and Vitamin D

The protective effect of calcium and vitamin D has been suggested by epidemiological studies that show an inverse correlation between their intake and the incidence of colorectal cancer [95,121].

The likely mechanism of action of calcium might reside in its ability to bind to bile acids, produced by a diet rich in red meat and animal fats, and to influence proliferation, differentiation, intracellular signals, and mechanisms of cellular apoptosis. Indeed, it has been shown that calcium intake induces a downregulation in epithelial-cell proliferation in the colonic mucosa of subjects at high risk of colon cancer [122], and can have a differential protective effect, depending on the genetic *Ki-ras* somatic mutations in human colon cancer [123].

The mechanism by which vitamin D level may alter cancer progression is explained by the action exerted by its ligation with vitamin D receptor (VDR) that seems to affect about 200 genes, some of which are co-activator, or co-repressor, and nuclear proteins involved in the regulation of cell proliferation, differentiation, and apoptosis [124].

As proof of such assumptions, experimental evidence has shown that vitamin D receptor knockout mice exhibited enhanced growth in response to predisposing factors and showed evidence of enhanced cancer development and growth [125].

Several clinical trials have proposed that the intake of calcium or vitamin D alone may be related to a reduction in adenoma recurrence and to a lower risk of colon cancer [9,95]. Consistent with these findings, a recent randomized clinical trial showed that calcium intake, either alone or in combination with vitamin D, might be related to lower risk of all cancers including CRC in postmenopausal women [126].

Despite these promising clinical results, however, a significant major obstacle in the routine use of these compounds as chemotherapeutic or chemopreventive agents [9] is represented by their dose-limiting hypercalcemic effects.

7.9 Selenium

Selenium is an essential trace element found in cereals, wheat, dairy products, meat, and fish and is a component of a series of compounds, such as some enzymes, and defined selenoproteins. A strong inverse association between selenium intake and colon cancer risk has been noted in several epidemiological studies and animal models [8,9]. The antioxidative and anti-inflammatory properties of selenoenzymes may explain the chemopreventive effect of selenium on CRC. In fact, recent studies highlight the correlation between genetic variants in selenoenzymes, abundantly expressed in the colon, and increased colorectal adenoma risk [127].

Although several biochemical pathways through which tumorigenesis is inhibited are currently acknowledged, the precise mechanisms by which selenium may prevent carcinogenesis are not yet known [128,129]. Experimental evidence has shown that several selenoprotein enzymes can increase DNA repair, activating the p53 tumor suppressor protein by a redox mechanism [128], and are important for maturation and activity of p53-mediated apoptosis [129].

Further experimental evidence has also helped to highlight the relationship between the metabolism of selenium and its chemopreventive effect on CRC, showing that selenium inhibits growth of colon cancer cells *in vivo* and *in vitro*, through the activation of AMPK via the downstream COX-2/PGE_2 pathway [130].

Selenium compounds used in cancer-prevention studies include the organic amino acid derivatives selenomethionine, selenocystein, and Se-methyl-selenocystein, inorganic salts selenite and selenate, as well as various synthetic inorganic compounds [130].

Several chemoprevention trials have been conducted using selenium supplementation in patients with sporadic colorectal adenomas and in healthy subjects, to investigate further whether selenium supplementation can reduce the cancer risk.

Combined analysis of pooled data from three randomized trials – the Wheat Bran Fiber Trial [131], the Polyp Prevention Trial [132], and the Polyp Prevention Study [133] – revealed that subjects with baseline blood selenium levels in the highest quartile had a significantly lower risk of adenoma recurrence when compared with those in the lowest quartile [134].

Primary and secondary prevention clinical trials that are currently in progress, such as the Selenium and Vitamin E Cancer Prevention Trial (SELECT), will provide further information not only about the inverse association observed between higher blood selenium concentration and adenoma or colorectal cancer risk, but also regarding some emerging evidence that selenium may affect not only cancer risk but also progression and metastasis [135,136].

7.10 Conclusion

Development of CRC is a multistep process, which results from complex and incompletely understood interactions between genes and the environment; therefore early detection and cancer prevention is certain to be a significant focus of translational research and intervention in the future, particularly since prevention of initiation requires lower doses of chemopreventive agents than effective inhibition or reversal of neoplastic progression.

At present, the chemopreventive action of various substances, such as NSAIDs, NO-NSAIDs, and folate, has been described, even if the ideal chemopreventive agent remains to be discovered, and several other questions regarding patient selection, optimal dosage, duration, and knowledge about the mechanism(s) by which these drugs act, need an answer [9].

A chemopreventive drug should have an efficacy of about 100% and an ideal safety profile; clearly, the risk versus benefit of any therapeutic intervention needs to be cautiously considered. For example, aspirin and NSAIDs, while appearing to be effective chemopreventive agents, have significant side-effects, such as gastrointestinal hemorrhage and stroke, which may limit their use for this indication; folate, ursodeoxycholic acid, and high-selenium yeast could represent potentially safer and cheaper alternatives, provided that they are shown to be effective in randomized controlled clinical trials. The effectiveness of chemoprevention could be improved through combinations of agents, limiting drug toxicity, and to be useful, any strategy of chemoprevention should be combined with screening and surveillance colonoscopy [137].

Further investigations of the metabolism of chemopreventive agents and of the enzyme polymorphisms involved in carcinogenesis are required, in order to conculsively identify which agents are most likely to be effective in particular genotype groups [8]. However, all the research and interest in pharmaceutical chemopreventive agents should not neglect the epidemiological evidence that lifestyle choices, such as regular physical exercise and a healthy diet will reduce an individual's risk of colorectal and other cancers. Indeed, it is estimated that up to 80% of colorectal cancers may be preventable by dietary change [138]. In conjunction with dietary measures, chemoprevention is currently being investigated in the population and in high-risk groups, since the achievement of this important goal may contribute to the conversion of CRC into a largely preventable disease [94,95].

References

1. Rustgi AK (2007) The genetics of hereditary colon cancer. Genes Dev 21:2525–2538.
2. Jemal A, Siegel R, Ward E, et al (2007) Cancer statistics. CA Cancer J Clin 57:43–66.
3. Bodmer WF (2006) Cancer genetics: colorectal cancer as a model. J Hum Genet 51:391–396.
4. Kinzler KW, Vogelstein B (1996) Lessons from hereditary colorectal cancer. Cell 87:159–170.
5. Johns LE, Houlston RS (2001) A systematic review and meta-analysis of familial colorectal cancer risk. Am J Gastroenterol 96:2992–3003.
6. Arnold CN, Goel A, Blum HE, et al (2005) Molecular pathogenesis of colorectal cancer: implications for molecular diagnosis. Cancer 104:2035–2047.
7. Telang NT, Li G, Katdare M (2006) Prevention of early-onset familial/hereditary colon cancer: new models and mechanistic biomarkers Int J Oncol 28:1523–1529.
8. Courtney EDJ, Melville DM, Leicester RJ (2004) Review article: chemoprevention of colorectal cancer. Aliment Pharmacol Ther 19:1–24.
9. Arber N, Levin B (2008) Chemoprevention of colorectal neoplasia: the potential for personalized medicine. Gastroenterology 134:1224–1237.
10. Dimitriadis A, Vincan E, Mohammed IM, et al (2001) Expression of WNT genes in human colon cancers. Cancer Lett 166:185–191.
11. Polakis P (2007) The many ways of WNT in cancer. Curr Opin Genet Dev 17: 45–51
12. Ken MC, Liu YI (2005) WNT signaling: complexity at the surface. J Cell Sci 119:395–402.
13. Eisenmann DM (2005) WNT signaling. WormBook 25:1–17.
14. Hajra KM, Fearon ER (2002) Cadherin and catenin alterations in human cancer. Genes Chromosomes Cancer 34:255–268.
15. Thorstensen L, Lothe RA (2003) The WNT signaling pathway and its role in human solid tumors. Atlas of Genetics and Cytogenetics in Oncology and Haematology. http://AtlasGeneticsOncology.org/Deep/WNTSignPathID20042.html (accessed 19 September 2008).
16. The WNT homepage. http://www.stanford.edu/~rnusse/wntwindow.html (accessed 19 September 2008).
17. Tetsu O, McCormick F (1999) Beta-catenin regulates expression of cyclin Dl in colon carcinoma cells. Nature 398:422–426.
18. Huber O, Korn R, McLaughlin J, et al (1996) Nuclear localization of beta-catenin by interaction with transcription factor LEF-1. Mech Dev 59:3–10.
19. He TC, Chan TA, Vogelstein B, et al (1999) PPARdelta is an APC-regulated target of nonsteroidal anti-inflammatory drugs. Cell 99:335–345.
20. He T-C, Sparks AB, Rago C, et al (1998) Identifcation of c-MYC as a target of the APC pathway. Science 281:1509–1512.
21. Mann B, Gelos M, Siedow A, et al (1999) Target genes of beta-catenin-T cell-factor/lymphoidenhancer-factor signaling in human colorectal carcinomas. Proc Natl Acad Sci U S A 96:1603–1608.
22. Palmirotta R, Curia MC, Esposito DL, et al (1995) Novel mutations and inactivation of both alleles of the APC gene in desmoid tumors. Hum Mol Genet 4:1979–1981.
23. Duval A, Gayet J, Zhou XP, et al (1999) Frequent frameshift mutations of the TCF-4 gene in colorectal cancers with microsatellite instability. Cancer Res 59:4213–4215.
24. Fukushima H, Yamamoto H, Itoh F, et al (2001) Frequent alterations of the β-catenin and TCF-4 genes, but not of the APC gene, in colon cancers with high-frequency microsatellite instability. J Exp Clin Cancer Res 20:553–559.
25. Palmirotta R, Matera S, Curia MC, et al (2004) Correlations between phenotype and microsatellite instability in HNPCC: implications for genetic testing. Fam Cancer 3:117–121.
26. Dihlmann S, von Knebel DM (2005) Wnt/beta-catenin-pathway as a molecular target for future anti-cancer therapeutics. Int J Cancer 113:515–524.
27. Schneikert J, Behrens J (2007) The canonical Wnt signalling pathway and its APC partner in colon cancer development. Gut 56:417–425.
28. Uthoff SM, Eichenberger MR, McAuliffe TL, et al (2001) Wingless-type frizzled protein receptor signaling and its

putative role in human colon cancer. Mol Carcinog 31:56–62.
29. Nojima M, Suzuki H, Toyota M, et al (2007) Frequent epigenetic inactivation of SFRP genes and constitutive activation of Wnt signaling in gastric cancer Oncogene 26:4699–4713.
30. Suzuki H, Watkins DN, Jair KW, et al (2004) Epigenetic inactivation of SFRP genes allows constitutive WNT signaling in colorectal cancer. Nat Genet 36:417–422.
31. Mao B, Wu W, Li Y, et al (2001) LDL-receptor-related protein 6 is a receptor for Dickkopf proteins. Nature 411(6835):321–325.
32. Aguilera O, Fraga MF, Ballestar E, et al (2006) Epigenetic inactivation of the Wnt antagonist DickkopF-1 (DKK-1) gene in human colorectal cancer. Oncogene 25:4116–4121.
33. Aguilera O, Peña C, García JM, et al (2007) The Wnt antagonist DICKKOPF-1 gene is induced by 1alpha,25-dihydroxyvitamin D3 associated to the differentiation of human colon cancer cells. Carcinogenesis 28:1877–1884.
34. Fuerer C, Nusse R, Ten Berge D (2008) Wnt signalling in development and disease. Max Delbrück Center for Molecular Medicine meeting on Wnt signaling in Development and Disease EMBO Rep 9:134–138.
35. Jaffe BM (1974) Prostaglandins and cancer: an update. Prostaglandins 6:453–461.
36. Bennett A, Del Tacca M (1975) Proceedings: prostaglandins in human colonic carcinoma. Gut 16:409.
37. Kune GA, Kune S, Watson LF (1988) Colorectal cancer risk, chronic illnesses, operations, and medications: Case control results from the Melbourne Colorectal Cancer Study. Cancer Res 48:4399–4404.
38. Yona D, Arber JN (2006) Coxibs and cancer prevention. Cardiovasc Pharmacol 47:S76–S81.
39. Fu JY, Masferrer JL, Seibert K, et al (1990) The induction and suppression of prostaglandin H2 synthase (cylooxygenase) in human monocytes. J Biol Chem 265:1727–1740.
40. Thun MJ, Henley SJ, Patrono C (2002) Nonsteroidal antiinflammatory drugs as anticancer agents: mechanistic, pharmacologic, and clinical issues. J Natl Cancer Inst 94:252–266.
41. Herschman HR (1996) Prostaglandin synthase 2. Biochim Biophys Acta 1299:125–140.
42. Kulmacz RJ (1998) Cellular regulation of prostaglandin H synthase catalysis. FEBS Lett 430:154–157.
43. Jones D, Carlton D, McIntyre T, et al (1993) Molecular cloning of human prostaglandin endoperoxide synthase type II and demonstration of expression in response to cytokines. J Biol Chem 268:9049–9054.
44. Eberhart CE, Coffey RJ, Radhika A, et al (1994) Up-regulation of cyclooxygenase 2 gene expression in human colorectal adenomas and adenocarcinomas. Gastroenterology 107:1183–1188.
45. Sheehan KM, Sheahan K, O'Dohoghue DP, et al (1999) The relationship between cyclooxygenase expression and colorectal cancer. JAMA 282:1254–1257.
46. FitzGerald GA, Patrono C (2001) The coxibs, selective inhibitors of cyclooxygenase-2. N Engl J Med 345:433–442.
47. Patrono C (2006) The PGH-synthase system and isozyme-selective inhibition. J Cardiovasc Pharmacol 47:S1–S6.
48. Peterson WL, Cryer B (1999) COX-1-sparing NSAIDs – is the enthusiasm justified? JAMA 282:1961–1963.
49. Willoughby DA, Moore AR, Colville-Nash PR (2000) COX-1, COX-2, and COX-3 and the future treatment of chronic inflammatory disease. Lancet 355:646–648.
50. Rigas B, Goldman IS, Levine L (1993) Altered eicosanoid levels in human colon cancer. J Lab Clin Med 122:518–523.
51. Guadagni F, Ferroni P, Palmirotta R, et al (2007) Nonsteroidal anti-inflammatory drugs in cancer prevention and therapy. Anticancer Res 27(5A):3147–3162.
52. Uefuji K, Ichikura T, Mochizuki H (2000) Cyclooxygenase-2 expression is related to prostaglandin biosynthesis and angiogenesis in human gastric cancer. Clin Cancer Res 6:135–138.
53. Grösch S, Jürgen Maier T, Schiffmann S, et al (2006) Cyclooxygenase-2 (COX-2)-independent anticarcinogenic effects of selective COX-2 inhibitors. J Natl Cancer Inst 98:736–747.
54. Tsujii M, Kawano S, Tsuji S, et al (1998) Cyclooxygenase regulates angiogenesis induced by colon cancer cells. Cell 93:705–716.
55. Ostrowski J, Wocial T, Skurzak H, et al (2003) Do altering in ornithine decarboxylase activity and gene expression contribute to antiproliferative properties of COX inhibitors? Br J Cancer 88:1143–1151.
56. Wei D, Wang L, He Y, et al (2004) Celecoxib inhibits vascular endothelial growth factor expression in and reduces angiogenesis and metastasis of human pancreatic cancer via suppression of Sp1 transcription factor activity. Cancer Res 64:2030–2038.
57. Pradono P, Tazawa R, Maemondo M, et al (2002) Gene transfer of thromboxane A(2) synthase and prostaglandin I(2) synthase antithetically altered tumor angiogenesis and tumor growth. Cancer Res 62:63–66.
58. Dandekar DS, Lopez M, Carey RI, et al (2005) Cyclooxygenase-2 inhibitor celecoxib augments chemotherapeutic drug-induced apoptosis by enhancing activation of caspase-3 and -9 in prostate cancer cells. Int J Cancer 115:484–492.
59. Jendrossek V, Handrick R, Belka C (2003) Celecoxib activates a novel mitochondrial apoptosis signaling pathway. Faseb J 17:1547–1549.
60. Zhang Z, DuBois RN (2000) Par-4, a proapoptotic gene, is regulated by NSAIDs in human colon carcinoma cells. Gastroenterology 18:1012–1017.
61. Chan TA, Morin PJ, Vogelstein B (1998) Mechanisms underlying nonsteroidal anti-inflammatory drug-mediated apoptosis. Proc Natl Acad Sci U S A 95:681–686.
62. Shureiqi I, Chen D, Lee JJ, et al (2000) 15 LOX-1: A novel molecular target of nonsteroidal anti-inflammatory drug induced apoptosis in colorectal cancer cells. J Natl Cancer Inst 92:1136–1142.
63. Shao J, Fujiwara T, Kadowaki Y, et al (2000) Overexpression of the wild-type p53 gene inhibits NF-kappa B activity and synergizes with aspirin to induce apoptosis in human colon cancer cells. Oncogene 19:726–736.
64. Husain SS, Szabo IL, Tamawski AS (2002) NSAID inhibition of GI cancer growth: clinical implications and molecular mechanisms of action. Am J Gastroenterol 97:542–553.
65. He T-C, Sparks AB, Rago C, et al (1998) Identification of c-myc as a target of the APC pathway. Science 281:1509–1512.
66. Lin HP, Kulp SK, Tseng PH, et al (2004) Growth inhibitory effects of celecoxib in human umbilical vein endothelial cells are mediated through G1 arrest via multiple signaling

mechanisms. Mol Cancer Ther 3:1671–1680.
67. Kulp SK, Yang YT, Hung CC, et al (2004) 3-phosphoinositidedependent protein kinase-1/Akt signaling represents a major cyclooxygenase-2-independent target for celecoxib in prostate cancer cells. Cancer Res 64:1444–1451.
68. Bombardier C, Laine L, Reicin A, et al (2000) Comparison of upper gastrointestinal toxicity of rofecoxib and naproxen in patients with rheumatoid arthritis. VIGOR Study Group. N Engl J Med 343:1520–1528.
69. Bresalier RS, Sandler RS, Quan H, et al (2005) Adenomatous Polyp Prevention on Vioxx (APPROVe) Trial Investigators: Cardiovascular events associated with rofecoxib in a colorectal adenoma chemoprevention trial. N Engl J Med 352:1092–1102.
70. Solomon SD, McMurray JJ, Pfeffer MA, et al (2005) Adenoma Prevention with Celecoxib (APC) Study Investigators: cardiovascular risk associated with celecoxib in a clinical trial for colorectal adenoma prevention. N Engl J Med 352:1071–1080.
71. Arber N, Eagle CJ, Spicak J, et al, PreSAP Trial Investigators (2006) Celecoxib for the prevention of colorectal adenomatous polyps. N Engl J Med 355:885–895.
72. Baron JA, Sandler RS, Bresalier RS, et al, APPROVe Trial Investigators (2006) A randomized trial of rofecoxib for the chemoprevention of colorectal adenomas. Gastroenterology 131:1674–1682.
73. Bertagnolli MM, Eagle CJ, Zauber AG, et al, APC Study Investigators (2006) Celecoxib for the prevention of sporadic colorectal adenomas. N Engl J Med 355:873–884.
74. Solomon SD, Pfeffer MA, McMurray JJ, et al, APC and PreSAP Trial Investigators (2006) Effect of celecoxib on cardiovascular events and blood pressure in two trials for the prevention of colorectal adenomas. Circulation 114:1028–1035.
75. Chan AT, Giovannucci EL, Meyerhardt JA, et al (2008) Aspirin dose and duration of use and risk of colorectal cancer in men. Gastroenterology 134:21–28.
76. Hisamuddin IM, Yang VW (2007) Genetic polymorphisms of human flavin-containing monooxygenase 3: implications for drug metabolism and clinical perspectives. Pharmacogenomics 8:635–643.
77. Bigler J, Whitton J, Lampe JW, et al (2001) CYP2C9 and UGT1A6 genotypes modulate the protective effect of aspirin on colon adenoma risk. Cancer Res 61:3566–3569.
78. Pilotto A, Seripa D, Franceschi M, et al (2007) Genetic susceptibility to nonsteroidal anti-inflammatory drug-related gastroduodenal bleeding: role of cytochrome P450 2C9 polymorphisms. Gastroenterology 133:465–471.
79. Arisawa T, Tahara T, Shibata T (2007) Association between genetic polymorphisms in the cyclooxygenase-1 gene promoter and peptic ulcers in Japan. Int J Mol Med 20:373–378.
80. Lee CR, Bottone FG Jr, Krahn JM, et al (2007) Identification and functional characterization of polymorphisms in human cyclooxygenase-1 (PTGS1) Pharmacogenet Genomics 17:145–160.
81. Cross JT, Poole EM, Ulrich CM (2008) A review of gene-drug interactions for nonsteroidal anti-inflammatory drug use in preventing colorectal neoplasia. Pharmacogenomics J 8:237–247.
82. Rigas B, Williams JL (2008) NO-donating NSAIDs and cancer: an overview with a note on whether NO is required for their action. Nitric Oxide 19:199–204.
83. Rigas B (2007) Novel agents for cancer prevention based on nitric oxide. Biochem Soc Trans 35:1364–1368.
84. Rigas B, Kashfi K (2004) Nitric-oxide-donating NSAIDs as agents for cancer prevention. Trends Mol Med 10(7):324–330.
85. Nath N, Kashfi K, Chen J, et al (2003) Nitric oxide-donating aspirin inhibits beta-catenin/T cell factor (TCF) signaling in SW480 colon cancer cells by disrupting the nuclear beta-catenin-TCF association. Proc Natl Acad Sci U S A 100:12584–12589.
86. Mackenzie GG, Rasheed S, Wertheim W, et al (2008) NO-Donating NSAIDs, PPARdelta, and Cancer: does PPARdelta contribute to colon carcinogenesis? PPAR Res 2008:919572.
87. Williams JL, Nath N, Chen J, et al (2003) Growth inhibition of human colon cancer cells by nitric oxide (NO)-donating aspirin is associated with cyclooxygenase-2 induction and β-catenin/T-cell factor signaling, nuclear factor-ÎB, and NO synthase 2 inhibition: implications for chemoprevention. Cancer Res 63:7613–7618.
88. Rigas B, Shiff SJ (2000) Is inhibition of cyclooxygenase required for the chemopreventive effect of NSAIDs in colon cancer? A model reconciling the current contradiction. Med Hypotheses 54:210–215.
89. Ouyang N, Williams JL, Rigas B (2008) NO-donating aspirin inhibits angiogenesis by suppressing VEGF expression in HT-29 human colon cancer mouse xenografts. Carcinogenesis 29:1794–1798.
90. Williams JL, Kashfi K, Ouyang N, et al (2004) NO-donating aspirin inhibits intestinal carcinogenesis in Min (APC(Min/+)) mice. Biochem Biophys Res Commun 313:784–788.
91. Kashfi K, Borgo S, Williams JL, et al (2005) Positional isomerism markedly affects the growth inhibition of colon cancer cells by nitric oxide-donating aspirin in vitro and in vivo. J Pharmacol Exp Ther 312:978–988.
92. Ouyang N, Williams JL, Tsioulias GJ, et al (2006) Nitric oxide-donating aspirin prevents pancreatic cancer in a hamster tumor model. Cancer Res 66:4503–4511.
93. Ulrich CM (2005) Nutrigenetics in cancer research–folate metabolism and colorectal cancer. J Nutr 135:2698–2702.
94. MacFarlane AJ, Stover PJ (2007) Convergence of genetic, nutritional and inflammatory factors in gastrointestinal cancer. Nutr Rev 65:S157–S166.
95. Ryan-Harshman M, Aldoori W (2007) Diet and colorectal cancer: review of the evidence. Can Fam Physician 53:1913–1920.
96. Giovannucci E, Stampfer MJ, Colditz GA, et al (1998) Multivitamin use, folate, and colon cancer in women in the Nurses' Health Study. Ann Intern Med 129:517–524.
97. Terry P, Jain M, Miller AB, Howe GR, et al (2002) Dietary intake of folic acid and colorectal cancer risk in a cohort of women. Int J Cancer 97:864–867.
98. La Vecchia C, Negri E, Pelucchi C, et al (2002) Dietary folate and colorectal cancer. Int J Cancer 102:545–547.
99. Fuchs CS, Willett WC, Colditz GA, et al (2002) The influence of folate and multivitamin use on the familial risk of colon cancer in women. Cancer Epidemiol Biomark Prev 11:227–234.

100. Logan RFA, Grainge MJ, Shepherd VC, et al (2008) UKCAP Trial Group.Aspirin and folic acid for the prevention of recurrent colorectal adenomas. Gastroenterology 134:29–38.
101. Cole BF, Baron JA, Sandler RS, et al (2007) Folic acid for the prevention of colorectal adenomas: a randomized clinical trial. JAMA 297:2351–2359.
102. Mokarram P, Naghibalhossaini F, Saberi Firoozi M (2008) Methylenetetrahydrofolate reductase C677T genotype affects promoter methylation of tumor-specific genes in sporadic colorectal cancer through an interaction with folate/vitamin B12 status. World J Gastroenterol 14:3662–3671.
103. Wallace HM, Fraser VA, Hughes A (2003) A perspective of polyamine metabolism. Biochem J 376:1–14.
104. Kingsnorth AN, Wallace HM (1985) Elevation of monoacetylated polyamines in human breast cancers. Eur J Cancer Clin Oncol 21:1057–1062.
105. Wallace HM, Duthie J, Evans DM, et al (2000) Alterations in polyamine catabolic enzymes in human breast cancer tissue. Clin Cancer Res 6:3657–3661.
106. Giardiello FM, Hamilton SR, Hylind LM, et al (1997) Ornithine decarboxylase and polyamines in familial adenomatous polyposis. Cancer Res 57:199–201.
107. Bello-Fernandez C, Packham G, Cleveland JL (1993) The ornithine decarboxylase gene is a transcriptional target of c-Myc. Proc Natl Acad Sci U S A 90:7804–7808.
108. Meyskens FL, Gerner EW, Emerson S, et al (1998) Effect of alphadifluoromethylornithine on rectal mucosal levels of polyamines in a randomized, double-blinded trial for colon cancer prevention. J Natl Cancer Inst 90:1212–1218.
109. Raul F (2007) Revival of 2-(difluoromethyl)ornithine (DFMO), an inhibitor of polyamine biosynthesis, as a cancer chemopreventive agent. Biochem Soc Trans 35:353–355.
110. Martinez ME, O'Brien TG, Fultz KE, et al (2003) Pronounced reduction in adenoma recurrence associated with aspirin use and a polymorphism in the ornithine decarboxylase gene. Proc Natl Acad Sci U S A 100:7859–7864.
111. Nair PP (1988) Role of bile acids and neutral sterols in carcinogenesis. Am J Clin Nutr 48(3 suppl):768–774.
112. Pérez-Holanda S, Rodrigo L, Viñas Salas J, et al (2007) Effect of ursodeoxycholic acid in an experimental colon cancer model. Rev Esp Enferm Dig 99:491–496.
113. Qiao D, Gaitonde SV, Qi W, et al (2001) Deoxycholic acid suppresses p53 by stimulating proteasome-mediated p53 protein degradation. Carcinogenesis 22:957–964.
114. Song S, Byrd JC, Koo JS, et al (2005) Bile acids induce MUC2 overexpression in human colon carcinoma cells. Cancer 103:1606–1614.
115. Khare S, Cerda S, Wali RK (2003) Ursodeoxycholic acid inhibits ras mutations, wild-type ras activation, and cyclooxygenase-2 expression in colon cancer. Cancer Res 63:3517–3523.
116. Khare S, Mustafi R, Cerda S, et al (2008) Ursodeoxycholic acid suppresses Cox-2 expression in colon cancer: roles of Ras, p38, and CCAAT/enhancer-binding protein. Nutr Cancer 60:389–400.
117. Alberts DS, Martínez ME, Hess LM, et al (2005) Phase III trial of ursodeoxycholic acid to prevent colorectal adenoma recurrence. J Natl Cancer Inst 97:846–853.
118. Brentnall TA (2003) Ursodiol: good drug makes good. Gastroenterology 124:1139–1140.
119. Hess LM, Krutzsch MF, Guillen J, et al (2004) Results of a phase I multiple-dose clinical study of ursodeoxycholic Acid. Cancer Epidemiol Biomarkers Prev 13:861–867.
120. Berkhout M, Roelofs HM, Friederich P, et al (2007) Ursodeoxycholic acid intervention in patients with familial adenomatous polyposis: a pilot study. Transl Res 150:147–149.
121. Peters U, McGlynn KA, Chatterjee N, et al (2001) Vitamin D, calcium, and vitamin D receptor polymorphism in colorectal adenomas. Cancer Epidemiol Biomarkers Prev 10:1267–1274.
122. Lipkin M, Newmark H (1985) Effect of added dietary calcium on colonic epithelial cell proliferation in subjects at high risk for familial colonic cancer. N Engl J Med 313:1381–1384.
123. Bautista D, Obrador A, Moreno V, et al (1997) Ki-ras mutation modifies the protective effect of dietary monounsaturated fat and calcium on sporadic colorectal cancer. Cancer Epidemiol Biomarkers Prev 6:57–61.
124. Carlberg C (2003) Current understanding of the function of the nuclear vitamin D receptor in response to its natural and synthetic ligands. Cancer Res 164:29–42.
125. Welsh JE (2004) Vitamin D and breast cancer: insights from animal models. Am J Clin Nutr 80(suppl): 1721S–12744S.
126. Lappe JM, Travers-Gustafson D, Davies KM, et al (2007) Vitamin D and calcium supplementation reduces cancer risk: results of a randomized trial. Am J Clin Nutr 85:1586–1591.
127. Peters U, Chatterjee N, Hayes RB, et al (2008) Variation in the selenoenzyme genes and risk of advanced distal colorectal adenoma. Cancer Epidemiol Biomarkers Prev 17:1144–1154.
128. Seo YR, Kelley MR, Smith ML (2002) Selenomethionine regulation of p53 by a ref1-dependent redox mechanism. Proc Natl Acad Sci U S A 99:14548–1153.
129. Anestål K, Arnér ESJ (2003) Rapid induction of cell death by selenium-compromised thioredoxin reductase 1 but not by the fully active enzyme containing selenocysteine. J Biol Chem 278:15966–15972.
130. Hwang JT, Kim YM, Surh YJ, et al (2006) Selenium regulates cyclooxygenase-2 and extracellular signal-regulated kinase signaling pathways by activating AMP-activated protein kinase in colon cancer cells. Cancer Res 66:10057–10063.
131. Alberts DS, Martinez ME, Roe DJ, et al (2000) Lack of effect of a high-fiber cereal supplement on the recurrence of colorectal adenomas. Phoenix Colon Cancer Prevention Physicians' Network. N Engl J Med 342:1156–1162.
132. Schatzkin A, Lanza E, Corle D, et al (2000) Lack of effect of a low-fat, high-fiber diet on the recurrence of colorectal adenomas. Polyp Prevention Trial Study Group. N Engl J Med 342:1149–1155.
133. Greenberg ER, Baron JA, Tosteson TD, et al (1994) A clinical trial of antioxidant vitamins to prevent colorectal adenoma. Polyp Prevention Study Group. N Engl J Med 331:141–147.
134. Jacobs ET, Jiang R, Alberts DS, et al (2004) Selenium and colorectal adenoma: results of a pooled analysis. J Natl Cancer Inst 96:1669–1675.

135. Xiang N, Zhao R, Song G, et al (2008) Selenite reactivates silenced genes by modifying DNA methylation and histones in prostate cancer cells. Carcinogenesis Aug 1 (Epub ahead of print).
136. Rayman MP (2005) Selenium in cancer prevention: a review of the evidence and mechanism of action. Proc Nutr Soc 64:527–542.
137. Ladabaum U, Chopra CL, Huang G, et al (2001) Aspirin as an adjunct to screening for prevention of sporadic colorectal cancer. A cost-effectiveness analysis. Ann Intern Med 135:769–781.
138. Cummings JH, Bingham SA (1998) Diet and the prevention of cancer. BMJ 317:1636–1640.

The Role of Imaging in Colonic Polyps and Polyposis

8

Riccardo Manfredi and Niccoló Faccioli

Abstract Radiology plays a critical role in the diagnosis and management of patients with colorectal polyps and cancer. Newer techniques, particularly virtual colonoscopy (VC), including computed tomography colonography (CTC), magnetic resonance colonography (MRC), and positron emission tomography (PET)/CTC may offer attractive alternatives for healthcare provider recommendation and patient use. With the exponential development in computer processing power, CT, MR, and PET/CTC offer numerous advantages over more traditional methods of radiological diagnosis, and provide essential information not only for initial diagnosis, but also for management, follow-up and detection of potential complications. Will CT, MR, and PET/CTC replace conventional colonoscopy in the future? We do not believe so at present.

Keywords Computed tomography • CT colonography • Double contrast barium enema • Magnetic resonance imaging • MR colonography • PET/CTC • Virtual colonoscopy

8.1 Introduction

Radiology plays a critical role in the diagnosis and management of patients with colorectal cancer. The double-contrast barium enema (DCBE) may be used for screening, especially for patients who are at higher than normal risk. It is also a method for diagnosing symptomatic colorectal cancer and for detecting complications such as obstruction or perforation.

Radiological examination of the colon can be performed with either single- or double-contrast technique. In most cases DCBE is superior for examination of the rectum and detection of small lesions [1]. In most cases, the role of radiology is to detect the presence of a polyp. A polyp is simply a protrusion of the mucosa into the bowel lumen. Therefore, it may be demonstrated as a radiolucent filling defect, as a contour defect, or as a ring shadow. The greatest and most frequent difficulty arises in distinguishing polyps from fecal residue. In general, fecal residue is mobile and is usually found on the dependent surface in the barium pool. In addition, several features may suggest that the filling defect represents a true polyp. Benign polyps appear as sessile soft tissue masses that protrude into the lumen of the bowel. The typical early colon cancer is a flat, sessile lesion that may produce a contour defect. A number of radiological criteria have been used for the detection of malignancy in colorectal polyps; the most important one is the size of the polyp. Malignant polyps tend to grow more quickly than benign polyps, although there is considerable overlap between the two groups. The presence of a long thin

R. Manfredi (✉)
Department of Radiology, University of Verona, Verona, Italy

G.G. Delaini, T. Skřička, G. Colucci (eds.), *Intestinal Polyps and Polyposis*,

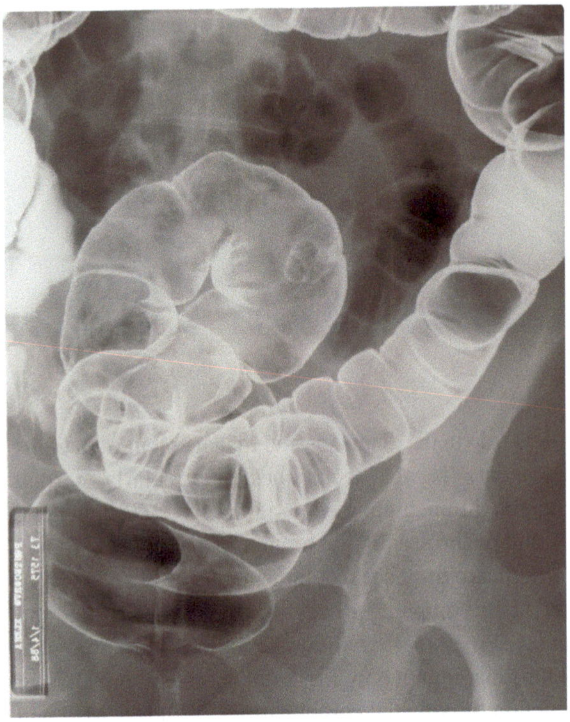

Fig. 8.1 Double-contrast barium enema examination of the sigmoid colon shows a polypoid mass vegetating within the bowel lumen

stalk is generally a sign of a benign polyp. If the head of the polyp is irregular or lobulated, the probability of malignancy is greater, although some benign polyps may have an irregular or lobulated surface (Fig. 8.1). The advanced lesions are generally annular or polypoid tumors seen as filling defects in the barium column, or as contour defects. Advanced cancers are often associated with "sentinel" polyps or additional polyps elsewhere in the colon. For this reason, whenever possible, the entire colon should be examined even when a carcinoma is encountered in the distal bowel. Carcinomas are particularly difficult to detect in patients with extensive diverticular disease of the sigmoid colon. If the radiological examination leaves any doubt regarding the presence of a carcinoma, virtual colonoscopy (VC) or flexible sigmoidoscopy should be recommended to confirm or exclude lesions in the sigmoid colon [1].

The most important complications of colorectal cancer include bleeding, bowel obstruction, and perforation with a pericolic abscess, or it may lead to a fistula to adjacent organs such as the stomach, duodenum, bladder, or vagina. The fistulous communication can be demonstrated by barium study or by the extracolonic presence of gas or contrast medium on computed tomography (CT) scans.

However, abdominal ultrasound is needed for assessment of distant metastases, and CT or magnetic resonance imaging (MRI) should be added for even more complete evaluation.

Patients who have had surgery for colorectal carcinoma undergo frequent postoperative examinations because of their relatively high risk of developing a second metachronous carcinoma. Ileocolic and colocolic anastomoses can be demonstrated in detail by DCBE examination. DCBE examination of the residual colon can usually be performed through colostomy [1].

Familial adenomatous polyposis syndrome (FAPS) is a relatively rare condition, but it is the most common of the polyposis syndromes [2]. FAPS is associated with the development of adenomas. The polyps involve all portions of the colon but may first appear distally. Thus, screening by procto-sigmoidoscopy (and colonoscopy or barium enema if adenomas are found) of all family members at risk for the disease is required.

The radiographic appearance of the colon in FAPS varies. Classically, innumerable small or moderate-sized filling defects carpet the entire colon. However, particularly in younger patients, the polyps may be more widely scattered. Correlation with colectomy specimens has shown that barium enemas significantly underestimate the number of polyps, especially in young patients whose polyps are usually less than 3 mm in diameter [3]. Carcinomas may present as a dominant polyp, a saddle lesion, or a typical infiltrating lesion. As in the general population, carcinomas are more commonly found in the left side of the colon.

Hamartomas of the Cronkhite–Canada, multiple hamartoma, and juvenile polyposis syndromes have similar histological features and share some radiological features [1–3].

Double-contrast examination of the alimentary tract is an accurate radiographic study for diagnosing these polyps. Enteroclysis is recommended for the diagnosis of small bowel polyps or obstruction. On CT scans, single and multiple Peutz–Jeghers polyps can be detected as soft tissue masses within the contrast-medium-filled intestinal loops.

Cross-sectional imaging modalities may also be

helpful for the diagnosis of extra-intestinal abnormalities. Cronkhite–Canada polyps are usually detected by this method, and are typically sessile in configuration.

The disease is best demonstrated radiographically with use of double-contrast techniques [1–3]. In the stomach, innumerable small to moderate-sized polyps carpet the mucosal surface, usually in its entirety. The small bowel, particularly the duodenum and terminal ileum, may contain multiple small polyps. The colon is diffusely involved but not to the extent seen in the stomach. Carpeting of the mucosa by polyps is less often seen in the colon. Despite the fact that the polyps are non-neoplastic, approximately 15% of reported patients have developed malignant neoplasms, including both gastric and colonic adenocarcinoma. The isolated juvenile polyp is the most common tumor of the colon in childhood. These patients most often present with painless rectal bleeding, and occasionally with a prolapsed rectal polyp; 60% of the polyps are located proximal to the rectum and sigmoid colon. The polyps are sessile or pedunculated, usually reddish, and often superficially ulcerated; they vary from 2 mm to more than 5 cm in diameter, and most patients have no family history of polyposis, so that the lesions presumably result from environmental factors or genetic mutations. Colonoscopy or DCBE should be performed for any child suspected of having a polyp.

Patients first diagnosed with juvenile polyposis of the colon should be further evaluated with DCBE or endoscopy for determining the presence of polyps elsewhere in the gastrointestinal tract. Because of the familial nature of the disease, management should be directed not only to the patient but also to the patient's family. The juvenile polyps are unevenly distributed throughout the gastrointestinal tract, with the small bowel and colon most severely involved [1–3].

When used appropriately, screening for CRC can reduce disease-related morbidity and mortality [4]. Current methods include fecal occult blood testing (FOBT), flexible sigmoidoscopy, DCBE, and conventional colonoscopy (CC); all are cost-effective techniques. Unfortunately, offering an array of options has not increased screening utilization, which continues to lag behind that of other common cancers [5]. Newer techniques, particularly VC, including computed tomography colonography (CTC), magnetic resonance colonography (MRC), and positron emission tomography (PET)/CTC may offer attractive alternatives for healthcare provider recommendation and patient use [6].

When used appropriately, screening for CRC can reduce disease-related morbidity and mortality [7]. Recent studies stress the fact that finding and resecting advanced adenomatous polyps, and thereby preventing cancer, is becoming a primary objective of screening programs. Several papers also show the potential of newly emerging methods of screening by imaging the colon with VC [8]. The sensitivity of VC for large adenomas and CRC appears to be high, although results vary by center and there is a steep learning curve.

VC is a new method for studying the colon. It consists of acquisition of CT, MR, and PET/CT images and can elaborate them with a workstation, creating endoluminal vision as good as traditional colonoscopy does, permitting the complete exploration of colonic lumen, and of tumoral stenosis [9]. Analysis of the differences between CT, MR, and PET/CT colonography shows that these techniques present both advantages and disadvantages, such as the impossibility of performing MR in patients with a pacemaker, or in claustrophobic patients, and the impossibility of performing CT with iodinated agents in patients with renal failure or with a history of adverse reactions. The increased use of these techniques is due to the high sensitivity of new-generation CT, MR, and PET/CT machines, the increased spatial resolution, specific software for digital cleaning of the colon, the introduction of high-end workstations, and the possibility of computer-assisted diagnosis (CAD) [10]. Therefore, it is desirable that the increasing spread of multidetector CT devices and future technical innovations should have the effect of increasing the use and experience of VC in various diagnostic centers, enabling increased use of VC as a screening tool [11].

Current CT techniques require meticulous bowel preparation and gas insufflation prior to the examination. With new multidetector CT scanners, the procedure requires a scan time of about 25–30 s, and sedation is not used. The advantages of CTC over CC include its safety, its ability to demonstrate the entire large bowel in almost all patients, even following incomplete endoscopy, and to accurately localize lesions, and examine the entire colon in patients with obstructing tumors. Additionally, CTC allows simultaneous preoperative tumor staging. There are few reported complications from CTC [12,13].

Screening for colorectal polyps is a controversially discussed indication for CTC. Sensitivity and specificity range widely and decrease with decreasing polyp size. Most frequently, the examination is well tolerated and assessed by patients to be more acceptable than CC [14]. CTC seems sufficiently sensitive and specific in the detection of large and medium polyps [15,16]. The sensitivity of CTC was heterogeneous but improved as polyp size increased (48% for detection of polyps <6 mm, 70% for polyps between 6 and 9 mm, and 85% for polyps >9 mm) [14,15]. Characteristics of the CTC scanner, including width of collimation, type of detector, and mode of imaging, explained some of this heterogeneity. In contrast, specificity was homogenous (92% for detection of polyps <6 mm, 93% for polyps from 6 to 9 mm, and 97% for polyps >9 mm) (Fig. 8.2). CTC is highly specific, but the range of reported sensitivities is wide [17]. Patient or scanner characteristics do not fully account for this variability, but collimation, type of scanner, and mode of imaging explain some of the discrepancy. This heterogeneity raises concerns about consistency of performance and about technical variability. These issues must be resolved before CTC can be advocated for generalized screening for CRC [18].

Because CRC has widely varying appearances in both endoscopy and CTC, familiarity with the gamut of morphologic appearances can help improve interpretation of the results [19,20]. The addition of intravenous contrast material to CTC can aid differentiation of true colonic masses from pseudolesions such as residual stool, and improves the depiction of enhancing masses that might otherwise be obscured by residual colonic fluid [21]. In contrast to staging of most other tumors, staging of CRC depends more on the depth of tumor invasion than on the size of the primary mass. The diverse appearances of colorectal cancers at two- and three-dimensional CTC include sessile, annular, ulcerated, necrotic, mucinous, invasive, and non-invasive lesions [22]. Imaging pitfalls that can simulate or obscure neoplasms are retained fecal material or fluid, incomplete distention, and

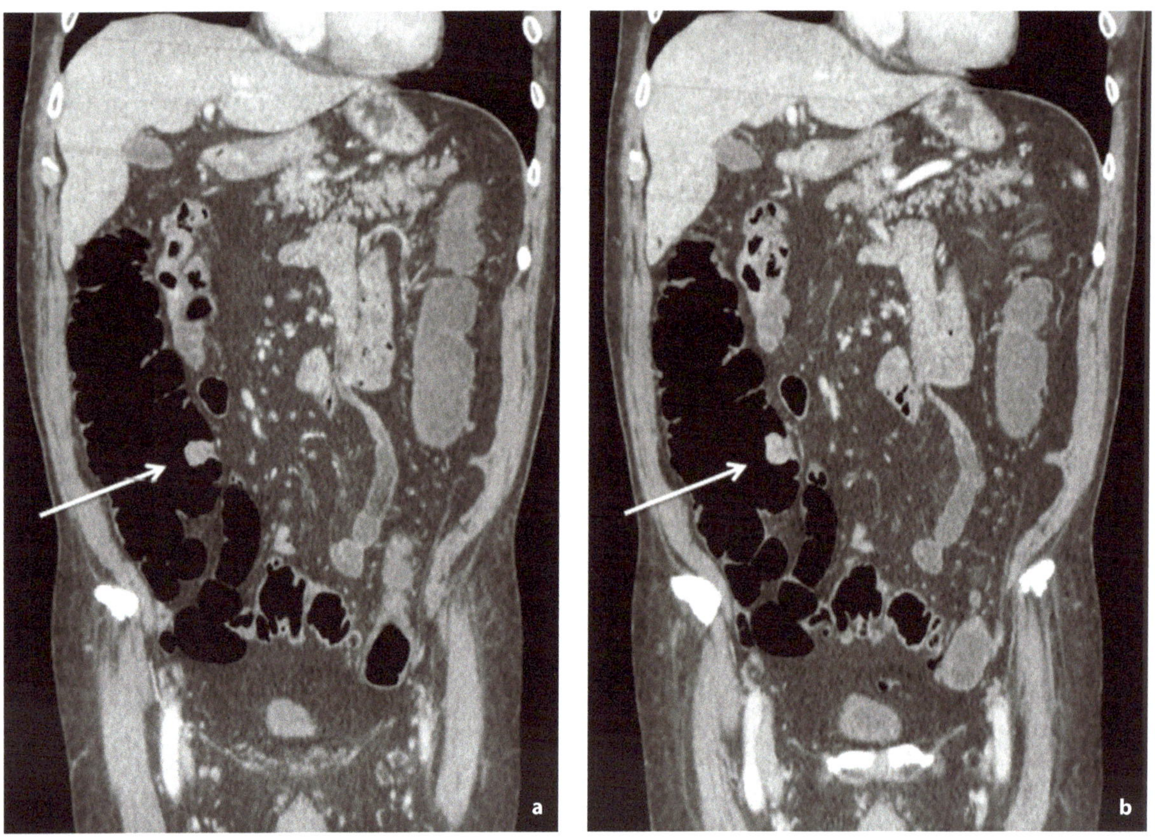

Fig. 8.2a,b Colonography of the right colon. The CT examination shows a polypoid lesion originating from the medial wall of the right colon, protruding within the bowel lumen (*arrow* in a); the stalk can also be depicted on CT colonography

advanced diverticulosis [23]. Contrast-enhanced CTC can simultaneously evaluate metastatic disease, local recurrence, and metachronous neoplasia in CRC, or in recurrent CRC. Contrast-enhanced CTC has the potential to detect local recurrence, metachronous disease, and distant metastases in patients with a history of invasive CRC. Suboptimal sigmoid distention can be seen on contrast-enhanced CTC, predominantly in patients with right hemicolectomies [24]. Contrast-enhanced CTC is a promising method for detecting local recurrence, metachronous disease, and distant metastases in patients with prior invasive CRC. The technique can also serve as a useful adjunct to colonoscopy by detecting local recurrences or metachronous disease that are endoscopically obscure, or by serving as a full structural colonic examination when endoscopy is incomplete [25].

8.2 Magnetic Resonance Colonography

Magnetic resonance colonography (MRC) has gained access into clinical routine as a means of assessing the large bowel. There are widely accepted indications for MRC, especially in patients with incomplete CC. Furthermore, virtual MRC is increasingly propagated as a screening tool, with particular advantages inherent to the non-invasive character of the procedure and the lack of exposure to ionizing radiation. Beyond a sufficiently high diagnostic accuracy, outstanding patient acceptance is a major advantage of MRC as a diagnostic modality [26]. A precondition for establishment of MRC as a diagnostic tool in secondary prevention of CRC is not only high diagnostic accuracy but also a good acceptance among patients [27].

Dark-lumen MRC has failed to detect all polyps smaller than 5 mm in diameter; these are generally not clinically relevant at the moment of their detection and thus can be kept under surveillance. However, MRC as a non-invasive imaging modality is a promising alternative to CC in the detection of clinically relevant polyps larger than 5 mm in diameter [28]. In patients at increased risk for CRC, specificity of MRC by using limited bowel preparation was high, but sensitivity was modest [29–31].

MRC is useful for detection of colonic pathology and assessment of the proximal colon in patients with colonic cancer after incomplete colonoscopy [32,33] (Fig. 8.3). A meta-analysis investigated MRC versus colonoscopy as a diagnostic investigation for CRC. This study suggested that MRC is an imaging technique with high discrimination for cases presenting with CRC. However, the exact diagnostic role of MRC needs to be clarified. Further evaluation is necessary to refine its applicability and diagnostic accuracy in comparison with other imaging methods such as CTC [34]. MRC could be useful in screening programs of patients at high risk for colon cancer. Patients with

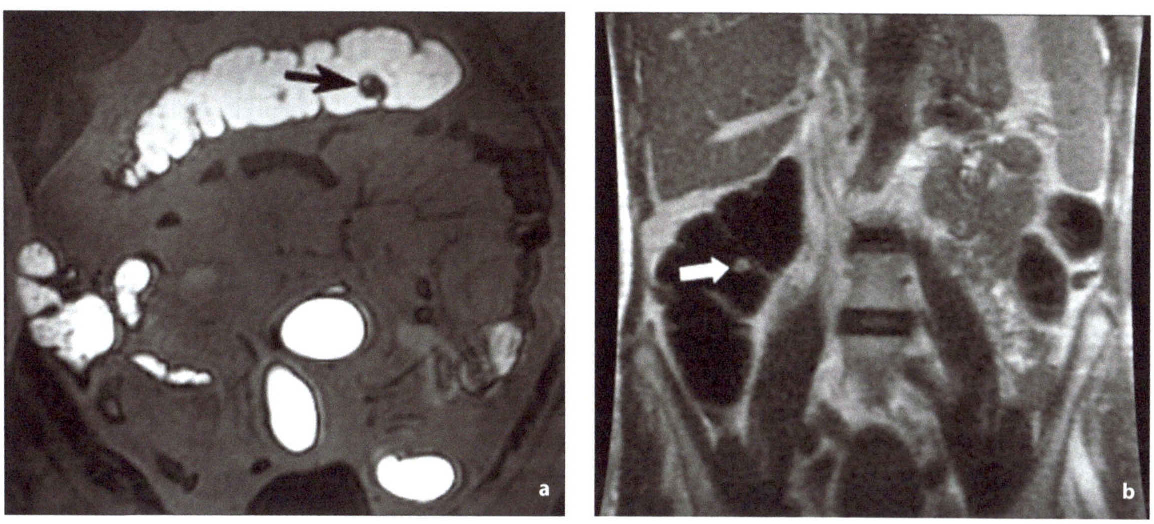

Fig. 8.3a,b MR colonography. **a**, Coronal true FISP (fast imaging with steady-state precision) image following the administration of polyethylene glycol shows a polypoid lesion within the bowel lumen (*arrow*). **b**, T1 weighted image on coronal plane showing the same lesion

MRC-detected endoluminal lesions must undergo CC for histological diagnosis [35].

8.3 Computed Tomography Colonography Versus Magnetic Resonance Colonography

A meta-regression technique was performed to compare the diagnostic accuracy of CTC and MRC, compared with CC for patients presenting with CRC. Overall sensitivity and specificity of CTC (both 95%) and MRC (both 95%) in detection of CRC were respectively similar. Meta-regression analysis showed no significant difference in the diagnostic accuracy of the two modalities. Both tests showed a large area under the summary receiver operating characteristic curve, with high diagnostic odds ratios. Factors that enhanced the overall accuracy of MRC were the use of intravenous contrast, fecal tagging, and exclusion of low-quality studies. This meta-analysis suggested that CTC and MRC have similar diagnostic accuracy in detecting CRC. Study quality, size and intravenous/intraluminal contrast agents affect diagnostic accuracies. For an exact comparison to be made, studies evaluating CTC, MRC, and CC in the same patient cohort would be necessary [36,37].

8.4 Virtual Colonoscopy Versus Conventional Colonoscopy

CTC is reliable for detecting lesions that are 6 mm or larger in size. It permits evaluation of the region proximal to an occlusive growth, which is often impossible with CC [37]. CTC is a good imaging tool for the exclusion of CRC in a population unfit for or unable to complete colonoscopy or barium enema, with reasonable sensitivity and specificity for detection of CRC [38]. MRC is a promising modality with high accuracy for detecting colorectal polyps larger than 5 mm in diameter. In inflammatory bowel disease (IBD), MRC can be used to assess disease activity, including spreading [39,40]. In detecting colonic lesions, MRC achieved a diagnostic accuracy similar to CC. However, MRC is minimally invasive, with no need for sedation or analgesics during investigation. There is a lower percentage of perforation risk, and all colon segments can be evaluated due to multisectional imaging availability; intramural, extra-intestinal components of colonic lesions, metastasis, and any additional lesions can be evaluated easily [41]. MRC is a feasible and useful method of evaluating the entire colon in patients with incomplete CC. The majority of patients find MRC less unpleasant than CC, and a majority would prefer MRC over CC for a future colon examination. MRC also appears to be less time consuming than CC with postprocedural monitoring, for patients and medical personnel [42,43]. MRC proved reliable in evaluating the majority of colonic segments that were inaccessible with CC [44]. The identification of additional disease at MRC underscores the need for a second diagnostic step in the setting of incomplete CC [44].

8.5 Complications and Limitations

Perforation of the colon and rectum is a rare complication of VC. Older age and underlying concomitant colonic disease were present in patients with perforation [45]. The cancer risks associated with the radiation exposure from VC are unlikely to be zero, but they are small. A best estimate for the absolute lifetime cancer risk associated with the radiation exposure using typical current scanner techniques is about 0.14% for paired VC scans for a 50-year-old, and about half that for a 70-year-old. These values could probably be reduced by factors of five or ten with optimized VC protocols [46].

There are many differences between the studies with high sensitivity (94%) [47]. Additional obstacles for implementation in prevention of CRC may be controversial results concerning patient acceptance, the large-scale use of ionizing radiation, difficulties in detecting flat adenomas, and extracolonic findings [48]. Flat lesions and small polyps are the other two main causes for missed lesions at eight multi-detector rows CTC [49,50]. Currently, CTC is less cost-effective than conventional endoscopy [51]. CTC and MRC present both advantages and disadvantages, such as the impossibility of performing MR in patients with a pacemaker or in claustrophobic patients, and the impossibility of performing CT with iodinated agents in patients with renal failure or with a history of adverse reactions [52,53]. Although application of PET and PET/CT in CRC diagnosis, staging and restaging has been widely accepted by oncologists, there are no data relating to PET/CTC in CRC screening. However, in high-risk patients and follow-up of CRC patients after treatment, the examination is still cost-effective. The dose of radiation exposures for the patient who accepts this examination is another major problem, which should be further studied in the near future [52,53].

8.6 Cost-Effectiveness Issues

CTC is an effective screening test for colorectal neoplasia. However, it is more expensive and generally less effective than CC. CTC can be reasonably cost-effective when its diagnostic accuracy is high, as with primary three-dimensional technology, as the costs are about 60% of those of CC. Overall, CTC technology will need to improve its accuracy and reliability to be a cost-effective screening option [54]. CRC screening is cost-saving in Italy, irrespective of the technique applied. CTC appears to be more cost-effective than flexible sigmoidoscopy, and it may also become a valid alternative to CC [55]. VC involves a CT or MR scan of the abdomen and pelvis to detect colorectal polyps and cancer. Both modalities have shown promising sensitivity in revealing larger polyps, in comparison with CC. Caution should be exercised in its clinical implementation, due to significant interobserver variation and individual learning curves. CTC can be performed cost-effectively compared to CC. CTC may be recommended in preference to DCBE after incomplete CC.

With the exponential development in computer processing power, CT, MR, and PET/CT colonography offer numerous advantages over more traditional methods of radiological diagnosis, and provide essential information not only for initial diagnosis, but also for management, follow-up, and detection of potential complications. Will CT, MR, and PET/CT colonography replace conventional colonoscopy in the future? We do not believe so at present. However, combined with several derivative techniques on the horizon involving stool DNA testing, computer-aided detection, and PET/MRI colonography, these techniques may further improve the specificity and sensitivity of imaging modalities in CRC screening and save the colonoscopy resource for those patients who need treatment.

References

1. Levine MS, Rubesin SE, Laufer I, et al (2000) Diagnosis of colorectal neoplasms at double-contrast barium enema examination. Radiology 216:11–18.
2. Gebert HF, Jagelman DG, McGannon E (1986) Familial polyposis coli. Am Fam Physician 33:127–137.
3. Bartram CI, Thornton A (1984) Colonic polyp patterns in familial polyposis. AJR Am J Roentgenol 142:305–308.
4. Low G, Tho LM, Leen E, et al (2008) The role of imaging in the pre-operative staging and post-operative follow-up of rectal cancer Surgeon 6:222–231.
5. Bromer MQ, Weinberg DS (2005) Screening for colorectal cancer now and the near future. Semin Oncol 32:3–10.
6. Burling D, Moore A, Taylor S, et al (2007) Virtual colonoscopy training and accreditation: a national survey of radiologist experience and attitudes in the UK. Clin Radiol 62:651–659.
7. Zimmerman RK, Nowalk MP, Tabbarah M, et al (2006) Predictors of colorectal cancer screening in diverse primary care practices. BMC Health Serv Res 6:116.
8. Saliangas K (2004) Screening for colorectal cancer. Tech Coloproctol 8(suppl 1):s10–s13.
9. Ajaj W, Goyen M (2007) MR imaging of the colon: technique, indications, results and limitations. Eur J Radiol 61:415–423.
10. Luo MY, Shan H, Yao LQ, et al (2004) Postprocessing techniques of CT colonography in detection of colorectal carcinoma. World J Gastroenterol 10: 1574–1577.
11. Gollub MJ, Schwartz LH, Akhurst T (2007) Update on colorectal cancer imaging. Radiol Clin North Am 45:85–118.
12. Limburg PJ, Fletcher JG (2006) Making sense of CT colonography related complication rates. Gastroenterology 131:2023–2024.
13. Sosna J, Sella T, Bar-Ziv J, et al (2006) Perforation of the colon and rectum—a newly recognized complication of CT colonography. Semin Ultrasound CT MR 27:161–165.
14. Graser A, Stieber P, Nagel D, et al (2008) Comparison of CT Colonography, Colonoscopy, Sigmoidoscopy, and Fecal Occult Blood Tests for the Detection of Advanced Adenoma in an Average Risk Population. Gut Oct 13 [Epub ahead of print].
15. Whitlock EP, Lin JS, Liles E, et al (2008) Screening for Colorectal Cancer: A Targeted, Updated Systematic Review for the U.S. Preventive Services Task Force. Ann Intern Med Oct 6 [Epub ahead of print].
16. Kim DH, Pickhardt PJ, Hoff G, et al (2007) Computed tomographic colonography for colorectal screening. Endoscopy 39:545–549.
17. Bazensky I, Shoobridge-Moran C, Yoder LH (2007) Colorectal cancer: an overview of the epidemiology, risk factors, symptoms, and screening guidelines. Medsurg Nurs 16:46–51.
18. Mulhall BP, Veerappan GR, Jackson JL (2005) Meta-analysis: computed tomographic colonography. Ann Intern Med 142:635–650.
19. Silva AC, Hara AK, Leighton JA, et al (2005) CT colonography with intravenous contrast material: varied appearances of colorectal carcinoma. Radiographics 25:1321–1334.

20. Yoshida H, Nappi J (2007) CAD in CT colonography without and with oral contrast agents: progress and challenges. Comput Med Imaging Graph 31:267–284.
21. Hjern F, Jonas E, Holmstrom B, et al (2007) CT colonography versus colonoscopy in the follow-up of patients after diverticulitis – a prospective, comparative study. Clin Radiol 62:645–650.
22. Mang T, Graser A, Schima W, et al (2007) CT colonography: techniques, indications, findings. Eur J Radiol 61:388–399.
23. Dachman AH, Dawson DO, Lefere P, et al (2007) Comparison of routine and unprepped CT colonography augmented by low fiber diet and stool tagging: a pilot study. Abdom Imaging 32:96–104.
24. Dachman AH, Lefere P, Gryspeerdt S, et al (2007) CT Colonography: Visualization Methods, Interpretation, and Pitfalls. Radiol Clin North Am 45:347–359.
25. Fletcher JG, Johnson CD, Krueger WR, et al (2002) Contrastenhanced CT colonography in recurrent colorectal carcinoma: feasibility of simultaneous evaluation for metastatic disease, local recurrence, and metachronous neoplasia in colorectal carcinoma. AJR Am J Roentgenol 178:283–290.
26. Lauenstein TC (2006) MR colonography: current status. Eur Radiol 16:1519–1526.
27. Kinner S, Lauenstein TC (2007) MR Colonography. Radiol Clin North Am 45:377–387.
28. Ajaj W, Ruehm SG, Gerken G, et al (2006) Strengths and weaknesses of dark-lumen MR colonography: clinical relevance of polyps smaller than 5 mm in diameter at the moment of their detection. J Magn Reson Imaging 24:1088–1094.
29. Bielen DJ, Bosmans HT, De Wever LL, et al (2005) Clinical validation of highresolution fast spin-echo MR colonography after colon distention with air. J Magn Reson Imaging 22:400–405.
30. Haykir R, Karakose S, Karabacakoglu A, et al (2006) Detection of colonic masses with MR colonography. Turk J Gastroenterol 17:191–197.
31. Luboldt W, Bauerfeind P, Wildermuth S, et al (2000) Colonic masses: detection with MR colonography. Radiology 2000; 216: 383–388.
32. Kinner S, Kuehle CA, Langhorst J, et al (2007) MR colonography with fecal tagging: do individual patient characteristics influence image quality? J Magn Reson Imaging 25:1007–1012.
33. Ajaj W, Lauenstein TC, Pelster G, et al (2005) MR colonography in patients with incomplete conventional colonoscopy. Radiology 234:452–459.
34. Purkayastha S, Tekkis PP, Athanasiou T, et al (2005) Magnetic resonance colonography versus colonoscopy as a diagnostic investigation for colorectal cancer: a meta-analysis. Clin Radiol 60:980–989.
35. Pappalardo G, Polettini E, Frattaroli FM, et al (2000) Magnetic resonance colonography versus conventional colonoscopy for the detection of colonic endoluminal lesions. Gastroenterology 119:300–304.
36. Purkayastha S, Athanasiou T, Tekkis PP, et al (2007) Magnetic resonance colonography vs computed tomography colonography for the diagnosis of colorectal cancer: an indirect comparison. Colorectal Dis 9:100–111.
37. Macari M, Bini EJ (2005) CT colonography: where have we been and where are we going? Radiology 237: 819–833.
38. Gallo TM, Galatola G, Laudi C, et al (2006) CT colonography: screening in individuals at high risk for colorectal cancer. Abdom Imaging 31:297–301.
39. Hartmann D, Bassler B, Schilling D, et al (2006) Colorectal polyps: detection with dark-lumen MR colonography versus conventional colonoscopy. Radiology 238:143–149.
40. Debatin JF, Lauenstein TC (2003) Virtual magnetic resonance colonography. Gut 5(suppl 4):iv17–iv22.
41. Haykir R, Karakose S, Karabacakoglu A, et al (2006) Three-dimensional MR and axial CT colonography versus conventional colonoscopy for detection of colon pathologies. World J Gastroenterol 12: 2345–2348.
42. Hartmann D, Bassler B, Schilling D, et al (2005) Incomplete conventional colonoscopy: magnetic resonance colonography in the evaluation of the proximal colon. Endoscopy 37:816–820.
43. Lauenstein TC, Goehde SC, Ruehm SG, et al (2002) MR colonography with barium-based fecal tagging: initial clinical experience. Radiology 223:248–254.
44. Rosman AS, Korsten MA (2007) Meta-analysis comparing CT colonography, air contrast barium enema, and colonoscopy. Am J Med 120:203–210.
45. Sosna J, Blachar A, Amitai M, et al (2006) Colonic perforation at CT colonography: assessment of risk in a multicenter large cohort. Radiology 239:457–458.
46. Brenner DJ, Georgsson MA (2005) Mass screening with CT colonography: should the radiation exposure be of concern? Gastroenterology 129:328–337.
47. Yoshida H, Nappi J (2007) CAD in CT colonography without and with oral contrast agents: Progress and challenges. Comput Med Imaging Graph 31:267–284.
48. Nio Y, Van Gelder RE, Stoker J (2006) Computed tomography colonography: current issues. Scand J Gastroenterol Suppl 243:139–145.
49. Park SH, Ha HK, Kim MJ, et al (2005) Falsenegative results at multi-detector row CT colonography: multivariate analysis of causes for missed lesions. Radiology 235:495–502.
50. Park SH, Lee SS, Choi EK, et al (2007) Flat colorectal neoplasms: definition, importance, and visualization on CT colonography. AJR Am J Roentgenol 188:953–959.
51. Vogt C, Cohnen M, Beck A, et al (2004) Detection of colorectal polyps by multislice CT colonography with ultra-low-dose technique: comparison with high-resolution video-colonoscopy. Gastrointest Endosc 60:201–209
52. Achiam MP, Bulow S, Rosenberg J (2002) CT- and MR colonography. Scand J Surg 91:322–327.
53. Hoppe H, Netzer P, Spreng A, et al (2004) Prospective comparison of contrast enhanced CT colonography and conventional colonoscopy for detection of colorectal neoplasms in a single institutional study using second-look colonoscopy with discrepant results. Am J Gastroenterol 99:1924–1935.
54. Vijan S, Hwang I , Inadomi J, et al (2007) The cost-effectiveness of CT colonography in screening for colorectal neoplasia. Am J Gastroenterol 102:380–390.
55. Hassan C, Zullo A, Laghi A, et al (2007) Colon cancer prevention in Italy: cost-effectiveness analysis with CT colonography and endoscopy. Dig Liver Dis 39:242–250.

Rectal Polyps and Early Rectal Cancer 9

Assessment by Three-Dimensional Endorectal Ultrasonography

Giulio Aniello Santoro, Sandro Magrini, Luciano Pellegrini, Giuseppe Gizzi and Giuseppe Di Falco

Abstract Accurate preoperative staging of rectal polyps and early invasive rectal cancer may determine the choice between a submucosal or a full-thickness excision, or even a radical resection. Imaging modalities used for local staging include computed tomography, two-dimensional endorectal ultrasonography (2D-ERUS), and magnetic resonance. The new technique of high-resolution three-dimensional ERUS, constructed from a synthesis of standard 2D cross-sectional images, promises to further improve the accuracy. This tool seems to offer the best information on the depth of tumoral submucosal invasion and presence of mesorectal lymph node metastases, and may guide surgical planning. It also has the advantage of being an office-based procedure and is well tolerated, with fast acquisition times and relatively low cost.

Keywords Endorectal ultrasonography • Mesorectal lymph node • Rectal cancer • Rectal polyps • Three-dimensional ultrasonography

9.1 Introduction

Benign rectal polyps are treated by endoscopic resection or local excision. The coexistence of malignancy (reported rate 2–10%) may change the therapeutic strategy [1–3]. It is important for the management of early invasive rectal cancer to classify the depth of submucosal invasion. When the depth of submucosal invasion is slight, there is no risk of regional lymph nodes metastases and cure can be achieved by local resection. However, if the depth of submucosal invasion is massive, the risk of lymph nodes metastases is substantial, and radical resection is necessary [4–8]. It follows that an accurate preoperative staging of rectal villous lesions is mandatory in therapeutic decision making.

Currently, conventional two-dimensional endorectal ultrasound (2D-ERUS) is the most sensitive technique for the assessment of early invasive rectal cancer. An accuracy rate between 81% and 92% has been reported, but series are small and are from single institutions [9–16]. This technique has some important limitations: interpretation of the images is highly operator dependent and is based on real-time examination. The recent advent of high-resolution three-dimensional (3D) ERUS, constructed from a synthesis of standard 2D cross-sectional images, and of 'volume render mode' (VRM), a technique to analyze information inside a 3D-volume by digitally enhancing individual voxels, promises to enhance the accuracy of

G.A. Santoro (✉)
Pelvic Floor Unit, Section of Anal Physiology and Ultrasound, Coloproctology Service, I° Department of Surgery, Regional Hospital, Treviso, Italy

G.G. Delaini, T. Skřička, G. Colucci (eds.), *Intestinal Polyps and Polyposis*,

ERUS in detecting the presence of tumor invasion in rectal polyps and in selecting appropriate candidates for local therapy [17–19].

This chapter is devoted to discussing the method for generating and using 3D-ERUS, particularly with regard to the advantages of this application in the preoperative staging of early invasive rectal tumors.

9.2 Equipment and Technique

The most widely used ERUS system is the B-K Medical scanner (ProFocus 2202, B-K Medical A/S, Mileparken 34, DK-2730 Herlev, Denmark) with a hand-held rotating endoprobe type 2050, which gives a 360° axial view of the rectal wall, and a built-in 3D automatic acquisition system (Fig. 9.1) [20]. The radial probe has a 270 mm metal shaft with a double crystal at its tip, frequency range from 6.0 to 16.0 MHz, and 90° degree scanning plane. It is rotated at 4–6 cycles/s to get radial scan of the rectum and surrounding structures. The probe is covered with a latex balloon that is filled with degassed water to maintain acoustic coupling between the transducer and the tissue [20]. It is important to eliminate all bubbles within the balloon to avoid artifacts that limit the overall utility of the study. The rectum can be of varying diameters, and therefore the volume of water in the balloon may have to be adjusted intermittently.

The acquisition of a 3D data volume and the underlying techniques are different from application to application. With the conventional 2D ultrasound, the screen resolution is measured in number of pixels (the display matrix), with each pixel having x- and y-plane only. A 3D model may be constructed from a synthesis of a high number of parallel transaxial 2D images (Fig. 9.2) [17]. Such reconstruction is possible by combining the ultrasound apparatus and the integrated computer technology with 3D software (BK3Di, B-K Medical, Herlev, Denmark). Adding the third dimension means that the pixel is transformed in a small 3D picture element called a voxel. Ideally, a voxel should be a cubic structure; however, the dimension in the z-plane is often slightly larger than that in the x- and y-planes. The depth of the voxel is critical to the resolution of the 3D image, and this depth is directly related to the spacing between two adjacent images. High-resolution 3D ultrasound acquires four to five transaxial images sampled per millimeter of acquisition length in the z-plane. This means that an acquisition based upon a sampling of transaxial images over a distance of 60 mm in the human body will result in a data volume block consisting of between 240 and 300 transaxial images. High-resolution data volumes will consist of typical voxel sizes around 0.15×0.15×0.2 mm. Because of this resolution in the longitudinal plane, which is close to the axial and transverse resolution of the 2D image, this technique ensures true dimensions of the 3D data cube in the reconstructed z-plane as well, and provides accurate distance, area, angle, and volume measurements [17]. The 2050 probe is designed so that no moving parts come in contact with human tissue. The transducer's 360° rotating head, the proximal–distal actuation mecha-

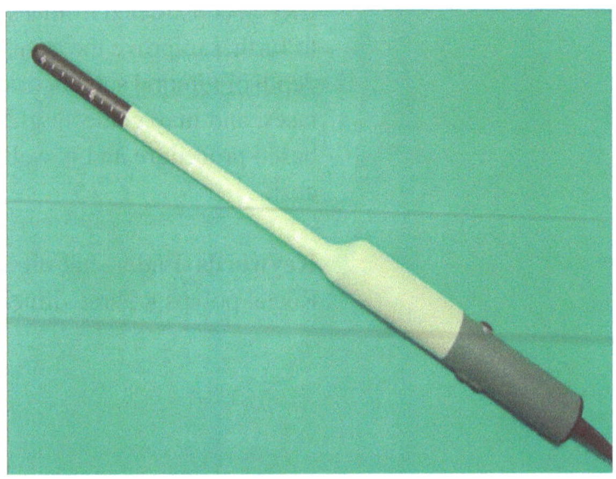

Fig. 9.1 B-K Medical anorectal probe type 2050

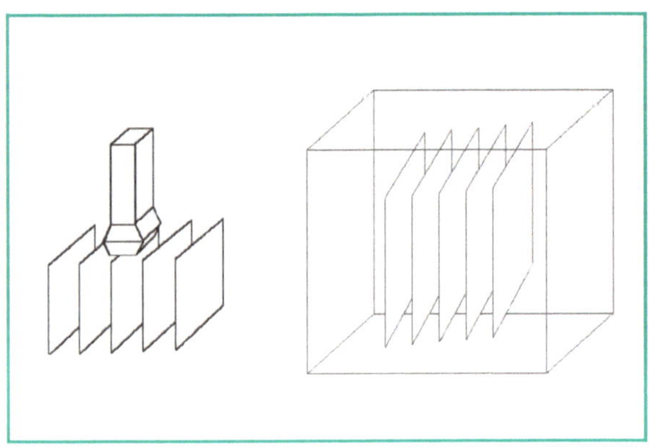

Fig. 9.2 Schematic model for acquisition of 3D anorectal endo image as a synthesis of a high number of parallel transaxial 2D images

nism, and the electronic mover are fully enclosed within the housing of the probe.

The ability to visualize information in the 3D image depends critically on the rendering technique [17]. Three basic types of technique are used:

1. *Surface render mode*: an operator or algorithm identifies the boundaries of the structures to create a wire-frame representation. It is the most commonly known version of *render mode* and it is extensively used by some medical centers in producing perhaps the very first images of an unborn baby's facial contours. Surface rendering techniques only give good results when a surface is available to render. These techniques fail when a strong surface cannot be found such as in the subtly layered structures within the anal canal and the rectal wall.
2. *Multiplane viewing techniques*: three perpendicular planes (axial, tranverse, and longitudinal) are displayed simultaneously and can be moved, rotated, tilted, and sliced to allow the operator to infinitely vary the different section parameters and visualize the lesion at different angles, and to get the most information out of the data. After data are acquired it is immediately possible to select coronal anterior–posterior or posterior–anterior as well as sagittal right–left views. The multiview function allows the operator to see up to six different and specialized views at once with real-time reconstruction (Fig. 9.3).
3. *Volume render mode*: this is a special feature that can be applied to high-resolution 3D data volume so information inside the cube is reconstructed to some extent. This technique uses a ray-tracing model as its basic operation. A ray or beam is projected from each point on the viewing screen (the display) back into and through the volume data. As the ray passes through the volume data it reaches the different elements (voxels) in the data set. Depending on the various render mode settings, the data from each voxel may be discarded, they may be used to modify the existing value of the ray, or they may be stored for reference to the next voxel and used in a filtering calculation. All of these calculations result in the current color or intensity of the ray being modified in some way. In normal VRM, the following four different post-processing display parameters can be used:
   (a) *opacity*: sets the relative transparency of the volume. The higher the value, the further into the volume the ray can travel before being terminated. Because of accumulated brightness as the ray traverses the volume, the net effect is to make the volume appear brighter as this control value is increased;
   (b) *luminance*: sets the inverse of the self-luminance value for the pixels, and should be used in conjunction with the opacity control for displaying certain voxel values for optimal visu-

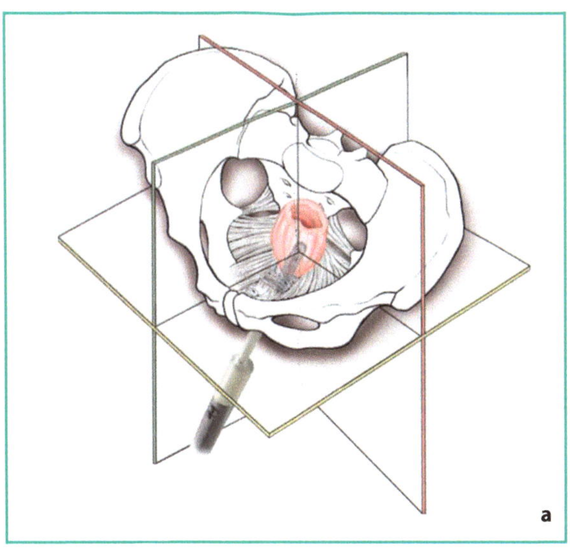

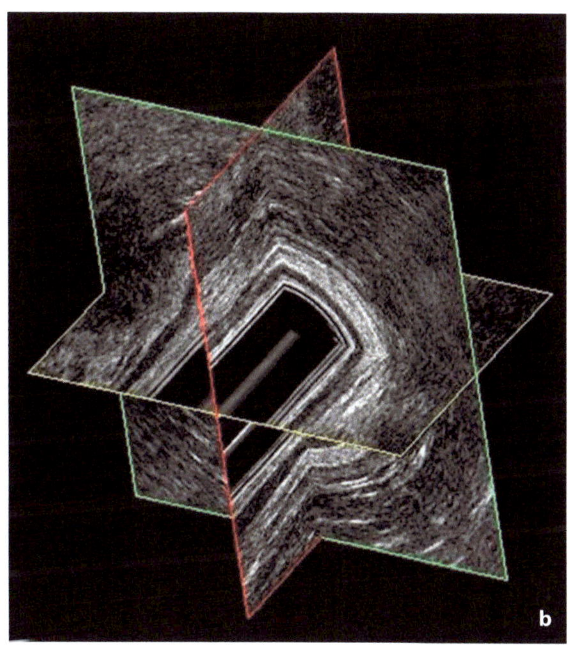

Fig. 9.3a,b Multiplane viewing technique. This function allows the operator to see different and specialized views at once. Reproduced from [20]

alization. The final image impression should be adjusted to the reader's requirements by setting the normal brightness and contrast controls;

(c) *thickness*: sets an upper limit to the penetration of the rays into the volume. This value is used in conjunction with the opacity parameter to determine when the ray traversal is terminated. Increasing the thickness setting allows deeper penetration, and the result is often a slightly smoother presentation together with a significant increase in the visual depth impression of an interface;

(d) *filter*: sets the lower threshold value for pixel intensities. Pixel values less than the filter value are not included in determining the intensity of the ray final value.

Endorectal ultrasound is usually performed with the patient in the left lateral decubitus position. An enema is administered 2 h before the examination. Initially, a digital examination should identify the size, fixation, morphology, and location of the tumor, if it is low enough. Proctoscopy with a dedicated rectosigmoidoscope (A.4522, Sapimed, Alessandria, Italy) is then performed (Sapimed, Alessandria, Italy) [21]. It allows visual examination of rectal lesions with exact determination of location with respect to both circumferential involvement of the rectal wall and the distance from the anal verge. Once the upper third of the rectum is reached, the endosonic probe is introduced through the rectosigmoidoscope. The presence of a double-graduated scale to measure the distance of the tip of the proctoscope and the tip of the probe from the anal verge, respectively, allows ascertainment of the correct positioning of both devices. The balloon is then filled, and the entire rectum down to the anal sphincter is evaluated while progressively withdrawing the conjoined probe and rectoscope. This is of extreme importance, as the lower border of a rectal lesion can differ significantly in the depth of invasion from the center or upper portions, and lymph nodes in the peri-rectal region are often just above the level of the tumor and will be missed if complete imaging is not obtained [21].

On the screen, the anterior aspect of the rectum will be superior (12 o'clock), right lateral will be left (9 o'clock), left lateral will be right (3 o'clock), and posterior will be inferior (6 o'clock) (just like the image on axial computed tomography (CT) scan). The tip of the ultrasound probe should be maintained in the center of the rectal lumen to gain optimal imaging of the rectal wall and peri-rectal structures.

9.3 Ultrasound Anatomy

On ultrasound, the normal rectal wall is 2–3 mm thick and is composed of a five-layer structure [22]. The first hyperechoic layer corresponds to the interface of the balloon with the rectal mucosal surface, the second hypoechoic layer to the mucosa and muscularis mucosa, the third hyperechoic layer to the submucosa, the fourth hypoechoic layer to the muscularis propria, and the fifth hyperechoic layer to the serosa or to the interface with the fibrofatty tissue surrounding the rectum (mesorectum) (Fig. 9.4). The mesorectum contains blood vessels, nerves, and lymphatics and has an inhomogeneous echo pattern. Very small, round to oval, hypoechoic lymph nodes should be distinguished from blood vessels, which also appear as circular hypoechoic structures. Three-dimensional ERUS offers a valuable supplement to conventional 2D-ERUS [20]. The five layers of the rectal wall are clearly illustrated in the coronal plane as well as in the

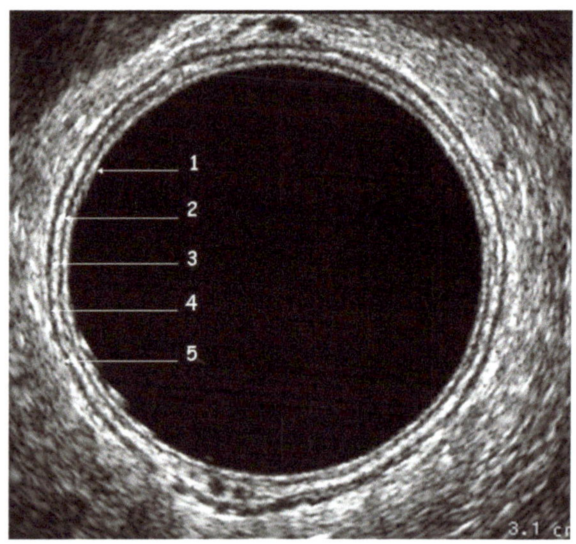

Fig. 9.4 Two-dimensional ultrasonographic five-layer structure of the normal rectal wall. The first hyperechoic layer (*1*) corresponds to the interface of the balloon with the rectal mucosal surface, the second hypoechoic layer (*2*) to the mucosa and muscolaris mucosa, the third hyperechoic layer (*3*) to the submucosa, the fourth hypoechoic layer (*4*) to the muscularis propria, and the fifth hyperechoic layer (*5*) to the serosa or to the interface with the mesorectum

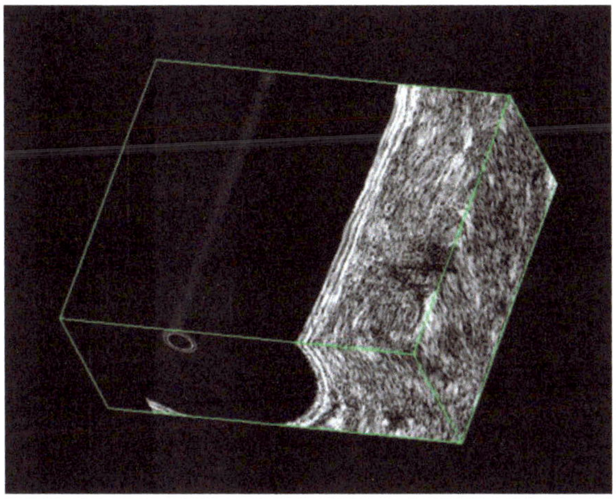

Fig. 9.5 Three-dimensional ultrasonographic five-layer structure of the normal rectal wall

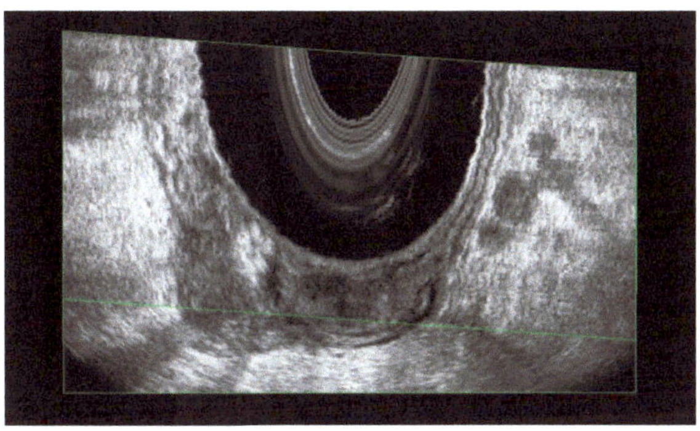

Fig. 9.6 Blood vessels can be followed longitudinally and distinguished from lymph nodes on reconstructed coronal plane

transaxial and sagittal planes (Fig. 9.5). Blood vessels can be followed longitudinally and distinguished from lymph nodes (Fig. 9.6).

9.4 Ultrasonographic Staging

Ultrasonographic criteria to determine the depth of tumor invasion in early rectal cancer, based on the classification proposed by Akasu et al [12], are as follows:

- *benign lesion (uT0)*: the mucosal layer is expanded but the submucosal layer remains intact around the entire breadth of the tumor (Fig. 9.7)
- *submucosal cancer (uT1)*: the hyperechoic submucosal layer is irregular or interrupted, consistent with tumor invasion. The depth of submucosal cancer invasion is classified in two subtypes: uT1-slight (*SM-s*: extent limited to the upper third of the third layer. The fourth hypoechoic layer of the muscularis propria appears intact) (Fig. 9.8) and uT1-massive (*SM-m*: tumor invasion extended to the middle or lower third of the third layer. The fourth hypoechoic layer is thickened consistent with peritumoral inflammation and desmoplastic reaction) (Fig. 9.9) [12,23]
- if a distinct break is seen in the submucosal layer, the muscularis propria has been invaded (*uT2 lesion*) (Fig. 9.10)
- *undetectable or benign lymph* nodes are classified as *uN–*
- *pathologic lymph nodes (uN+)* appear as circular or slightly oval-shaped structures, often with an irregular border, and with an echogenicity similar to the primary tumor (Fig. 9.11) [24,25].

9.5 Accuracy

Sonographic evaluation of a villous rectal lesion is helpful in determining the presence of tumor invasion. The presence of an intact hyperechoic submucosal interface indicates lack of tumor invasion into the submucosa. Heintz et al [26] reported that ERUS cannot differentiate between villous adenoma and invasive cancer because neither the muscularis mucosa nor the submucosa is sonographically definable, and that the first hypoechoic layer corresponds anatomically with the mucosa and the submucosa. They suggested that uT0 and uT1 tumors, which manifest as a broadening of the first hypoechoic layer, should be classified together. Instead Adams and Wong [14] considered the first hypoechoic layer as the mucosa and muscularis mucosa, and the middle hyperechoic layer as the submucosa. Consequently, for such authors, lesions that expand the inner hypoechoic layer and are surrounded by a uniform middle hyperechoic layer represent villous adenoma, and lesions that expand the inner hypoechoic layer and have distinct echo defects of the middle hyperechoic layer are considered uT1 tumors.

According to the literature, 2D-ERUS is the most accurate technique for determining the depth of tumor invasion in early-stage rectal cancer [27,28]. In a

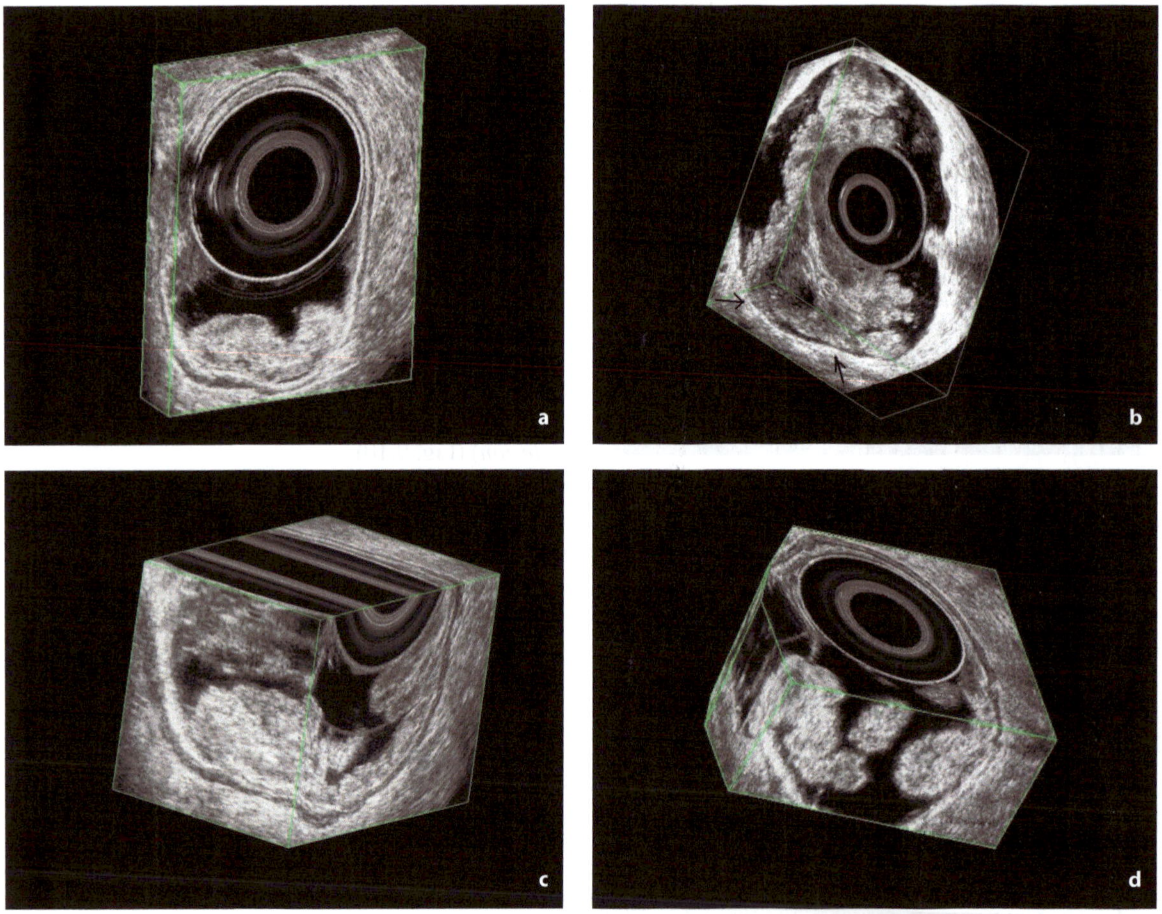

Figure 9.7a–d uT0 rectal lesions. The mucosal layer is expanded and the presence of an intact hyperechoic submucosal interface indicates lack of tumor invasion into the submucosa (**b**, *arrows*)

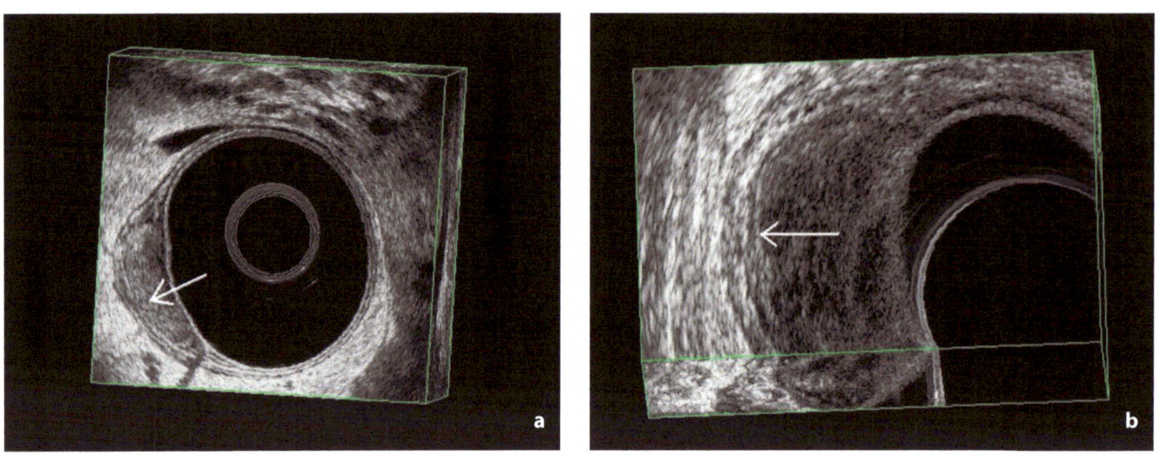

Fig. 9.8a,b uT1 rectal tumors with slight submucosal invasion [SM-s: extent limited to the upper third of the third layer (*arrows*)]. The fourth hypoechoic layer of the muscolaris propria appears intact

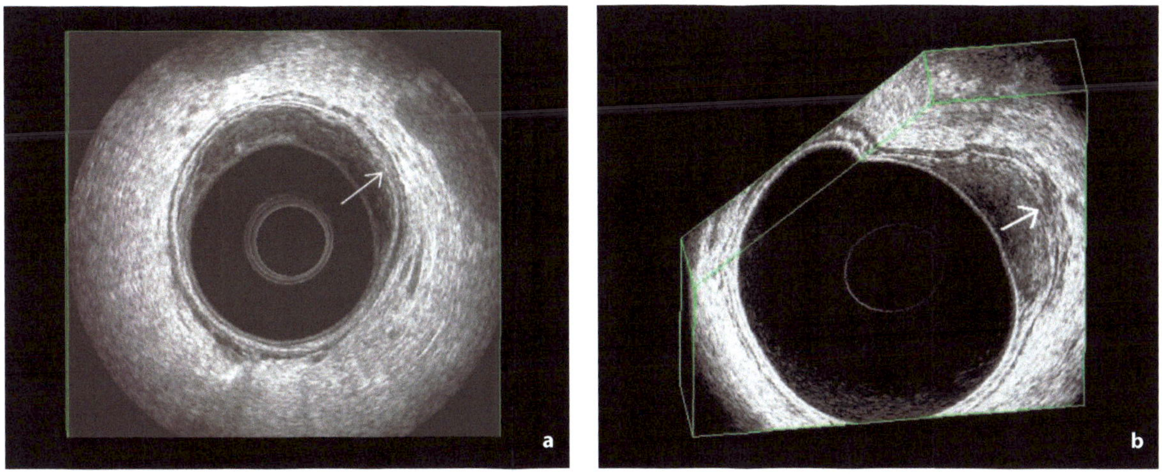

Fig. 9.9a,b uT1 rectal tumors with massive submucosal invasion [SM-m: tumor invasion extended to the middle or lower third of the third layer (*arrows*)]. The fourth hypoechoic layer is thickened consistent with peritumoral inflammation and desmoplastic reaction

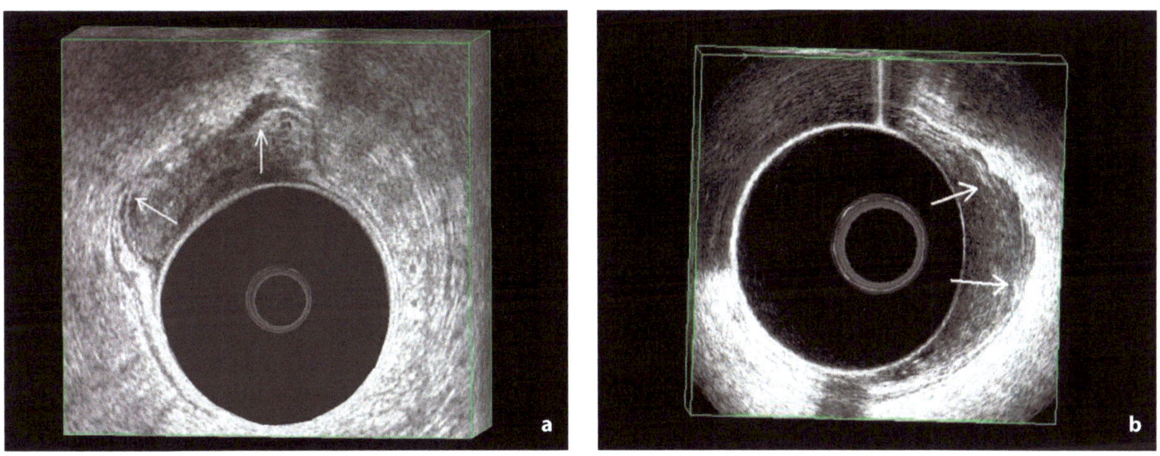

Fig. 9.10a,b uT2 rectal tumors. A distinct break is seen in the submucosal layer and the muscolaris propria is thickened (*arrows*). The surrounding hyperechoic layer corresponding to the serosa or perirectal fat remains intact

systematic literature review, Worrell et al [10] reported that ERUS correctly established a cancer diagnosis in 81% of 62 biopsy-negative rectal adenomas that had focal carcinoma on histopathology. In another study from the Cleveland Clinic Florida [13], the final pathology results confirmed the preoperative ERUS diagnosis of non-invasive villous rectal tumors in 26 out of 27 patients. Doornebosch et al [29] recently reported that ERUS is very reliable in diagnosing tubolovillous adenoma (sensitivity 89%; specificity 86%), and therapeutic decision making regarding local excision versus radical surgery based on ERUS is valid. By adding ERUS to preoperative biopsies, the rate of missed carcinomas was reduced from 21% to 3% ($P<0.001$).

Technical difficulties associated with scanning villous adenoma can be due to very large exophytic lesions that tend to attenuate rectal layers or to produce fixed artifacts over one part of the rectal wall, obscuring the image [30]. In large carpeting lesions, careful evaluation of the entire tumor is necessary to determine that a small area of invasion has not been overlooked. Konishi et al [9] reported that the overall accuracy of ERUS-based evaluation of tumor invasion depth was 60% in villous lesions and 91% in non-villous lesions. Snare biopsy of lesions before referral

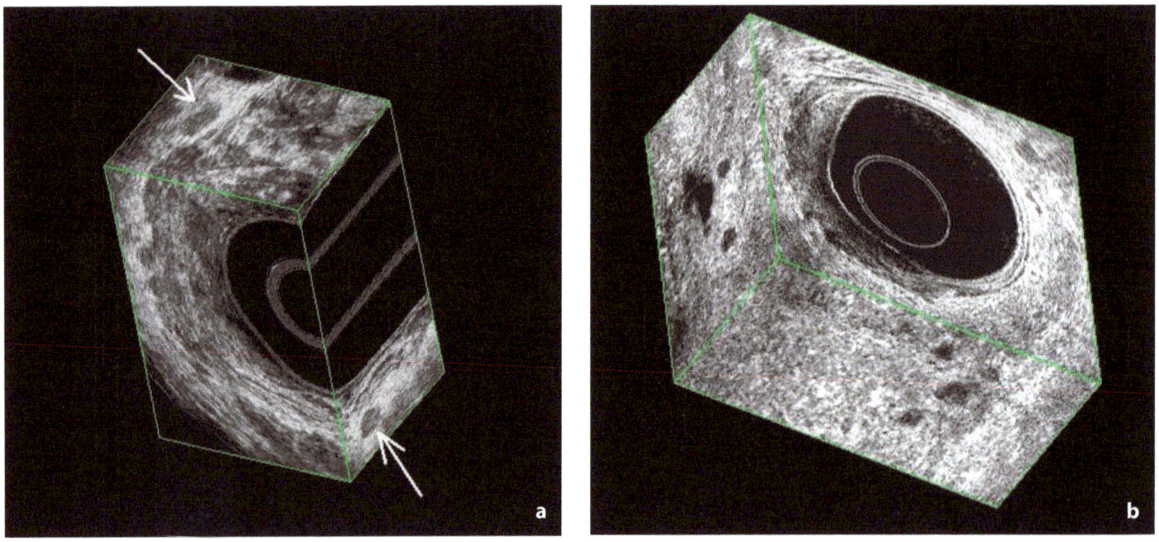

Fig. 9.11a,b Sonograms of enlarged, hypoechoic, malignant-appearing lymph nodes of the mesorectum (a, *arrows*)

to ERUS produces a burn artifact, which can also lead to overstaging [30].

If a tumor arises in a polyp it is important to determine whether the stalk is invaded. Differences in classification are reported between Western [31] and Japanese [4] pathologists. In 1985, Haggitt et al [31] divided the depth of invasion into four levels:

- *level 0*: carcinoma *in situ* or intramucosal carcinoma
- *level 1*: carcinoma invading through the muscularis mucosa into the submucosa but limited to the head of the polyp
- *level 2*: carcinoma invading the level of the neck of the adenoma
- *level 3*: carcinoma invading any part of the stalk
- *level 4*: carcinoma invading into the submucosa of the bowel wall below the stalk of the polyp.

By definition, all sessile polyps with invasive adenocarcinoma are in level 4. They studied 129 patients with pTis to pT1 colorectal tumors and found that level 4 invasion was a statistically significant factor ($P < 0.001$) predicting positive nodes. Similar results were reported by Nivatvongs et al [8] on 151 patients with pT1 colorectal tumors undergoing bowel resection in which invasion into the submucosa of the bowel wall at the base of the stalk (level 4) was the single most significant risk factor for positive nodes. Seitz et al [1] suggested that Haggitt's classification applies well for pedunculated polyps; however, it should not be used for malignant sessile polyps. Suzuki et al [32] determined the risk of lymph node metastases in 65 patients with Haggitt's level 4 invasion into the submucosa. Lymph node metastasis was noted in 11 (16.9%) of the 65 patients; however, the width of submucosal invasion was significantly greater in node-positive than in node-negative patients ($P = 0.001$). When 5-mm-wide submucosal invasion was used as an indicator for intestinal resection, 37 patients were found to have indications for bowel resection, and 11 (29.7%) of the 37 patients had lymph node metastases. The positive predictive value increased from 17% to 30% when the width of submucosal invasion was added to Haggitt's level 4 as an indicator for bowel resection. Kudo [4] was the first to differentiate three different types of early invasive cancers:

- *SM-1 tumor*, invading the superior third of the submucosa
- *SM-2 tumor*, invading the superficial two-thirds of the submucosa
- *SM-3 tumor*, invading the deep one-third of the submucosa.

The group type SM-1 was divided in three subtypes: type SM-1a (invasion <1/4 of the submucosa), type SM-1b (invasion <1/2 of the submucosa) and type SM-1c (invasion >1/2 of the submucosa). Kikuchi et al [7] found that the risk of lymph node metastases was 0% for SM-1 lesions, 10% for SM-2 lesions, and 25% for SM-3 lesions ($P < 0.001$). In their study, tumoral invasion of the deep third of the submucosa (SM-3) was the only independent risk factor for lymph node metastasis. Akasu et al [12] proposed a classification of the depth of submucosal cancer in two groups:

- *SM-slight (SM-s)*: extent limited to the upper third of the submucosa
- *SM-massive (SM-m)*: tumor invasion extended to the middle or lower third of the submucosa.

In their series of 154 patients with early-stage rectal cancer, sensitivity, specificity, and overall accuracy rates of preoperative ERUS for detection of slight and massive submucosal invasion were 99/74/96% and 98/88/97%, respectively. Incidences of lymph node metastases in pT1-slight and pT1-massive were 0% and 22%, respectively. They suggested that patients with massive submucosal invasion should be best treated by radical surgery. Another study from the Mayo Clinic confirmed these data [5]. Among patients with T1 carcinoma in the middle or lower third of the rectum, the multivariate risk factor for long-term cancer-free survival was invasion into the lower third of the submucosa. For lesions with SM3 invasion, the oncologic resection group had lower rates of distant metastases and better survival compared with patients who underwent local excision. Therefore, a decision whether to perform radical surgery, or local excision, or polypectomy should be based principally on assessment of invasion depth. Akahoshi et al [33] improved the accuracy of ERUS by using a high-frequency (12 MHz) ultrasound catheter probe. The depth of invasion was correctly assessed in 87% (46/53) of pT1 tumors. Starck et al [11] reported their experience on the adoption of high-multifrequency probes. Their conclusion was that endosonography reliably distinguished benign from early invasive rectal lesions.

Metastatic involvement of the mesorectal lymph node is a major independent prognostic factor. It has been observed that the presence of >3 nodes is associated with a poor prognosis [34,35]. Moreover, identification of a metastatic peri-rectal lymph node is important, as early T1 lesions with mesorectal node involvement are not suitable for local excision. The nodal disease not addressed by local extirpation may eventually progress to clinical recurrence, usually located deep in the mucosa and locally advanced at the time of diagnosis [36]. In this setting, ERUS nodal assessment is particularly important. Sonographic evaluation of lymph node metastases is somewhat less accurate than depth of wall invasion [6]. The criteria used to identify metastatic lymph nodes in most of the studies are echogenicity, border demarcation, and node diameter [37,38]. Normal, non-enlarged peri-rectal nodes are not usually seen on ERUS. Inflamed, enlarged lymph nodes appear hyperechoic, with ill-defined borders. Much of the sound energy is reflected because the lymphatic tissue architecture is intact. In contrast, metastatic lymph nodes that have been replaced with tumor appear hypoechoic with an echogenicity similar to the primary lesion. Malignant lymph nodes tend to be circular rather than oval, have discrete borders, and are most commonly found adjacent to the primary tumor or in the mesorectum proximal to the tumor. The sonographic features of lymph nodes generally can be distributed into four groups [6]:

1. if lymph nodes are not visible by ultrasound, the probability of lymph node metastases is low
2. hyperechoic lymph nodes are often benign and result from non-specific inflammatory changes
3. hypoechoic lymph nodes larger than 5 mm are highly suggestive for lymph node metastases
4. lymph nodes larger than 5 mm that are visible with mixed echogenic patterns should be considered metastatic.

Overstaging and understaging can occur during assessment of lymph node involvement [24]. At conventional 2D-ERUS, the cross-sectional appearance of blood vessels in the perirectal fat may be confused with positive lymph nodes. Three-dimensional reconstruction allows differentiation of vessels from lymph nodes by following their branching pattern. On size characteristics alone, sonographically detected nodes in the mesorectum greater than 5 mm in diameter have 50–70% chance of being involved, whereas those smaller than 4 mm have less than 20% chance [34]. Enlarged nodes, however, may be benign and reactive, whereas small nodes, which are difficult to identify, may be infiltrated. Up to 20% of patients have involved nodes of less than 3 mm, limiting the accuracy of the technique. The size of nodal metastasis is proportional to pT stage. This explains the relationship between nodal staging accuracy and T stage: <50% accuracy for pT1 lesions and >80% for pT3 lesions [36]. Even with an improved understanding of the ultrasonographic characteristics of malignant lymph node, micrometastases and granulomatous inflammation are impossible to detect. In these cases ERUS-guided needle biopsy or ERUS-guided fine-needle aspiration biopsy may improve diagnostic accuracy [39].

Endorectal ultrasound also has a good accuracy in the diagnosis of recurrence after local excision of early rectal cancer. In a series of 108 patients treated by local excision, de Anda et al [40] reported that ERUS identified one-third (10 patients out of 32) of asymptomatic local recurrences that were missed by digital or proctoscopic examination.

New software technology has allowed a series of 2D images to be assembled, giving a 3D-representation of the rectum and mesorectum [17]. After a 3D data set has been acquired, it is immediately possible to select coronal anterior–posterior or posterior–anterior as well as sagittal right–left views, together with any oblique image plane. The 3D image can be rotated, tilted, and sliced to allow the operator to infinitely vary the different section parameters, visualize the lesion at different angles, and measure accurately distance, area, angle, and volume. By using a combination of the different postprocessing display parameters, the 3D image can be rendered to provide better visualization performance when there are not large differences in the signal levels of pathologic structures compared with surrounding tissues [17]. Several studies have shown important benefits of 3D high-resolution ERUS in terms of better parietal staging [17–19,21]. In a recent comparative study on 86 patients with rectal cancer examined with 3D-ERUS, 2D-ERUS and CT, Kim et al [41] demonstrated that 3D-ERUS was greatly superior to the other modalities in both T staging and N staging. In this series, the accuracy of 3D-ERUS for T and N staging was 91% and 90%, respectively. In a study from Memorial Sloan-Kettering Cancer Center [42], the use of the new implemented 3D-ERUS appeared to facilitate the understanding of the spatial relations between different structures, compared to 2D-ERUS and MRI. However, no definitive conclusions over the new diagnostic tool were drawn. Vyslouzil et al [43] reported an accuracy of 100% in the pT1 stage using 3D-ERUS, and concluded that preoperative 3D ultrasonographic staging plays a decisive role in selecting patients who are suitable for local resection. In a preliminary study [23], we reported that the accuracy of 3D-ERUS was significantly superior to 2D-ERUS in determining the presence of submucosal invasion in 89 patients with rectal villous lesions (85 versus 62.5%, respectively; $P = 0.022$). Moreover, 3D-ERUS with render mode provided better delineation of the superficial invasion of submucosa (SM-s lesions) compared to 2D-ERUS (83.3 versus 54.1%, respectively; $P = 0.029$). To provide a higher-resolution image of the different layers of the rectal wall, render mode was used with high opacity, high thickness, and normal filter and luminance setting, adjusted with high contrast and brightness. Compared with normal mode, VRM with this setting offered a clear view of the submucosal layer and helped to differentiate slight from massive submucosal invasion.

9.6 Conclusion

The great transformation in the management of rectal polyps and early rectal cancer has increased the importance of accurate preoperative staging for therapeutic decision making. Among the different imaging modalities, 3D-ERUS has been evolving as the best procedure, due to enhanced spatial resolution, providing detailed information on early tumoral invasion into the rectal wall and the presence of mesorectal lymph node metastases. It also has the advantage of being an office-based procedure, which is well tolerated, with fast acquisition times and relatively low cost.

References

1. Seitz U, Bohnacker S, Seewald S, et al (2004) Is endoscopic polypectomy an adequate therapy for malignant colorectal adenomas? Presentation of 114 patients and review of the literature. Dis Colon Rectum 47:1789–1797.
2. Dell'Abate P, Iosca A, Galimberti A, et al (2001) Endoscopic treatment of colorectal benign-appearing lesions 3 cm or larger: techniques and outcome. Dis Colon Rectum 44:112–118.
3. Tsuda S, Veress B, Toth E, Fork FT (2002) Flat and depressed colorectal tumors in a southern Swedish population: a prospective chromoendoscopic and histopathological study. Gut 51:550–555.
4. Kudo S (1993) Endoscopic mucosal resection of flat and depressed types of early colorectal cancer. Endoscopy 25:455–461.
5. Nascimbeni R, Nivatvongs S, Larson DR, Burgart LJ (2004) Long-term survival after local excision for T1 carcinoma of the rectum. Dis Colon Rectum 47:1773–1779.
6. Nascimbeni R, Burgart RJ, Nivatvongs S, Larson DR (2002) Risk of lymph node metastasis in T1 carcinoma of the colon and rectum. Dis Colon Rectum 45:200–206.
7. Kikuchi R, Takano M, Takagi K, et al (1995) Management of early invasive colorectal cancer: risk of recurrence and clinical guidelines. Dis Colon Rectum 38:710–717.
8. Nivatvongs S, Rojanasakul A, Reiman HM, et al (1991) The risk of lymph node metastasis in colorectal polyps with

invasive adenocarcinoma. Dis Colon Rectum 34:323–328.
9. Konishi K, Akita Y, Kaneko K, et al (2003) Evaluation of endoscopic ultrasonography in colorectal villous lesions. Int J Colorectal Dis 18:19–24.
10. Worrell S, Horvath K, Blakemore T, Flum D (2004) Endorectal ultrasound detection of focal carcinoma within rectal adenomas. Am J Surg 187:625–629.
11. Starck M, Bohe M, Simanaitis M, Valentin L (2003) Rectal endosonography can distinguish benign rectal lesions and invasive early rectal cancers. Colorectal Dis 5:246–250.
12. Akasu T, Kondo H, Moriya Y, et al (2000) Endorectal ultrasonography and treatment of early stage rectal cancer. World J Surg 24:1061–1068.
13. Pikarsky A, Wexner S, Lebensart P, et al (2000) The use of rectal ultrasound for the correct diagnosis and treatment of rectal villous tumors. Am J Surg 179:261–265.
14. Adams WJ, Wong WD (1995) Endorectal ultrasonic detection of malignancy within rectal villous lesions. Dis Colon Rectum 38:1093–1096.
15. Glaser F, Schlag P, Herfarth CH (1990) Endorectal ultrasonography for the assessment of invasion of rectal tumors and lymph node involvement. Br J Surg 77:883–887.
16. Kim HJ, Wong WD (2000) Role of endorectal ultrasound in the conservative management of rectal cancers. Semin Surg Oncol 19:358–366.
17. Santoro GA, Fortling B (2007) The advantages of volume rendering in three-dimensional endosonography of the anorectum. Dis Colon Rectum 50:359–368.
18. Kim JC, Cho YK, Kim SY, et al (2002) Comparative study of three-dimensional and conventional endorectal ultrasonography used in rectal cancer staging. Surg Endosc 16:1280–1285.
19. Giovannini M, Bories E, Pesenti C, et al (2006) Three-dimensional endorectal ultrasound using a new freehand software program: results in 35 patients with rectal cancer. Endoscopy 38:339–343.
20. Santoro GA, Fortling B (2006) New technical developments in endoanal and endorectal ultrasonography. In: Santoro GA, Di Falco G (eds) Benign anorectal diseases. Springer-Verlag Italia, Milan, pp 13–26.
21. Santoro GA, D'Elia A, Battistella G, Di Falco G (2007) The use of dedicated rectosigmoidoscope for ultrasound staging of tumors of the upper and middle third of the rectum. Colorectal Dis 9:61–66.
22. Santoro GA, Di Falco G (2004) Surgical and endosonographic anatomy of the rectum. In: Santoro GA, Di Falco G (eds) Atlas of endoanal and endorectal ultrasonography. Springer-Verlag Italia, Milan, pp 37–41.
23. Santoro GA, Bara Egan D, Di Falco G (2004) Three-dimensional endorectal ultrasonography in the evaluation of early invasive rectal cancer. Colorectal Dis 6(suppl 2):0–20.
24. Santoro GA, Di Falco G (2004) Endorectal ultrasound in the preoperative staging of rectal cancer. In: Santoro GA, Di Falco G (eds) Atlas of endoanal and endorectal ultrasonography. Springer-Verlag Italia, Milan, pp 49–95.
25. Hildebrandt U, Feifel G (1985) Preoperative staging of rectal cancer by intrarectal ultrasound. Dis Colon Rectum 28:42–46.
26. Heintz A, Buess G, Frank K, et al (1989) Endoluminal ultrasonic examination of sessile polyps and early carcinomas of the rectum. Surg Endosc 3:92–95.
27. Sailer M, Leppert R, Kraemer M, et al (1997). The value of endorectal ultrasound in the assessment of adenomas, T1- and T2-carcinomas. Int J Colorectal Dis 12:214–219.
28. Nesbakken A, Løvig T, Lunde OC, Nygaard K (2003) Staging of rectal carcinoma with transrectal ultrasonography. Scand J Surg 92:125–129.
29. Doornebosch PG, Bronkhorst PJB, Hop WCJ, et al (2008) The role of endorectal ultrasound in therapeutic decision-making for local vs. transabdominal resection of rectal tumors. Dis Colon Rectum 51:38–42.
30. Kim J, Yu CS, Jung HY, et al (2001) Source of errors in the evaluation of early rectal cancer by endoluminal ultrasonography. Dis Colon Rectum 44:1302–1309.
31. Haggitt RC, Glotzbach RE, Soffer EE, Wruble LD (1985) Prognostic factors in colorectal carcinomas arising in adenomas: implications for lesions removed by endoscopic polypectomy. Gastroenterology 89:328–336.
32. Suzuki T, Sadahiro S, Mukoyama S, et al (2003) Risk of lymph node and distant metastases in patients with early invasive colorectal cancer classified as Haggitt's level 4 invasion: image analysis of submucosal layer invasion. Dis Colon Rectum 46:203–208.
33. Akahoshi K, Yoshinaga S, Soejima A, et al (2001) Transit endoscopic ultrasound of colorectal cancer using 12MHz catheter probe. Br J Radiol 74:1017–1022.
34. Dworak O (1989) Number and size of lymph nodes and node metastases in rectal carcinomas. Surg Endosc 3:96–99.
35. Monig SP, Baldus SE, Zirbes TK, et al (1999) Lymph node size and metastatic infiltration in colon cancer. Ann Surg Oncol 6:579–581.
36. Landmann RG, Wong WD, Hoepfl J, et al (2007) Limitations of early rectal cancer nodal staging may explain failure after local excision. Dis Colon Rectum 50:1520–1525.
37. Hulsmans FJ, Bosma A, Mulder PJ, et al (1992) Perirectal lymph nodes in rectal cancer: in vitro correlation of sonographic parameters and histologic findings. Radiology 184:553.
38. Katsura Y, Yamada K, Ishizawa T, et al (1992) Endorectal ultrasonography for the assessment of wall invasion and lymph node metastasis in rectal cancer. Dis Colon Rectum 35:362–368.
39. Milsom JW, Czyrko C, Hull TL, et al (1994) Preoperative biopsy of pararectal nodes in rectal cancer using endoluminal ultrasonogarphy. Dis Colon Rectum 37:364–368.
40. de Anda EH, Lee S-H, Finne CO, et al (2004) Endorectal ultrasound in the follow-up of rectal cancer patients treated by local excision or radical surgery. Dis Colon Rectum 47:818–824
41. Kim JC, Kim HC, Yu CS (2006) Efficacy of 3-dimensional endorectal ultrasonography compared with conventional ultrasonography and computed tomography in preoperative rectal cancer staging. Am J Surg 192:89–97.
42. Schaffzin DM, Wong WD (2004) Endorectal ultrasound in the preoperative evaluation of rectal cancer. Clin Colorectal Cancer 4:124–132.
43. Vyslouzil K, Cwiertka K, Zboril P, et al (2007) Endorectal ultrasonography in rectal cancer staging and indication for local surgery. Hepatogastroenterolgy 54:1102–1106.

What is What?

Clinical Vignettes of Multiple Polyps

10

Johann Pfeifer

Abstract Colorectal polypoid lesions are the most common pathology found during endoscopy. The clinical challenge nowadays is to differentiate during endoscopy between neoplastic and non-neoplastic lesions, in order to choose the correct treatment modality. The aim of this chapter is to describe the clinical features, appearances, and frequencies of the various colorectal (polypoid) lesions and how to differentiate between them. New, technically advanced methods such as narrow-band imaging (NBI), magnifying chromoendoscopy (CE), and autofluorescence imaging (AFI) are described, along with their current and possible future roles in colonoscopy to diagnose polyps "histologically" before the pathologist has even seen a specimen.

Keywords Adenoma • Autofluorescence imaging (AFI) • Chromoendoscopy • Colon polyp • Computed virtual chromoendoscopy (CVC) • Early colorectal cancer • Flat adenoma • Fujinon intelligent color enhancement (FICE) • Hyperplastic polyp • Laser scanning confocal microscopy (LCM) • Narrow-band imaging (NBI)

10.1 Introduction

Colorectal polypoid lesions are the most common pathology found during endoscopy. They may be single or multiple, small or large, pedunculated or sessile. As colorectal adenomatous polyps can potentially advance to colorectal cancer, their endoscopic removal by polypectomy or endoscopic mucosal resection (EMR) is mandatory. In contrast, hyperplastic polyps and other non-neoplastic lesions without neoplastic potential are often seen during colonoscopy. Removal of these lesions brings no advantage, but consumes time and money and entails the risk of complications. The clinical challenge nowadays is to differentiate during endoscopy between neoplastic and non-neoplastic lesions, to choose the correct treatment modality.

The aim of this chapter is to describe the clinical features, appearances, and frequencies of the various colorectal (polypoid) lesions and how to differentiate between them. New, technically advanced methods such as narrow-band imaging (NBI), magnifying chromoendoscopy (CE), and autofluorescence imaging (AFI) are described along with their current and possible future roles in colonoscopy to diagnose polyps "histologically" before the pathologist has even seen a specimen.

10.2 Classification

Colorectal polypoid lesions can be classified according to their morphology or histological type including

J. Pfeifer (✉)
Department of General Surgery, Medical University of Graz, Graz, Austria

G.G. Delaini, T. Skřička, G. Colucci (eds.), *Intestinal Polyps and Polyposis*,

Table 10.1 Morphological classification of polyps in the colorectum

Characteristic	Description
Location	Rectum, sigmoid, descending colon, etc Right-sided/left-sided
Size	Small: ≤0.5 cm Medium: 0.5–1.0 cm Large: >1.0 cm
Surface	Smooth (not involved) Irregular
Number of lesions	Single Multiple
Presentation	Lobulated, berrylike (typical for tubular adenomas) Shaggy, velvety (typical for villous adenomas) Mixed form
Fixation	Sessile (lacking a stalk) Pedunculated (with a stalk) Flat
Appearance	Bleeding Non-bleeding

Table 10.2 Histological classification of polyps in the colorectum

Tumor type	Description
Epithelial tumors (or tumor-like lesions)	Adenoma Hyperplastic (metaplastic) polyp Sessile serrated adenoma (SSA) Juvenile polyp Inflammatory polyp Inflammatory fibroid polyp Peutz–Jeghers polyp Adenocarcinoma Hereditary non-polyposis colorectal cancer (HNPCC) Endocrine cell or carcinoid tumor
Polyposis syndromes	Familial adenomatous polyposis (FAP) Juvenile polyposis Peutz–Jeghers syndrome Cowden's syndrome Bananyan–Riley–Ruvalcaba syndrome Cronkhite–Canada syndrome
Non-epithelial tumors Tumors of lymphoid tissue	Benign lymphoid polyp Benign lymphoid polyposis Malignant lymphoma of the large intestine
Tumors of connective tissue	Gastrointestinal stromal tumors (GIST) Stromal tumors in the rectum (Leiomyomatous polyp)
Tumors of adipose tissue	Lipoma Lipohyperplasia of the ileocecal valve
Tumors of vascular tissue	Hemangioma Lymphangioma
Neurogenic tumors	Benign neurofibroma GIST

benignity or malignancy (Tables 10.1 and 10.2). Gross morphological classification for colorectal lesions most often follows the Paris workshop guidelines (Table 10.3) [1]. For grading of dysplasia and early colorectal cancer, the Vienna classification is most often used (Table 10.4) [2].

Table 10.3 Paris classification of colorectal lesions (gross morphology)

Endoscopic appearance	Paris class	Picture	Description
Protruded lesions	Ip	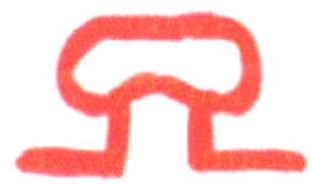	Pedunculated polyps
	Ips		Subpedunculated polyps
	Is		Sessile polyps
Flat elevated lesions	0-IIa		Flat elevation of mucosa
	0-IIa/c		Flat elevation with central depression
Flat lesions	0-IIb		Flat mucosal change
	0-IIc		Mucosal depression
	0-IIc/IIa	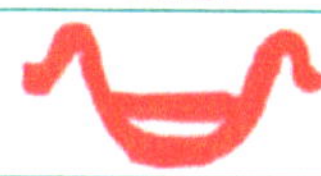	Mucosal depression with raised edge

Table 10.4 Vienna classification of gastrointestinal epithelial neoplasia

Category	Description
1	Negative for neoplasia/dysplasia
2	Indefinitive for neoplasia/dysplasia
3	Non-invasive low-grade neoplasia (low-grade adenoma/dysplasia)
4	Non-invasive high grade neoplasia
4.1	High-grade adenoma/dysplasia
4.2	Non-invasive carcinoma (carcinoma *in situ*)[a]
4.3	Suspicion of invasive carcinoma
5	Invasive neoplasia
5.1	Intramucosal carcinoma[b]
5.2	Submucosal carcinoma or beyond

[a]Non-invasive indicates absence of evident invasion
[b]Intramucosal indicates invasion into the lamina propria or muscularis mucosae

10.3 Clinical Appearance

10.3.1 Epithelial Tumors

A large variety of neoplasms may occur in the colorectum – a reflection of the complexity of this organ and its cellular elements. The most frequently diagnosed tumors arise from the surface epithelium and are called polyps. For the purpose of cancer prevention it is of utmost importance to differentiate between neoplastic and non-neoplastic lesions.

10.3.1.1 Adenomas

Historically, the adenoma has been considered to be the most important type of epithelial polyp in the colorectum. Adenomas represent a family of mucosal neoplasms, which show some diversity in their appearance but share certain essential acquired genetic characteristics. Adenomas are common in a population with high colorectal cancer incidence; when the population develops from a low-risk to a high-risk group, the frequency of adenomas and colorectal cancer usually also increases [3]. Within the colon itself, there appears to have been a "shift to the right" over time, in terms of segmental location of neoplasia [4]. The basis for this shift is unknown. Interestingly, women appear more prone to right-sided cancer than men [5].

Adenomas may be encountered upon endoscopy, in sizes that range from very small, 1–2 mm protrusions, to polyps 10 to 100 times larger. They may be pedunculated, sessile, or flat. Small adenomas are usually sessile and slightly redder than the surrounding mucosa. With increasing size, the adenoma usually becomes pedunculated, and the head is darker red and is broken into lobules with intercommunicating clefts. This darkening is due to a combination of increased vascularity and the different light-scattering properties of neoplastic, mucin-depleted epithelium [6].

A typical tubular adenoma has a lobulated berry-like surface, a villous adenoma a shaggy or velvety appearance sometimes described as "cauliflower-like". Often the surface of a villous adenoma is soft and friable. Adenomas are classified according to their histologically proven percentage of components: up to 20% villous component is classified as a tubular adenoma (branching glands), 20–80% as a tubulo-villous, and >80% as a villous adenoma (villiform means that the glands extend straight down from the surface of the polyp, creating villous-like projections from its surface).

The earliest identifiable morphological change in the growing adenoma is that the adenomatous epithelium, growing along the basement membrane, replaces part or all of a single crypt. At a later stage there is irregular branching of crypts close to the mucosal surface. We believe that villous formation may be the result of continuous growth leading to crypt elongation with compression of the intervening stroma [7]. Surface growth without crypt elongation may account for the appearance of "flat adenomas" [8], which are usually about 1 cm in diameter.

Size and histological type influence the change to a malignant tumor. All adenomas are, by definition, dysplastic. Tubular adenomas, which are the majority of adenomatous polyps, have a low risk of becoming malignant, whereas villous adenomas, which are more often sessile, have a higher risk of becoming malignant. Basically, with a polyp size of ≤1 cm, the cancer risk is approximately 1%, if the size is 1–2 cm it is up to 10%, and for size >2 cm the cancer risk is 20–50% [9].

10.3.1.2 Hyperplastic (Metaplastic) Polyps

Hyperplastic polyps are the most commonly encountered polyps in the adult colorectum. By definition, hyperplastic polyps are a metaplastic proliferation of differentiated colonic epithelium with no malignant potential. Their frequency increases with age, especially after 40 years. This kind of polyp is more common in men than in women, and more often diagnosed in Westernized populations. The polyp is often very small (<5 mm) and dome shaped; it is pale or, more commonly, the same colour as the surrounding mucosa, (most often) sessile, and situated upon mucosal folds; it is clinically asymptomatic and detected incidentally during colonoscopy. Hyperplastic polyps are unique among gastrointestinal polyps as being limited to the large intestine and appendix; the most common location is the rectosigmoid. Giant hyperplastic polyps are an exceedingly rare presentation. As these small polyps are also often seen in patients with concomitant adenomas, the use of hyperplastic polyps during sigmoidoscopy as biomarkers for further colonoscopic evaluation has been suggested [10–13]. This suggestion is, however, controversial [10–13]. Hyperplastic polyps are generally considered to be harmless and non-neoplastic; polypectomy should be done only to distinguish these lesions histologically from adenomas.

10.3.1.3 Sessile Serrated Adenomas

Current evidence suggests that small left-sided hyperplastic polyps carry very little risk of malignant transformation. However, multiple metaplastic polyps and larger right-sided metaplastic polyps seem to have an increased cancer risk when they transform into "serrated adenomas" (SSAs) [14]. It seems that this kind of polyp has an alternative carcinogenesis pathway [15,16]. Macroscopically, the serrated adenoma is usually between 5 and 20 mm in diameter, sessile, and either protuberant or flat. Magnification of the surface may show star-shaped or asteroid crypt openings as in "normal" hyperplastic polyps. These lesions are more common in women, with an average age at diagnosis of about 70 years.

In clinical practice there is still some discussion as how to diagnose and classify serrated adenomas. As clinicians, we distinguish between hyperplastic polyps, admixed polyps and serrated adenoma. However, the following issues confront the pathologist and are included in diagnostic considerations:

- large hyperplastic polyps, particularly when multiple in the proximal colon

- hyperplastic polyps with foci of dysplasia
- large hyperplastic polyps with enterocytic (eosinophilic) change
- admixed polyps, partly hyperplastic, partly traditional adenomas
- admixed polyps, partly hyperplastic, partly serrated adenoma
- classical serrated adenoma, unequivocally adenomatous but with serrated configuration
- tubulovillous or villous adenoma with slight or focal serration
- dysplastic epithelium with serration, abundant eosinophilic cytoplasm, and enlarged vesicular nuclei containing prominent nucleoli [17].

Among the pathological parameters, crypt branching, crypt dilatation, and horizontal crypts are more frequent in SSA than in traditional serrated adenoma (TSA) (P<0.001) [18]. SSAs are larger than TSAs (12.6±7.3 versus 9.8±6.9 mm, P=0.005), more likely to be flat (P=0.006), and more frequently located in the proximal colorectum (P=0.012). There are no significant differences in age, sex, and body mass index between TSA and SSA [18]; however, intraobserver agreement for SSA, TSA and hyperplastic polyps (HP) among pathologists is only moderate ($\kappa = 0.58$) [19].

10.3.1.4 Juvenile Polyps

The juvenile polyp is a non-neoplastic epithelial polyp composed of tissues indigenous to the site of origin, but arranged in a random manner. Thus, juvenile polyps are classified as hamartomas. They are very common, occurring in about 2% of children, but might also be seen in adults [20]. Typically, there are only a few polyps. With >10 juvenile polyps, we speak of juvenile polyposis; approximately one-third of cases have a hereditary etiology and the remainder are sporadic. In hereditary juvenile polyposis, the polyps may be found throughout the whole gastrointestinal tract, and in contrast to the sporadic form, polyps will continue to form throughout life.

The polyps can range from several millimeters to several centimeters, and may be sessile or, more often, pedunculated. The typical lesion found during endoscopy is 1.0–1.5 cm in size, with an intensely erythematous, friable, eroded, ulcerated surface. Sometimes the surface is nodular rather than ulcerated, making this lesion macroscopically indistinguishable from a tubular adenoma. The cut surface shows grossly visible cystic spaces containing gray-to-yellow mucoid material. Clinically, the most frequent presentation is painless rectal bleeding at the age of 9 to 10 years.

10.3.1.5 Inflammatory Polyps

Non-neoplastic proliferations of either mucosa or granulation tissue may be due to various injuries to the colorectal epithelium. The most frequent causes are ulcerative colitis, Crohn's disease, and ischemic bowel disease. Usually there are multiple polyps with chronic inflammation of varying extent in the vicinity of the lesion. Gross appearance varies from finger-like structures to rounded masses. Radiologists tend to call inflammatory polyps pseudopolyps (to describe the remaining normal mucosa islands in a severely ulcerated colon segment on a double-contrast enema). A special form is the "cap polyp", found mainly in the rectosigmoid region and caused by mucosal prolapse and ischemia as well as diverticular disease. The polyps are dark red, sessile, and usually located on the crest of a mucosal fold.

10.3.1.6 Peutz–Jeghers Polyps

In about half of the cases in Peutz–Jeghers syndrome, one or more small polyps are found in the colon and rectum. They are hamartomas with a characteristic histological appearance of an excessive and redundant muscularis mucosae covered by a non-neoplastic epithelium and lamina propria. Upon gross examination, the polyps vary in size but may also be several centimeters in size, sessile, or pedunculated, and may have a lobulated or irregular surface. Macroscopically, they may mimic adenomas, demanding histological investigations.

10.3.1.7 Endocrine Cell or Carcinoid Tumors

Endocrine tumors are rare (<1% of all colorectal neoplasms). Most arise in the rectum, while colonic forms are concentrated in the right colon. Right-sided endocrine cell tumors usually present as bulky polypoid or ulcerated masses, and are indistinguishable from carcinomas; in the rectum, bulky tumors are rare. Most often a small nodule less then 1 cm is discovered by chance. The cut surface is pink or tan.

10.3.2 Non-Epithelial Tumors

Non-epithelial tumors are rare.

10.3.2.1 Lymphoid Polyps

A localized hyperplasia of mucosal and submucosal lymphoid tissue may present as a single polyp or, less commonly, as multiple polyps. These polyps are most frequently seen in the lower rectum; in the colon they are exceptional. Usually they are smooth, round tumors, and mainly sessile. They are slightly more common in men in the third or fourth decade.

10.3.2.2 Gastrointestinal Stromal Tumors

Despite the preponderance of connective tissue in the large bowel, these tumors are relatively unusual; gastrointestinal stromal tumors (GISTs) occur there much less frequently than in the stomach and small intestine. The ascending and transverse colon are the most common sites for colonic GISTs, which usually present as very large masses (>5 cm, often even more than 10 cm). One very special feature of colonic GISTs is that they may grow as confluent nodules in a longitudinal fashion, with a dominant mass and multinodular thickening of the adjacent colonic wall [21]. Smooth muscle tumors occur mainly in the rectum. These incidental findings may present to the endoscopist as a small submucosal nodule or elevation. Beside these leiomyomatous polyps arising from the muscularis mucosae, deep intramural stroma tumors of the rectum have also been described. Local recurrence is a characteristic feature of the latter [22].

10.3.2.3 Lipomas

Although lipomas are the most common benign, non-epithelial tumors found in the colon, they are nonetheless very infrequent (0.035–4% of all colon tumors) [23]. These tumors are asymptomatic and found incidentally, though they can be very large: the largest reported colonic lipoma measured 16 cm [24]. The typical appearance is a smooth, yellowish, sessile polyp, although some lipomas may also be pedunculated. If mucosa can be grasped and pulled away from the submucosa ("tenting sign"), or if repeated biopsies of the polyp head reveal fatty tissue ("naked fat sign"), the diagnosis is definitive. However, if the surface is necrotic or ulcerated, lipomas may mimic malignant polyps [25].

10.4 New Investigational Methods

Standard conventional colonoscopy offers no reliable discrimination between neoplastic and non-neoplastic colorectal lesions, as the clinical impression of small colonic polyps does not correlate well with histology. Sensitivity and specificity are 87.4% and 65%, while the positive predictive value, negative predictive value, and accuracy of clinical impression for the detection of neoplastic polyps are 76.0%, 80.2%, and 73.4%, respectively [26]. However, conventional (non-magnifying) colonoscopy after spraying with indigo carmine dye achieves a sensitivity for adenomatous polyps as high as 91% (423/467 adenomas correctly identified) and a specificity of 82% (153/187 non-neoplastic lesions correctly predicted [27].

The issue of whether a polyp detected only upon endoscopy is non-neoplastic, or is neoplastic and so a possible precursor of a colorectal cancer, and the description of flat adenomas in 1985 by Muto from Japan caused considerable controversy [28]. Conventional thinking was that all adenomas are stalked or sessile. The interesting and dangerous issue of flat adenomas was pursued by subsequent papers that also reported the existence of small (<1 cm) flat adenocarcinomas [29]. Flat adenomas can be minimally elevated (IIa), flush with the mucosa (IIb), or depressed below the level of the mucosal surface (IIc). Flat adenomas IIb and IIc account for only 10% of all flat adenomas and only 2–3% of adenomas overall, but these lesions have a high incidence of high-grade dysplasia or early invasive cancer [30]. Conversely, type IIa lesions have no increased risk of severe dysplasia or cancer, and grow slowly [31]. Good bowel preparation is essential for the detection of flat adenomas, and with the acetic acid spray technique, adenomas covered with mucus can be effectively identified. Acetic acid is inexpensive and safe; it eliminates the mucus layer and causes an aceto-white reaction in the mucosal layer. These interactions highlight the pit pattern image in magnification colonoscopy [32].

In the West, it is common practice to diagnose colorectal cancer only when invasion through the muscularis mucosae has been demonstrated. This policy reflects the view that a diagnosis of cancer

Table 10.5 Kudo's classification for colonic neoplasms with high-magnification endoscopy

Classification	Surface pattern	Suspected growth type
Type I	Regular round crypt	Non-neoplastic
Type II	Stellar or papillary crypt	
Type III	Tubular and roundish crypts	Adenoma (neoplastic)
Type IV	Sulcus, branch, or gyrus-like crypts	
Type V	Irregular or severely distorted crypts	Invasive carcinoma

should equate with the acquisition of metastatic potential. Because colorectal neoplasia that is limited to the mucosa does not metastasize, the diagnostic label of cancer is not only unwarranted but could also lead to unnecessary surgery. This means that when a Western pathologist diagnoses "severe dysplasia", a Japanese pathologist might diagnose the same sample as "mucosal carcinoma" [33]. Whereas Western pathologists place major emphasis on the presence of invasion, Japanese pathologists rely on architecture and cytology to a much greater degree. While there is relatively little disagreement in polypectomy specimens between Western and Japanese pathologists, discrepancy mainly exists in diagnosing dysplasia in flat adenomas [34]. The Vienna classification has been introduced to facilitate exchange of scientific knowledge and discussions between pathologists in Japan and Europe (Table 10.4) [2].

Furthermore, these scientific questions stimulated the industry to develop new and better investigational tools. Beside conventional colonoscopy, the following tools alone or in combination are currently available: high-magnification (zoom) endoscopy, chromoendoscopy (CE), narrow-band imaging (NBI), computed virtual chromoendoscopy (CVC), autofluorescence imaging (AFI), and confocal laser scanning endomicroscopy (LCM). From an endoscopist's perspective, the attraction of using advanced imaging techniques is improved accuracy of screening for early superficial neoplasia.

10.4.1 High-Magnification Colonoscopy

The first magnifying endoscopes were introduced as early as 1978 [35]. Nishizawa identified the different characteristic patterns of crypt orifices by observing adenomatous polyps with magnifying colonoscopy in the 1980s [36]. In 1993, the Olympus 200Z series colonoscope was introduced, permitting *in vivo* magnification of up to 100× normal. With this technology, it is now possible to examine the detailed morphology and colonic crypt patterns similarly to *in vitro* stereomicroscopy. In 1994, Kudo introduced a classification for colonic neoplasms with high-magnification endoscopy, with the aim of distinguishing benign and potentially malignant lesions (Table 10.5) [37].

10.4.2 High-Magnification Chromoscopic Colonoscopy

A simple and well-known method is the application of dye (e.g. indigo carmine, methylene blue, Lugol's solution, etc), either injected directly down the biopsy channel or sprayed extensively over the colon surface using special catheters to improve the detection rate by increasing contrast [38]. Indigo carmine is usually used as it is not absorbed and so pools in depressions and mucosal ridges. The disadvantage of CE is that the colorectum must be perfectly clean, and application and inspection of the dye-sprayed and coloured mucosa is time consuming. Combination with high-magnification (HMCC) is popular: Kudo's classification can be used to differentiate between neoplastic and non-neoplastic growth patterns on the surface epithelium (Table 10.4) [37,39].

10.4.3 Narrow-Band Imaging

Recently, another technique, narrow-band imaging, has been proposed to study the capillary pattern of a suspected lesion [40]. When light is directed onto tissue, there can be four different outcomes: reflection, scattering, absorption, and fluorescence.

The NBI system utilizes short and limited wavelength within the hemoglobin absorption band, such that blood vessels can be demonstrated with adequate contrast. In principle, the NBI system is based on modification of the spectral features, with each optical filter narrowing a bandwidth of spectral transmittance [41]. The central wavelengths of trichromatic optical filters used are 500 nm, 445 nm, and 415 nm, and each has a bandwidth of 30 nm; these features correspond to a penetration depth into the mucosa of 240 μm, 200 μm and 170 μm, respectively. Capillary vessels are usually visualized as dark complexes when the 415 nm wavelength is used in which the blue light is mostly absorbed by hemoglobin [41]. The layered, structured gastrointestinal mucosa has a thickness of 700–800 μm; diagnosing neovascularization by neoangiogenesis is an important feature in malignant tumors [42]. Probably the most convenient feature of the NBI system is that with just one click of a special button on the colonoscope, the user can switch from conventional white light endoscopy (WLE) to the NBI system. This is a major advantage compared to the CE method. Essentially, low- and high-magnification NBI are able to distinguish neoplastic from non-neoplastic colorectal lesions. The accuracy of NBI is better than conventional colonoscopy and equivalent to chromoendoscopy [43].

CE mainly uses indigo carmine, which has proved to be effective in increasing the rate of colorectal adenoma detection in patients with hereditary non-polyposis colorectal cancer (HNPCC). NBI seems to be more effective than high-definition WLE in patients with a high cancer risk [44,45]. In patients with average cancer risk, NBI is similar to CE, whereby both are not really more effective than white light colonoscopy [46].

10.4.4 Computed Virtual Chromoendoscopy

Computed virtual chromoendoscopy with the Fujinon intelligent color enhancement (FICE) system is a new dyeless imaging technique that enhances mucosal and vascular patterns. The CVC imaging technique is based on narrowing the bandwidth of the conventional endoscopic image using spectral estimation technology [47]. CVC technology enhances the vascular network as well as the pit pattern. The NBI system is based on optical filters within the light source of a videoendoscope system that selects light in short and limited wavelengths within the haemoglobin absorption band, such that blood vessels can be demonstrated with adequate contrast [48]. FICE is based on the same physical principle as NBI, but due to a special computed spectral estimation technique, optical filters are not needed. FICE with low magnification shows a sensitivity of 89.9% and a specificity of 73.8%; FICE with high magnification has a sensitivity of 96.6% and a specificity of 80.3%. These results are better than with standard colonoscopy and low and high magnification, but not different from results obtained with conventional chromoendoscopy [49].

10.4.5 Autofluorescence Imaging

Autofluorescence imaging (AFI) is another novel technique. The idea is to use short-wavelength light (blue; 390–470 nm) to excite endogenous tissue fluorophores, which emit fluorescent light of longer wavelength (green; 540–560 nm). Normal mucosa is highlighted green, neoplastic tissue purple without administration of exogenous fluorophores. Most false-positive results occur when inflamed mucosa is evaluated [50]. The 415 nm image channel analyzes the fine surface architecture of the mucosa and the superficial capillary network; the 540 nm image channel analyzes the collecting vessels more in depth. Superficial and deep details are superimposed in a single image, enhancing the visibility of flat lesions and displaying subepithelial capillaries in brown and veins in the submucosa in cyan.

Autofluorescence is abnormal in neoplastic tissue because of:

- the increase in the nuclear–cytoplasmic ratio, which leads to reduction of autofluorescence in neoplasia, as nuclei show no autofluorescence compared with cytoplasm
- loss of collagen, since collagen in submucosa is the strongest fluorophore, and with thickened neoplastic mucosa, the submucosal collagen shows no autofluorescence
- neovascularization, resulting in increased hemoglobin concentration that absorbs autofluorescent light [51].

A randomized study from the Netherlands comparing AFI and conventional colonoscopy for the detection of colorectal adenomas did not show a statistically significant difference (73% versus 70%) [52]. However, detection of neoplasia in patients with ulcerative colitis has improved, and decreases the yield of random biopsies [53]. All diagnostic modalities that rely on morphological imaging also involve a learning effect if the investigator is to apply a method effectively. This is also true for AFI [54]. At the moment, the future role of AFI is not clearly defined [51].

10.4.6 Confocal Laser Scanning Endomicroscopy

Laser scanning confocal microscopy is another interesting newly developed technology, which for the first time permits cellular resolution imaging of the surface and subsurface mucosa down to a depth of 250 μm. The components of the confocal laser microscope are mounted in the tip of a conventional colonoscope, so that this technology can be used beside standard videoendoscopy. During LCM, a 488 nm laser delivers an excitation wavelength with a maximum laser output of <1 mW at the surface of the mucosa. The optical slide thickness is 7 μm; the image collection rate is 0.8–1.6 frames/s. To increase contrast, intravenous fluorescein sodium is injected, and topical flurophores such as acriflavin hydrochloride are sprayed onto the mucosal surface immediately before an image is obtained [55]. At the moment, however, it is uncertain whether endomicroscopy offers any advantage over chromoendoscopy and targeted biopsy alone.

10.5 Practical Approach

For practical reasons, a step-by-step methodology in endoscopy for detection of lesions should be followed:

- cleanliness of mucosal surface: good preparation, jet wash, antifoam agents
- detection of a suspected area: withdrawal time >8 min
- characterization of the lesion: gross morphology (indigo carmine), microvascular surface (NBI), surface microarchitecture of the superficial epithelium (NBI, chromoendoscopy – pit pattern) (Figs. 10.1–10.4). Analysis of microcirculation at low magnification with NBI (without chromoendoscopy). Gross morphology with standard vision and chromoendoscopy. Severe vascular alterations are evaluated with zoom and chromoendoscopy
- classification of the lesion and treatment decision.

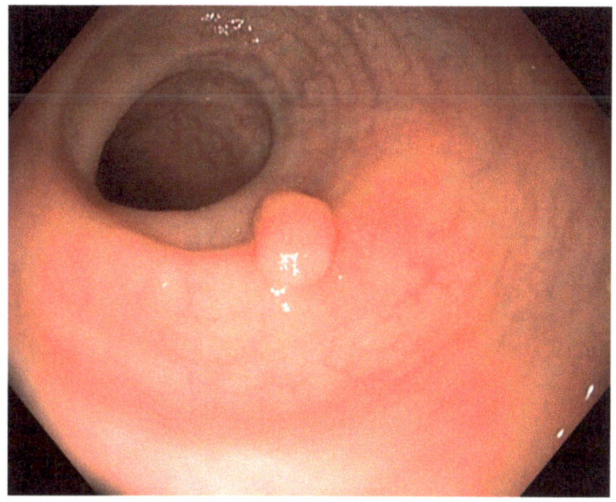

Fig. 10.1 Adenomatous polyp

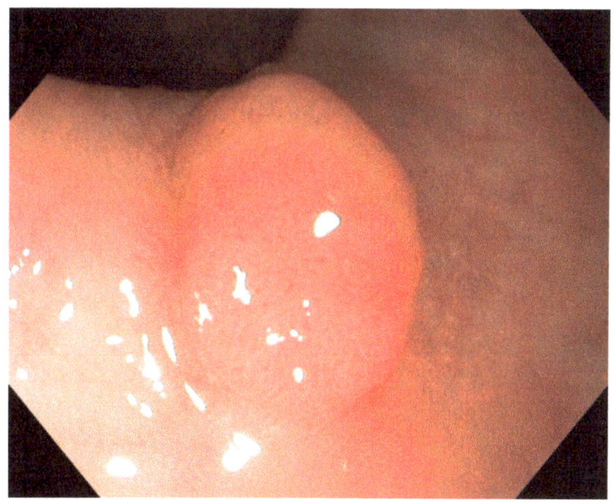

Fig. 10.2 The same adenomatous polyp as in Figure 10.1 (zoom 1.5×)

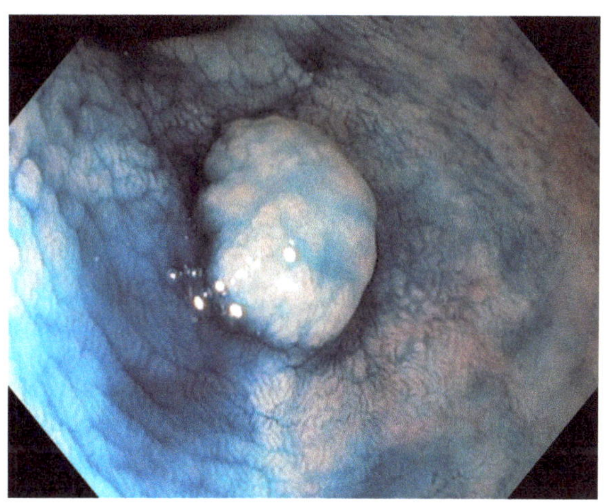

Fig. 10.3 The same adenomatous polyp as in Figure 10.1 with chromoendoscopy (indigo carmine)

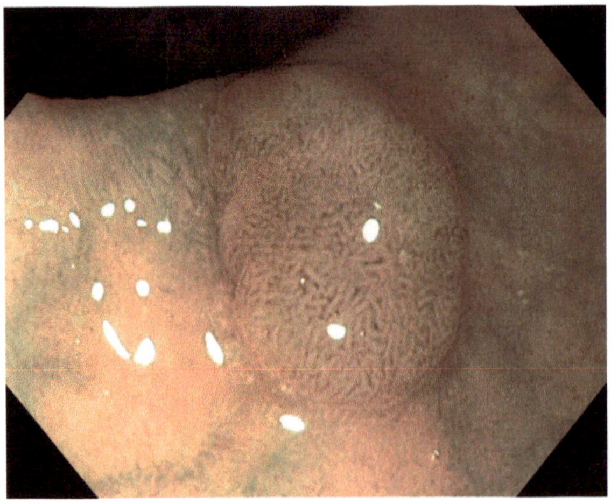

Fig. 10.4 The same adenomatous polyp as in Figure 10.1 (zoom 1.5×) and NBI

10.6 Future Colonoscopy

At the moment, colonoscopes equipped with optical magnification are costly and therefore not widely used [56]. However, if these new technologies prove to be superior in daily practice, economics will not stand in the way of their general availability and use.

We can expect that in the future, screening colonoscopies will be largely replaced by painless virtual colonoscopy [57]. However, if a lesion is suspicious and potentially malignant, newer technologies like high-resolution HMCC, NBI, and AFI will probably be used. The new technologies seem to have the advantage of timely recognition of malignant transformation of the mucosa, especially in longstanding colitis. Only time will tell whether LCM can finally replace the pathologist's diagnosis made with a biopsy under the microscope.

10.7 Conclusion

Understanding and recognizing different forms of polypoid lesions and their neoplastic potential is of utmost importance to the endoscopist as well as to the colorectal surgeon. New technologies are currently being investigated with regard to their practicability and cost-effectiveness. The future will provide answers in this exciting evolving field of early colorectal cancer detection.

References

1. Paris Workshop Participants (2002) The Paris endoscopic classification of superficial neoplastic lesions: esophagus, stomach, and colon. Gastrointest Endosc 58:S3–43.
2. Schlemper RJ, Riddell RH, Kato Y, et al (2000) The Vienna classification of gastrointestinal epithelial neoplasia. Gut 47:251–255.
3. Clark JC, Collan Y, Eide TJ, et al (1985) Prevalence of polyps in an autopsy series from areas with varying incidence of large bowel cancer. Int J cancer 36:179–186.
4. Bufil JA (1990) Colorectal Cancer: evidence for distinct genetic categories based on proximal or distal tumor location. Ann Intern Med 113:779–788.
5. Devesa SS, Chow WH (1993) Variation in colorectal cancer incidence in the United States by subsite of origin. Cancer 71:3819–3826.
6. Biemer-Hüttmann AE, Walsh MD, McGuckin MA, et al (1999) Immunohistochemical staining patterns of MUC1, MUC2, MUC4, and MUC5AC mucins in hyperplastic polyps, serrated adenomas, and traditional adenomas of the colorectum. J Histochem Cytochem 47:1039–1048.
7. O'Brian MJ, Winawer SJ, Zauber AG et al (1990) The National Polyp Study: determinants of high grade dysplasia in colorectal adenomas. Gastroenterology 98:371–379.
8. Fearon ER, Hamilton SR, Vogelstein B (1987) Clonal analysis of human colorectal tumours. Science 238:193–196.
9. Nusko G, Mansmann U, Altendorf-Hofmann A, et al (1997) Risk of invasive carcinoma in colorectal adenomas assessed by size and site. Int J Colorectal Dis 12:267–271.
10. Naveau S, Brajer S, Bedossa P, et al (1991) Hyperplastic colonic polyps as a marker for adenomas colonic polyps. Eur J Gastroenterol Hepatol 3:57–61.
11. Kellokumpu I, Kyllönen L (1991) Multiple adenomas and synchronous hyperplastic polyps: predictors of metachronous colorectal adenomas. Ann Chir Gynecol 80:30–35.
12. Opelka FG, Timmcke AE, Gathright JB, et al (1992) Diminutive colonic polyps: an indication for colonoscopy. Dis Colon Rectum 35:178–181.
13. Imperiale TF, Wagner DR, Lin CY, et al (2000) Risk of advanced proximal neoplasms in asymptomatic adults according to the distal colorectal findings. N Engl J Med 343:169–174.
14. Longacre TA, Fenoglio-Preiser CM (1990) Mixed hyperplasia adenomatous polyps/serrated adenomas. Am J Surg Pathol 14:524–537.
15. Baker K, Zhang Y, Jin C, Jass JR (2004) Proximal versus distal hyperplastic polyps of the colorectum: different lesions or a biological spectrum? J Clin Pathol 57:1089–1093.
16. Jass RJ (2004) Hyperplastic polyps and colorectal cancer: is there a link? Clin Gastroenterol Hepatol 2:1–8.
17. Day DW, Jass JR, Price AB, et al (2003) Epithelial tumours of the large intestine. In: Morson BC, Dawson AM (eds) Morson and Dawson's Gastrointestinal Pathology, 4th edn. Blackwell Science, Oxford, pp 551–609.
18. Lee S, Chang H, Kim T et al (2008) Clinicopathologic findings of colorectal traditional and sessile serrated adenomas in Korea: a multicenter study. Digestion 77:178–183.
19. Farris A, Misdraji J, Srivastava A, et al (2008) Sessile

Adenoma. Challenging discrimination from other serrated colonic polyps. Am J Surg Pathol 32:30–35.
20. Merg A, Howe JR (2004) Genetic conditions associated with intestinal juvenile polyps. Am J Med Genet 129C:44–55.
21. Appelman HD (1998) Mesenchymal tumours of the gastrointestinal tract. In: Ming S, Glodman (eds) Pathology of the gastrointestinal tract. Williams & Wilkins, Baltimore, pp 361–364.
22. Haque S, Dean PJ (1992) Stromal neoplasms of the rectum and anal canal. Hum Pathol 23:762–767.
23. Milkes DE, Soetikno RM (2007) Other benign and malignant colonic tumors. In: Wexner SD, Stollman N (eds) Diseases of the colon. Informa Healthcare, New York, pp 517–542.
24. Alponat A, Kok KY, Goh PM, Ngoi SS (1996) Intermittent subacute intestinal obstruction due to a giant lipoma of the colon: a case report. Am J Surg 62:918–921.
25. Ryan J, Martin JE, Pollock DJ (1989) Fatty tumours of the large intestine: a clinicopathological review of 13 cases. Br J Surg 76:793–796.
26. Lawrance IC, Sherrington C, Murray K (2006) Poor correlation between clinical impression, the small colonic polyp and their neoplastic risk. J Gastroenterol Hepatol 21:563–568.
27. Sonwalker S, Rotimi O, Rembacken BJ (2006) Characterization of colonic polyps at conventional (nonmagnifying) colonoscopy after spraying with 0.2% indigo carmine dye. Endoscopy 38:1218–1223.
28. Muto T, Kamiya J, Sawada T, et al (1985) Small "flat adenoma" of the large bowel with special reference to its clinicopathologic features. Dis Colon Rectum 28:847–851.
29. Balanchandar G, Trowers EA (1999) Early colorectal cancer in a flat adcnoma. J Natl Med Assoc 91:631–632.
30. Kudo S, Kashida H, Tamura T, et al (2000) Colonoscopic diagnosis and management of nonpolypoid early colorectal cancer. World J Surg 24:1081–1090.
31. Rozen P, Brazowski E (2003) Flat colorectal neoplasia: identification, pathogenesis and clinical significance. Dig Liver Dis 35:135–137.
32. Togashi K, Hewett DG, Whitaker DA, et al (2006) The use of acetic acid in magnification chromocolonoscopy for pit pattern analysis in small polyps. Endoscopy 38:612–616.
33. Schlemper RJ, Kato Y, Stolte M (2001) Review of histological classifications of gastrointestinal epithelial neoplasia: differences in diagnosis of early carcinomas between Japanese and Western pathologists. J Gastroenterol 36:445–456.
34. Jass JR (2000) Histopathology of early colorectal cancer. World J Surg 24:1016–1021.
35. Tada M, Misaki FF, Kawai K (1978) A new approach to the observation of minute changes of the colonic mucosa by means of magnifying colonoscopy, type CF-MB-M (Olympus). Gastrointest Endosc 24:146–147.
36. Nishizawa M, Okata T, Sato F, et al (1980) A clinicopathological study of minute polypoid lesions of the colon based on magnifying fiber-colonoscopy and microscopy. Endoscopy 12:124–129.
37. Kudo S, Hirota S, Nakajuama T, et al (1994) Colorectal tumors and pit pattern. J Clin Pathol 47:880–885.
38. Kiesslich R, van Bergh M, Hahn M, Hermann G, Jung M (2001) Chromoendoscopy with indigocarmine improves the detection of adenomatous and nonadenomatous lesions in the colon. Endoscopy 33:1001–1006.
39. Kudo S, Rubio CA, Teixeira CR, et al (2001) Pit pattern in colorectal neoplasia: endoscopic magnifying view. Endoscopy 33:367–373.
40. Machida H, Sano Y, Hamamoto Y, et al (2004) Narrow-band imaging in the diagnosis of colorectal mucosal lesions: a pilot study. Endoscopy 36:1094–1098.
41. Sano Y, Kobayashi M, Hamamoto Y, et al (2001) New diagnostic method based on color imaging using narrow-band imaging (NBI) system for gastrointestinal tract. Gastrointest Endosc 53:AB125.
42. Risau W (1997) Mechanisms of angiogenesis. Nature 386:671–674.
43. Chiu H, Chang C, Chen C, et al (2007) A prospective comparative study of narrow-band imaging, chromoendoscopy, and conventional colonoscopy in the diagnosis of colorectal neoplasia. Gut 56:373–379.
44. East JE, Suzuki N, Stavrinidis M, et al (2007) Narrow-band imaging for colonoscopic surveillance in HNPCC syndrome: a back-to-back study. Gastrointest Endosc 65:AB126.
45. East JE, Suzuki N, Stavrinidis M, et al (2007) Narrow-band imaging improves adenoma detection in patients with high risk for adenomas: a randomized trial. Gastrointest Endosc 65:AB116.
46. Repici A, Hervoso CM, Preatoni P, et al (2007) A prospective study comparison of NBI versus chromoendoscopy with indigo carmine versus white-light colonoscopy in the detection of adenomatous colonic lesions. Gastrointest Endosc 65:AB351.
47. Pohl J, May A, Rabenstein T, et al (2007) Computed virtual chromoendoscopy: a new tool for enhancing tissue surface structures. Endoscopy 39:80–83.
48. Gono K, Yamazaki K, Dogucji N, et al (2003) Endoscopic observation of tissue by narrow band illumination. Opt Rev 10:1–5.
49. Pohl J, Nguyen-Tat M, Pech O, et al (2008) Computed virtual chromoendoscopy for classification of small colorectal lesions: a prospective comparative study. Am J Gastroenterol 103:562–569.
50. Ochsenkuhn T, Tillack C, Stepp H, et al (2006) Low frequency of colorectal dysplasia in patients with long-standing inflammatory bowel disease colitis: detection by fluorescence endoscopy. Endoscopy 38:477–482.
51. Ragunath K (2007) Autofluorescence endoscopy – not much gain after all? Endoscopy 39:1021–1022.
52. van den Broek F, Curvers WL, Hardwick JC, et al (2007) Interobserver variability and accuracy of colonic pit pattern by narrow-band imaging and the additional value of autofluorescence characteristics. Gastrointest Endosc 65:AB349.
53. van den Broek FJ, Fockens P, van Eeden S, et al (2008) Endoscopic tri-modal imaging for surveillance in ulcerative colitis: randomised comparison of high-resolution endoscopy and autofluorescence imaging for neoplasia detection and evaluation of narrow-band imaging for classification of lesions. Gut 57:1083–1089.
54. Kato M, Kaise M, Yonezawa J, et al (2007) Autofluorescence endoscopy versus conventional white light endoscopy for the detection of superficial gastric neoplasia: a prospec-

tive comparative study. Endoscopy 39:937–941.
55. Smith L, Tiffin N, Thomson M, et al (2008) Chromoscopic endomicroscopy: in vivo cellular resolution imaging of the colorectum. J Gastroenterol Hepatol 23:1009–1023.
56. Lambert R, Saito H, Saito Y (2007) High-resolution endoscopy and early gastrointestinal cancer . . . dawn in the East. Endoscopy 39:232–237.
57. Summerton S, Little E, Cappell MS (2008) CT colonography: current status and future promise. Gastroenterol Clin North Am 37:161–189.

Diagnosis and Treatment of Upper Gastrointestinal Polyps in Polyposis

11

Michele Comberlato and Federico Martin

Abstract Familial adenomatous polyposis is a hereditary autosomal dominant disease; the *APC* gene is responsible. The multifactorial regulation of this gene expresses many different clinical features and almost all patients present with polyps in the foregut, so gastroenterologists and surgeons should carefully check patient's upper gastrointestinal tract as well. The diagnosis of polyps in the stomach and, mostly, in the duodenum is of crucial relevance due the high risk of developing a cancer. A better understanding of the natural history of the disease, together with recent developments in operative endoscopy and surgery, make it possible to offer these patients better interdisciplinary management.

Keywords Chromoendoscopy • Complications • Diagnosis • Duodenectomy, Endoscopy • Excision • Gastrectomy • Mucosectomy • Polypectomy • Upper gastrointestinal tract • Staining • Submucosal dissection

11.1 Introduction

Polyps, usually small ones, represent an incidental diagnosis in the upper gastrointestinal (GI) tract, as they are detected in no more than 2% of endoscopic procedures, which, in the majority of cases, are performed for unrelated indications [1].

The esophagus is very seldom involved in polyposis, but the stomach and particularly the duodenum are widely affected in patients with familial adenomatous polyposis (FAP).

German colleagues described gastric polyps for the first time in FAP in 1895 [2], and duodenal polyps in 1904 [3], and the first diagnosis of duodenal carcinoma associated with polyposis was published in 1935 [4].

Until the 1970s, only sporadic cases were reported, but since then, the wide diffusion of flexible endoscopy has allowed a widespread knowledge of the relevance of a correct diagnosis of polyposis in the upper GI tract [5].

The upper GI tract represents the most common extracolonic site of malignancy in FAP, and it is therefore obvious that any endoscopic examination performed in this group of patients should offer the maximal diagnostic potential. Duodenal adenomas and carcinomas are very uncommon in the general population, but the majority of patients affected by FAP present with a duodenal site of disease [6].

A major breakthrough of the last decade has been the start of surveillance programs in the upper GI tract as soon as a diagnosis of FAP has been made, and not just in adulthood, or, worse, only after a prophylactic colectomy [7].

M. Comberlato (✉)
Department of Gastroentrology and Digestive Endoscopy, Central Hospital, Bolzano, Italy

G.G. Delaini, T. Skřička, G. Colucci (eds.), *Intestinal Polyps and Polyposis*,

11.2 The Pattern of Disease

Gastric and duodenal polyps appear as rounded, small, and whitish lesions that are sessile or nodular and superficial, although sometimes they can reach a diameter of 2–3 cm. In cases where the diagnosis is late, the polyposis may be so diffuse that large flat lesions have formed, or the polyps have developed as a carpet along the mucosa (Figs 11.1–11.4) [8]. The gold standard of the diagnostic procedures in this context is always flexible endoscopy.

Presentation of the most recent technical evolution in the field of foregut endoscopy goes beyond the scope of our paper, but it is of crucial relevance to remember the great impact of the more recent methods for performing an accurate examination of the gastric and duodenal mucosa. The new technique of dying the mucosa and performing a real chromoendoscopy, with standard or high-resolution and magnifying video endoscopes, allows us to recognize very early even small, rounded lesions, that would otherwise be undetectable; this should now be the standard procedure for endoscopic surveillance in this group of patients [9].

The vast majority of gastric polyps are represented by fundic gland polyps, usually located in the fundus and the body of the stomach, and detected in up to 50% of the screened population. The histology of these small lesions demonstrates cystic dilation of fundic glands, usually without dysplasia, but with sporadic reports of cancerization [10–12].

Gastric adenomas occur in a minority of patients, and are usually situated in the antrum and as a single small lesion. The evolution of these rare lesions is without risk, as they do not proceed to a real malignant lesion [13].

There are no data in the literature to suggest any treatment for small, rounded fundic gland polyps, but where there is clear evidence of an adenoma, it seems reasonable to treat the lesion as an adenoma, and remove it endoscopically with either a polypectomy or a mucosectomy. Alternative endoscopic therapies such as argon plasma coagulation or Nd:YAG laser could be employed in cases of diffuse small adenomas.

Duodenal lesions are similar to gastric ones, grouped in the second or third section of the duodenum, with a regular duodenal bulb; they can be found in 50–90% of cases [14].

When the polyps are located in the second part of the duodenum, they tend to cluster near the papilla or precisely on it. At histological examination they appear as standard adenomas with the usual degree of possible dysplasia [14]. Duodenal adenomas are detected in 20–80% of patients, with differences partly explained by the different endoscopes used to screen the patients with either a forward or side view. The

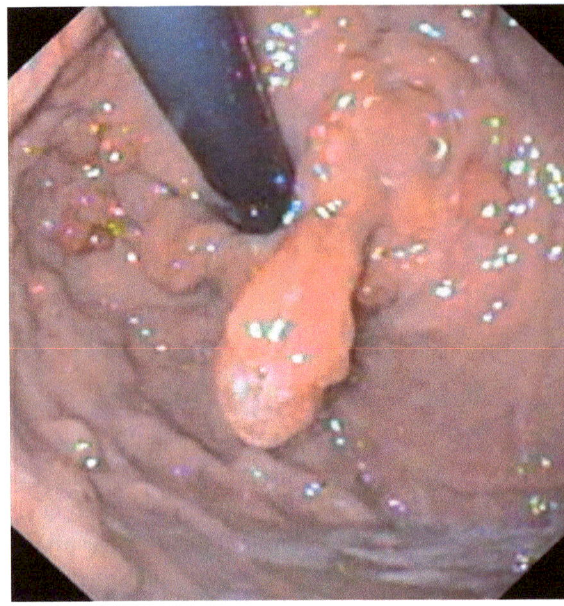

Fig. 11.1 Fundic gland polyps

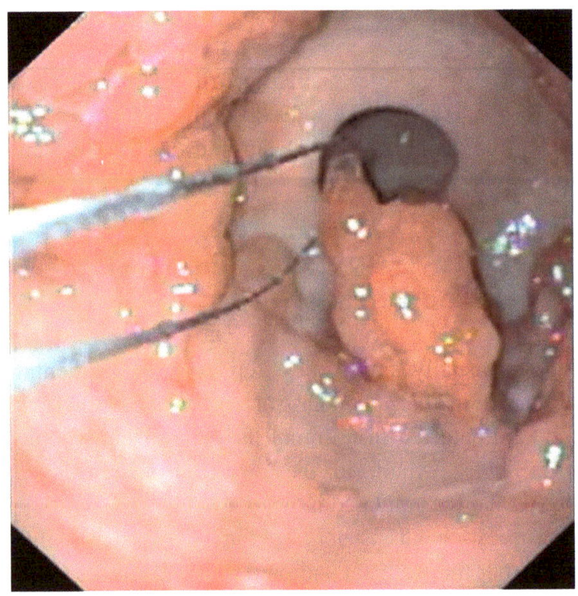

Fig. 11.2 Antral polyps

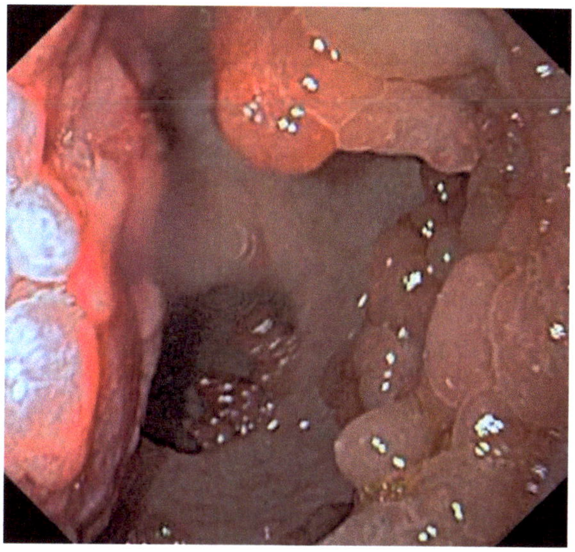

Fig. 11.3 Duodenal polyps

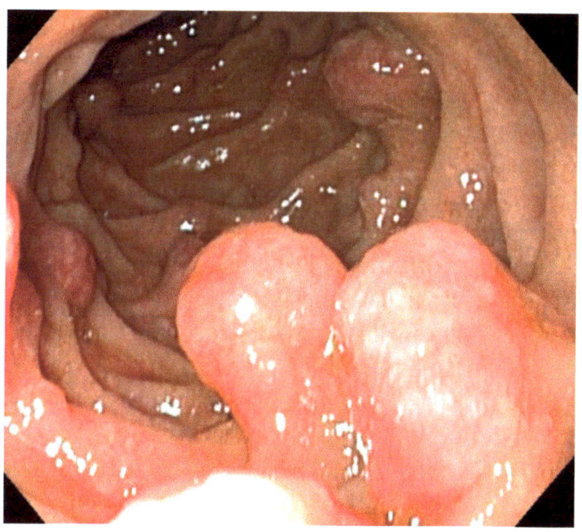

Fig. 11.4 Duodenal polyps

most accurate examinations are achieved with the side-viewing scope and multiple large biopsies [8,15].

The relevance of duodenal adenomas is clearly explained by the fact that duodenal adenocarcinoma is the major cause of death in FAP, after cancer of the colon and rectum [16]. Furthermore, if we analyze the impact of duodenal lesions on cancer risk, the epidemiological data are very impressive, with a 100–300-fold major risk of cancer in comparison with the general population, and an overall cumulative risk of 4–10% [17,18].

Gastric polyps are simply described as they appear during a standard foregut endoscopy, but, in contrast, duodenal polyps were classified by Spigelman in 1989, and this coding is still valid and routinely used worldwide [5] (Table 11.1).

The Spigelman classification allows to score duodenal polyps according to four general criteria, with stage 0 to I identifying no or minimal disease, and stage IV associated with more aggressive disease, often accompanied by severe dysplasia.

Long-term cohort studies report very slow progression of disease with development of malignancy in between 7% and 36% of patients enrolled in Spigelman stage IV [18,19].

Considering the many factors involved in the different evolution of the disease, age seems to be one of the major risk factors for malignancy. It is essential to make an effort to clearly identify the risk situations in which there should be very close surveillance, to allow intervention before invasive carcinoma develops and while disease is still in an early, theoretically curable, stage.

There are three very important surveillance studies in this field reported in the literature, which are summarized in Table 11.2 [18–20].

Table 11.1 Spigelman classification for duodenal polyps in FAP

Criterion	1 point	2 points	3 points
Polyp number	1–4	5–20	>20
Polyp size (mm)	1–4	5–10	>10
Histology	Tubular	Tubulo-villous	Villous
Dysplasia	Mild[a]	Moderate[a]	Severe[b]

Stage 0: 0 points; Stage I: 1–4 points; Stage II: 5–6 points; Stage III: 7–8 points; Stage IV: 9–12 points

[a]A low degree of dysplasia according to current classification

[b]A high degree of dysplasia

Table 11.2 Risk of progression of duodenal polyps in FAP

Author	Groves et al, 2002 [19]	Saurin et al, 2004 [20]	Bulow et al, 2004 [18]
Patients, *n*	99	35	368
Age, years	42	37	25
Sex, male %	55	57	49
Follow-up, years	10	4	7.6
Spigelman IV, %			
At start	9.6	14	7
At end	14	35	15
Duodenal cancer	6[a]	0	4[b]

[a]Spigelman stage in previous endoscopies: II, III, IV, IV, IV, IV
[b]Spigelman stage in previous endoscopies: II, III, IV, IV

The conclusions of these studies are confirmed by a recent paper from Vasen et al. about the clinical management of FAP [21]:

- disease progression in terms of number, size, and histology is very slow
- the cancer risk is related to the Spigelman stage at entry into the surveillance program
- the evolution from adenoma to cancer may take more than 20 years
- standard current protocols detect duodenal polyposis at an early Spigelman stage.

11.3 Pharmacological Therapy of Duodenal Polyposis

The first drug that proved to be effective in the treatment of FAP was sulindac, which, in long-term therapy, was able to reduce the number of colorectal adenomas by more than 50%, before and after colectomy [22–24]. Unfortunately, the drug has no effect in the prevention of polyposis.

Twenty years ago, a new class of non-steroidal anti-inflammatory drugs (NSAIDs) became available, the COX-2 inhibitors, and celecoxib was successfully tested in the treatment of polyposis, with a significant action in treating duodenal adenomas as well [25].

Unfortunately, the reported undesirable cardiovascular effects with rofecoxib stopped many trials. A recent meta-analysis on 11 published studies confirmed the safety of celecoxib at a dose of 200 mg per day, but the standard doses prescribed for polyposis are higher [26].

Therefore, at present, sulindac appears to be the only safe available drug for colorectal polyposis, although celecoxib could be considered in a tailored therapy for patients at high risk for progression of the disease, but without cardiovascular risk factors.

11.4 Endoscopic Management of Duodenal Polyposis

As in other fields, the good management of duodenal polyposis in FAP is the result of a close relationship between the gastroenterologist and the surgeon, because in all situations a shared plan of therapy and follow-up is recommended.

At present there are no data in the literature comparing treatment and surveillance versus no surveillance, and the following considerations are based on the classical Spigelman's theorem.

It is widely accepted that duodenal polyps Spigelman stage 0–II have a minimal risk of progression, in contrast with the high risk of malignancy in the III–IV stage, up to 36% [18,19].

The first consideration when starting to examine the possible therapeutic options is that there is no consensus about the optimal way to treat these patients.

The two options for treatment are endoscopy and surgery. It seems easy, but which and when?

We will start with endoscopy.

The recent technological improvement in flexible digestive endoscopy has allowed a broadening of the choice of the treatment and a multimodal approach to duodenal polyposis. Small polyps could be coagulated with forceps or destroyed with argon plasma coagulation, but the real battle is not with the small polyps Spigelman 0–II. Larger polyps could be snared, using the technique of endoscopic mucosal resection to warrant a more radical excision, and cutting to the submucosa; the resection could be completed with coagulation of the polyp's edge (Figs 11.5–11.8).

However, snaring is not enough if the outcome of the procedure is not surveyed, and a recent review by Brosens et al analyzed 11 papers published on this topic [27]. The recurrence of adenoma after polypectomy is high, up to 50% with a mean 17% incidence of complications: perforation, haemorrhage, or pancreatitis.

Thus, an endoscopic therapeutic approach appears not to be indicated for small non-advanced polyps, and only larger lesions, bigger than 1 cm, or adenomas with high-grade dysplasia should be endoscopically resected.

In these cases, the endoscopist should be highly skilled in performing an endoscopic mucosal resec-

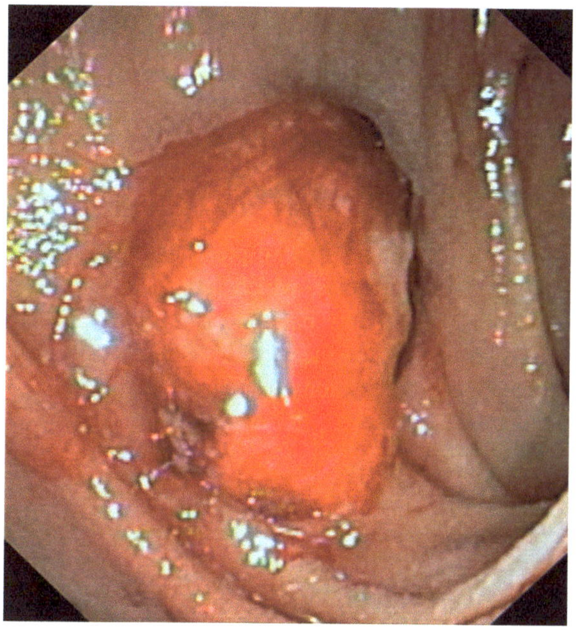

Fig. 11.5 Duodenal adenoma

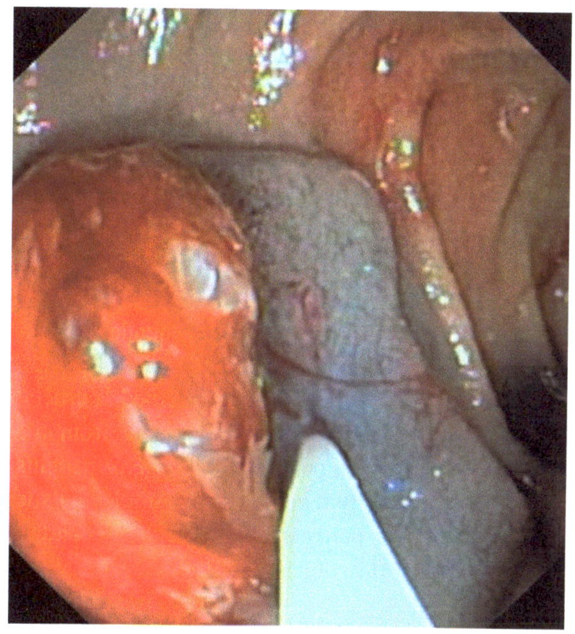

Fig. 11.6 Submucosal saline injection

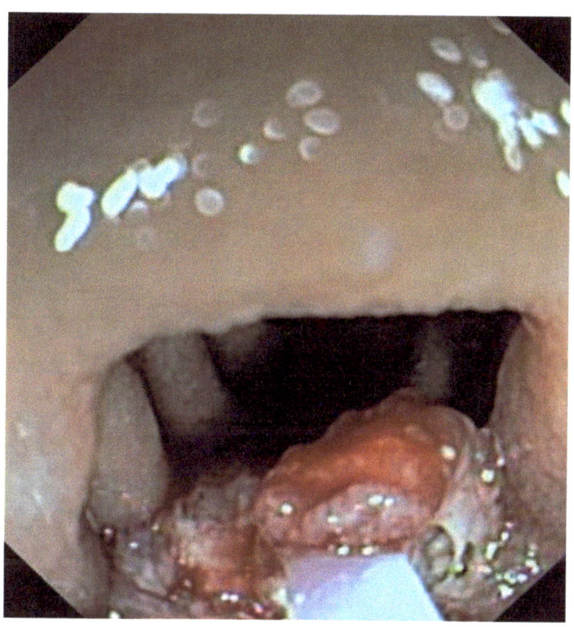

Fig. 11.7 Endoscopic mucosal resection

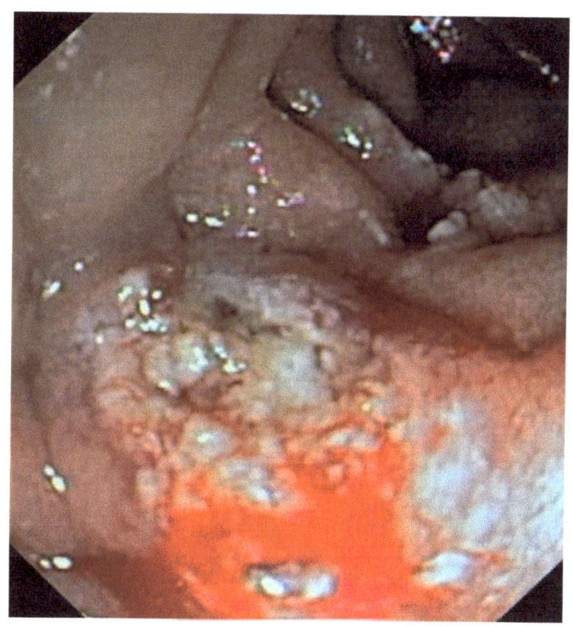

Fig. 11.8 Base of mucosectomy

Table 11.3 Suggested program of endoscopic surveillance

Spigelman stage	Endoscopic control interval
0–I	5 years
II	3 years
III	1–2 years
IV	Consider surgery

tion, which is preceded by submucosal injection of saline or volume expander solutions in order to minimize the risk of perforation; the surgeon should also be skilled in the use of endoclips for treating bleeding or small perforations [28].

Therefore, the real value of endoscopic treatment of duodenal polyps Spigelman's class II or more is still debated, and while the delay in major surgical procedures could be a clear advantage, this is limited by a significant morbidity and even mortality.

Table 11.3 summarizes the recommended scheduled procedures for the upper GI tract surveillance.

As far as duodenal polyps are concerned, a crucial topic to consider is the adenomas arising from the papilla of Vater, which occur in 50–100% of patients [29]; the periampullary region is the most relevant site for malignancy in patients after colectomy, and it develops in up to 10% of cases [20].

Traditionally the first-line therapy was surgery, with either duodenocefalopancreasectomy or transduodenal excision. Due to the fact that only small series or case reports have been published, it is difficult to establish a rule in the procedure for endoscopic ampullectomy. Up today, the endoscopic resection of the adenoma represents the ideal treatment, but it also presents a high rate of complications and should therefore be reserved to endoscopists with proven experience in therapeutic endoscopy and only used in patients with clear clinical indications.

Norton and colleagues reported immediate postprocedure complications in 7 out of 26 patients (4 mild pancreatitis, 2 focal bleeding stopped with adrenaline (epinephrine) injection, and 1 perforation successfully managed conservatively) [30].

Regarding the technique, is very important to perform a clear cholangiopancreatography before the resection and to inject some drops of dye into the pancreatic duct, to facilitate post-resection cannulation and insertion of a short stent for prevention of pancreatitis.

All snare ampullectomies have to be performed without submucosal injection, in order to perform a full-thickness resection of the adenoma and of the deeper sphincter musculature. This allows rapid and easier access to the pancreatic duct [31].

A crucial point of debate in this topic is the setting of the electrosurgical generator. As in other fields, fixed rules are lacking, but the authors agree with the view that advises setting the generator on endocut blended energy with more cutting than coagulating power to avoid deep thermal injury and major risk of secondary complications.

11.5 Jejuno-ileal Polyps

When jejuno-ileal polyps are present, they are very small and the diagnosis, which is made with either capsule [32] or double balloon enteroscopy [33], is usually fortuitous. As there is no evidence of cancer risk, there is no indication for polypectomy [34].

11.6 Surgery and Upper GI Tract Polyposis

Surgical treatment of duodenal adenomatosis should be performed before biopsy shows invasive adenocarcinoma. There is also a general agreement that prophylactic surgery should be reserved for patients with advanced disease (Spigelman stage III–IV) considering that the risk of developing cancer in this group of patients is about 7–35% compared with 5% of the overall population with FAP [35].

Surgical options include local and resective interventions.

Ampullectomy and duodenotomy with polypectomy are followed by a high recurrence rate and, in our opinion, should be reserved for young patients in order to postpone major surgery.

Major surgical procedures (the classical Whipple operation or pylorus-preserving pancreatoduodenectomy) are probably the best options for adults with advanced Spigelman stage. However, the recurrence rate after extensive surgery is not zero. In a paper published by Vasen and colleagues, this figure is about 30% in patients treated by pylorus-preserving pancreatoduodenectomy, and 18% in a group that underwent the Whipple procedure [36].

References

1. Dekker W (1990) Clinical relevance of gastric and duodenal polyps. Scand J Gastroentrol 178:7–12.
2. Hauser G (1895) Über Polyposis intestinalis adenomatosa und deren Beziehungen zur Krebsentwicklung. Deutsche Arch Klin Med 55:429–438.
3. Funkestein O (1904) Über Polyposis intestinalis. Z Klin Med Berl 55: 236–248.
4. Cabot J (1935) Case records of the Massachusetts General Hospital N 21061. N Engl J Med 212:263–267.
5. Spigelman AD, Williams CB, Talbot IC, et al (1989) Upper gastrointestinal cancer in patients with familial adenomatous polyposis. Lancet 2:783–785.
6. Vasen HF, Bulow S, Myrhoj T, et al (1997) Decision analysis in the management of duodenal adenomatosis in familial adenomatous polyposis. Gut 40:716–719.
7. Vasen HF, Bulow S and the Leeds Castle Polyposis Group (1999) Guidelines for the surveillance and management of familial adenomatous polyposis (FAP): a world wide survey among 41 registries. Colorectal Dis 1:214–221.
8. Iida M, Aoyagi K, Fujimura Y, et al (1996) Non polypoid adenomas of the duodenum in patients with familial adenomatous polyposis. Gastrointest Endosc 44:305–308.
9. Bruno MJ (2003) Magnification endoscopy, high resolution endoscopy and chromoendoscopy: towards a better optical diagnosis. Gut 52(S4):7–11.
10. Sarre RG, Frost AG, Jagelman DG, et al (1987) Gastric and duodenal polyps in familial adenomatous polyposis; a prospective study of the nature and prevalence of upper gastrointestinal polyps. Gut 28:306–314.
11. Domizio P, Talbot IC, Spigelman AD, et al (1990) Upper gastrointestinal pathology in familial adenomatous polyposis: results from a prospective study of 102 patients. J Clin Pathol 43:738–743.
12. HofgartnerWT, Thorp M, Ramus MW, et al (1999) Gastric adenocarcinoma associated with fundic gland polyps in a patient with attenuated familial adenomatous polyposis. Am J Gastroenterol 94:2275–2281.
13. Jagelman DG, DeCosse JJ, Bussey HJ (1988) Upper gastrointestinal cancer in familial adenomatous polyposis. Lancet 1:1149–1151
14. Debinski HS, Spigelman AD, Hatfield A, et al (1995) Upper intestinal surveillance in familial adenomatous polyposis. Eur J Cancer 31A:1149–1153.
15. Bulow S, Alm T, Fausa O, et al (1995) Duodenal adenomatosis in familial adenomatous polyposis. DAF Project Group. Int J Colorectal Dis10:43–46.
16. Trimbath Jd, Giardiello FM (2002) Review article: genetic testing and counselling for hereditary colorectal cancer. Aliment Pharmacol Ther 16:1843–1857.
17. Bjork J, Akerbrant H, Iselius L, et al (2001) Periampullary adenomas and adenocarcinomas in familial adenomatous polyposis: cumulative risk and *APC* gene mutations. Gastroenterology 121:1127–1135.
18. Bulow S, Bjork J, Christensen IJ, et al (2004) Duodenal adenomatosis in familial adenomatous polyposis. Gut 53:381–386.
19. Groves CJ, Saunders BP, Spigelman AD, et al (2002) Duodenal cancer in patients with familial adenomatous polyposis (FAP): results of a 10 years prospective study. Gut 50:636–641.
20. Saurin JC, Gutknecht C, Napoleon B, et al (2004) Surveillance of duodenal adenomas in familial adenomatous polyposis reveals high cumulative risk of advanced disease. J Clin Oncol 22:493–498.
21. Vasen HFA, Moeslein G, Alonso A, et al (2008) Guidelines for the clinical management of familial adenomatous polyposis (FAP). Gut 57:704–713.
22. Waddel WR, Loughry RW (1983) Sulindac for polyposis of the colon. J Surg Oncol 24:83–87.
23. Cruz-Correa M, Hylind LM, Romans KE, et al (2002) Long term treatment with sulindac in familial adenomatous polyposis: a prospective cohort study. Gastroenterology 122:641–645.
24. Giardiello FM, Yang VW, Hylind LM, et al (2002) Primary chemoprevention of familial adenomatous polyposis with sulindac. N Engl J Med 346:1054–1059.
25. Phillips RK, Wallace MH, Lynch PM, et al (2002) A randomized double blind , placebo controlled study of celecoxib, a selective cyclooxygenase-2 inhibitor, on duodenal polyposis in familial adenomatous polyposis. Gut 50:857–860.
26. McGettigan P, Henri D (2006) Cardiovascular risk and inhibition of cyclooxygenase: a systematic review of the observational studies of selective and non selective inhibitors of cyclooxygenase-2. JAMA 296:1633–1644.
27. Brosens LA, Keller JJ, Offerhaus GJ, et al (2005) Prevention and management of duodenal polyps in familial adenomatous polyposis. Gut 54:1034–1043.
28. Ahmadi A, Draganov P (2008) Endoscopic mucosal resection in the upper gastrointestinal tract. World J Gastroenterol 14):1984–1989.
29. Bleau BL, Gostout CJ (1996) Endoscopic treatment of ampullary adenomas in familial adenomatous polyposis. J Clin Gastroenterol 22:237–241.
30. Norton ID, Gostout CJ, Baron TH, et al (2002) Safety and outcome of endoscopic snare excision of the major duodenal papilla. Gastrointest Endosc 56:239–243.
31. Harewod GC, Pochron NL, Gostout CJ (2005) Prospective randomized controlled trial of prophylactic pancreatic stent placement for endoscopic snare excision of the duodenal ampulla. Gastrointest Endosc 62:367–370.
32. Iaquinto G, Fornasarig M, Quaia M (2008) Capsule endoscopy is useful and safe for small-bowel surveillance in familial adenomatous polyposis. Gastrointest Endosc 67:61–67.
33. Moenkemueller K, Weigt J, Treiber G, et al (2006) Diagnostic and therapeutic impact of double-balloon enteroscopy. Endoscopy 38:67–72.
34. Offerhaus GJ, Giardiello FM, Krush AJ, et al (1992) The risk of upper gastrointestinal cancer in familial adenomatous polyposis. Gastroenterology 102:1980–1982.
35. Bulow S, Bjork J, Christensen IJ, et al (2004) Duodenal adenomatosis in familial adenomatous polyposis. Gut 53:381–386.
36. de Vos tot Nederveen Cappel WH, Järvinen HJ, Björk J, et al (2003) Worlwide survey among polyposis registries of surgical management of severe duodenal adenomatosis in familial adenomatous polyposis. Br J Surg 90:705–710.

Lower Gastrointestinal Endoscopy for Polyps and Polyposis

12

Guido Missale, Gianpaolo Cengia, Dario Moneghini, Luigi Minelli, Gian Paolo Lancini, Domenico Della Casa, Michele Ghedi and Renzo Cestari

Abstract Colonoscopy has become the leading method to explore the entire colon, and is currently considered the gold standard for colorectal cancer screening. Improvements in technology have provided specific diagnostic capability, and the treatment of dysplastic and neoplastic superficial lesions is now achievable in the majority of patients, by adopting sophisticated resection techniques. Endoscopic treatment of polyps must be performed in order to both minimize the risks of the procedure and optimize the completeness of the removal, thereby reducing recurrence; therefore operators must be skilled and continuously trained, in order to perform local treatment by either endoscopic mucosal resection (EMR) or endoscopic submucosal dissection (ESD). In this way, endoscopic resection can be considered a safe and effective alternative to surgery for the treatment of colorectal polyps.

Keywords Colonoscopy • Colonic and rectal polyps • Endoscopic mucosal resection • Endoscopic submucosal dissection • Polypectomy • Techniques

12.1 Colonoscopy

Endoscopy has become the ideal method for rectal, colonic, and distal ileum mucosal evaluation, and has replaced the double-contrast barium enema totally [1]; moreover, the introduction of videoendoscopes has led to colonoscopy being considered the "gold standard" in colorectal cancer screening, due to the high diagnostic yield combined with the range of therapeutic options available [2].

Diagnostic accuracy and safety of colonoscopy depends greatly on the following factors.

- The quality of the *bowel preparation*: iron oral intake should be preliminarily suspended (3–4 days), and a restricted dietary regimen (low-residue foods) is recommended for 3–5 days before colonoscopy [3]. The optimal laxative is not standardized yet [4], although polyethyleneglycol (PEG) solutions and sodium phosphate (SP) are preferred among colonoscopists. PEG is a non-absorbable solution and significant fluid or electrolyte shifts are therefore avoided; it is generally safer, although limitations are the high volume needed (4 liters) and the poor palatability (in 5–15% of subjects) [5]. SP is a low-volume hyperosmotic solution which draws plasma water into the bowel lumen to promote colonic cleansing [6,7]. Since significant fluid and electrolyte shifts can occur, SP must be diluted to prevent emesis, and significant oral fluid intake is required to prevent dehydration, especially in children, elderly patients, small intes-

R. Cestari (✉)
University of Brescia, Digestive Endoscopy Unit, A.O. Spedali Civili, Brescia, Italy

tine disorders, poor gut motility, renal or liver insufficiency, or congestive heart or liver failure [8–10].

- *Informed consent* is the precondition to colonoscopy, due to the high likelihood of operative procedures, even in asymptomatic subjects. Patients must be informed about all aspects of the examination, such as technical (scope progression, air distension, manual abdominal compression, expected duration time), pharmacological (need for conscious sedation and related driving hazards; co-medications), and clinical (indications, symptoms and disease correlation, comorbidities) factors. Information should also include the diagnostic capability of the technique; the use of specific diagnostics methods (biopsy, magnifying/chromoendoscopy), the potential for endoscopic treatment, the type of lesions detected, the risk for cancer progression, and the rationale of removal must be explained. The complication rate (bleeding, perforation) must be declared, even for diagnostic exploration, and the choices for management of complications should be explained (endoscopic, angiographic, or surgical option). The postoperative course, the risk of delayed complications and symptoms (rectal bleeding, fever, abdominal pain), and the likelihood of hospitalization must also be explained.
- *Co-medications*: use of antiplatelets drugs must be suspended prior to the procedure (4 days), as well as anticoagulants, because of the risks of bleeding [11].
- *Antibiotic prophylaxis* is usually not necessary, except in patients in high-risk categories [12].
- *Sedation* with a combination of meperidine and midazolam or propofol (anesthesiologist-assisted) improves colonoscopy outcomes (such as cecal intubation rate), and allows patient position changes, with an overall low incidence of complications [13]. All sedated patients should have continuous monitoring of blood pressure, pulse, and oxygen saturation [14].
- *Digital rectal examination* is mandatory to rule out anorectal diseases, and anal lubrication facilitates scope insertion.
- The patient lies in the left lateral position for the insertion and left colon exploration, whereas a supine position can be helpful for crossing the transverse colon; right lateral decubitus may facilitate passage through the hepatic flexure towards the cecum. Manual abdominal compression can prevent loop formation, improve scope advancement, and minimize pain.

The quality of colonoscopy is defined by the following parameters [15,16]:

- caecal intubation ≥90%, and in screening cases ≥95% with documentation in endoscopic reports (100%) and photography when available
- withdrawal time should average at least 6–10 min
- polyp detection rate (≥25% in males and ≥15% in females older than 50 years)
- documentation of the quality of bowel preparation (goal: 100%).

Concerning *complications*, quality-improvement targets are [15,16]:

- the percentage of cases with informed consent (goal: 100%)
- the incidence of minor sedation reactions, such as unplanned reversal of sedation (goal: ≤1 in 100)
- the incidence of more serious adverse reactions (mask ventilation or endotracheal intubation) (goal ≤1 in 300)
- the incidence of perforation by type (goal <1 per 1000; for screening <1 per 2000)
- the incidence of post-polypectomy bleeding (immediate and delayed) (goal <1 per 100).

12.2 Instruments and Diagnostic Techniques

Since the first introduction of fiberoptic models, technology advance has led to considerable improvements in the performance of endoscopic instruments. *Standard electronic scopes* with distal charge-coupled device (CCD) provided high-resolution images (up to 200,000 pixels) and better luminal visualization (130–140° vision angle) across angulations and within the folds; recently, *high-definition* (HD) *scopes* (> 850,000 pixels), using optical or electronic magnification up to 80 times, can visualize early adenomatous lesions, aberrant crypt foci (ACF), and dyscromic mucosal areas that are not otherwise visible [17].

Optical chromoendoscopy, using surface contrast dyes and vital colorant staining, combined with magnification, can now be used to improve recognition of early lesions, to target biopsies, to visualize glandular opening (pit pattern), and for histological definition of parietal invasion (Table 12.1) [18,19]. Additional tech-

Table 12.1 The pit pattern classification system

System type	Pattern
I	Normal
II	Non neoplastic • Star-like • Onion-like
III	Neoplastic (tubular) • L (large) • S (small)
IV	Neoplastic (adenoma)
V	Neoplastic (carcinoma) • I (irregular) • N (non-structured)

niques available for *electronic chromoendoscopy* are *narrow-band imaging* (NBI) which provides, by optical RGB (red, green, blue) filters, visualization of pit patterns and the superficial vascularization [20], and *Fuji intelligent chromoendoscopy* (FICE), which can be applied to acquired images [21]. Moreover, NBI can differentiate the vascular pattern (according to the intensity of vessel network clarity around the pits) into a three-point scale, as follows: (1) weaker (paler) than surrounding mucosa (non-neoplastic); (2) similar to surrounding mucosa (non neoplastic); (3) stronger (darker) (neoplastic). Predictive values of vascularity, obtained both by NBI and optical chromoendoscopy are comparable for differentiating between dysplastic and neoplastic changes (sensitivity 96.8%, specificity 81.1%) [22,23]. Furthermore, NBI does not require either mucosal cleansing or staining solutions, and is user-friendly as there is a manual switch on the handle of the scope. Current limitations concern the inter-observer variability in imaging interpretation, the need for new processors, and a detailed cost analysis.

12.3 Polyps

Colorectal polyps are very different in size, superficial extension, and aspect; moreover, these features may be not related to the histological type of the lesion.

According to the *Paris classification of superficial gastrointestinal neoplasia* (Table 12.2), polyps can be morphologically differentiated into *polypoid* (protruding type) and *non-polypoid* (non-protruding type) [24]. Laterally spreading tumors (LSTs) are polyps greater than 10 mm in diameter with a low vertical axis, that extend laterally along the lumen and can be classified into G-type (granular, uniform or mixed) and F-type (flat surface non-granular) according to the predominant aspect [25]. The likelihood of malignancy and submucosal invasion can be estimated for polypoid lesions, according to the size of the head and stalk, as well as surface depressed areas [26]. Moreover, correlation between pits and histological analysis is well demonstrated with overall diagnostic accuracy of 75% for non-neoplastic (types I and II), 94% for adenomatous (types III and IV), and 85% in polyps suspected for invasive cancer (type V); sensitivity ranges from 42% to 98% to 82% respectively, whereas specificity is 99%, 52%, and 99% respectively [27]. Furthermore, pit patterns associate to submucosal invasion (0% in types II–IIIL, 3.9% in IIIS, 3.8% in IV, and 21.1% in VI, up to 65.6% in VN group – 100% for depressed, type IIc lesions) [18]. According to morphology (flat, elevated, depressed), lymph node involvement has proven to be related to submucosal invasion (ranging from 0.76%, 2%, to 29.3% respectively) and to pit pattern (from 0% to 4% for IIIS–IIIL and IV, up to 41% for type V) [27].

12.4 Polypectomy

12.4.1 Instruments – Accessories

Electrosurgical generators (ESGs) allow polypectomy by energy delivery for endoscopic accessories through monopolar (snares, hot-biopsy forceps, argon beam coagulation) or bipolar circuit (hemostatic proce-

Table 12.2 The Paris classification of superficial gastrointestinal neoplasia

Type	Morphology	Classification group
0-I	Pedunculated	0–Ip
	Sessile	0–Is
0-II	Slightly elevated	0-IIa
	Completely flat	0-IIb
	Depressed without ulcer	0-IIc
0-III	Excavated or ulcerated	0-III
Mixed	0-IIa + 0-IIc	0-IIc + 0-IIa

dures). ESGs provide two principal therapeutic effects: cutting (by >200 V continuous current), and coagulation (by interrupted or <200 V continuous current); the type and intensity of the current has been proven to influence the risk and timing of bleeding after polypectomy, since cutting or blended current expose the tissue to the danger of immediate bleeding (elevated tissue necrotic effect), whereas the use of pure low-voltage coagulation current is more likely to result in delayed hemorrhage [28].

Forceps (single-use or reusable) are very different in cup shape and size (standard, jumbo) and include standard "cold" and "hot" devices, which may join both tissue sampling and electrocautery; hot forceps are generally considered the easiest cautery instrument to resect small polyps and the hardest for controlling thermal injury [29].

Snares usually consist of a monopolar wire loop to surround the target tissue, which is then transected by both mechanical and electrosurgical cutting as the loop is withdrawn into the sheath. All the snares are designed for electrosurgical use, but either hot or cold techniques (small or mini snares) can be used with any device. Both single-use and reusable snares are available. Wires can be classified as monofilament (stiffer) or braided [30]. Differences in snare size (from 1 to 6 cm in diameter) and shape (oval, hexagonal or crescent) facilitate the capture of the lesion; moreover, rotator wires allow orientation of the loop; recently, needle-tip snares (small caliber) have been introduced to ameliorate positioning and grasping the base of polyps [31].

Polyp retrieval devices: polyp and fragments retrieval is essential to achieve histological analysis, and several endoscopic devices may be useful for this purpose, according to size, consistence, site, and number of polypectomies. Currently, snares, tricuspidal forceps, or Dormia baskets may be useful to recover small, pedunculated polyps; otherwise, after piecemeal resection, fragments can be removed by filtered suction traps directly through the operative channel. The most powerful tools for polyp recovery seem to be retrieval nets (assembled over a metallic wire) that are different in size and length, which provide safe capture and optimal preservation of the specimen for histology [32,33].

Injection needles consist of an outer sheath (plastic, Teflon, or stainless steel) and an inner core needle (23–25G), different in length (from 200 to 240 cm) and in needle exposure (prefittable by screw set on the handle), to avoid extracolonic injection and peritoneal contamination. The caliber of the needle has a role in submucosal injection of hypertonic, viscous solutions (such as sodium hyaluronate, glycerol, or fibrin glue) [34].

Plastic transparent caps (similar to endoscopic ligation devices) allow tissue to be suctioned into the chamber and excised with a snare to facilitate resection of sessile polyps. For endoscopic mucosal resection (EMR) and endoscopic submucosal dissection (ESD) techniques, caps are available in a variety of sizes and shapes (symmetric or oblique), and addition of a distal circular rim at the distal cap's tip permits housing of a prefitted snare [35].

Tattooing: when colonic resection is required (lesion flat or small), or during surveillance, localization can be optimized by submucosal four-quadrant tattooing with sterile India ink, although x-ray detection by endoscopically placed clips and intra-operative colonoscopy have been adopted. Ink tattooing has been shown to be the ideal marker, satisfying all requirements (long duration, serosal visibility, neither systemic nor visceral injuries); no long-term adverse effects from this technique are reported (more than 1000 procedures performed). The absence of bleb is most commonly due to ink leakage along the needle tract and spilling into the lumen, or deep penetration in the wall, which can result in intraperitoneal spraying [36].

Hemostatic devices: post-polypectomy hemorrhage occurs in 0.3–2.25% of patients, both immediately after or up to three weeks later. Bleeding during EMR or ESD cannot be considered as a complication, but as an aspect of the technique, and is usually managed at the same time [37]. Several devices are currently available to stop bleeding, and the endoscopist's skill and confidence are essential for the choice of the optimal tool.

For mild bleeding from small polyps, *"hot" biopsy forceps* can be adopted. *Metallic clips* provide hemostasis through vessel compression around the polypoid stalk or base. Orientation can be achieved by rotational mechanism, and positioning can be modified by re-opening the catheter device. Clips are multi-angle shaped, made from stainless steel ribbons of different lengths (short/standard/long), size, number (2 or 3), and angulations (90°/130°); both mono- or multiple-clip deployment systems are available [38,39]. *Detachable nylon loops* can be placed over pedunculated polyps to prevent or to stop bleeding after stalk resection. After exit out of the Teflon sheath (1950 mm long and 2.5 mm in diameter) and placement, the loop is then tightened with advancement of a silicon-rubber stopper; once the desired closure extent is obtained, it should then be released; application can be difficult for large polyps and because of wire floppiness, leading to difficulties in polypectomy [40].

Argon plasma coagulation (APC) applies high-frequency monopolar current to target tissues through ionized argon gas (plasma), which is an excellent electrical conductor, and an electrical spark at the tip of the probe leads to current flow through the gas to the nearest available tissue. Several flexible probes are available, different in length, diameter, and direction of the flushing gas (forward or tangential). The main advantage of this coagulation method is the non-contact mode, which results in lower wall penetration (compared to standard laser), providing minimized thermal injuries. Factors affecting the depth of coagulation effect are the duration of the energy burst and the power setting; in the right colon, power settings of 40–45 W are appropriate, whereas from the transverse colon to the rectum, higher power settings can be applied (60 W maximum). Current indications for APC are ablation of residual sessile polyps after piecemeal polypectomy and hemostasis for minimal spurting during resection [41].

Endoscopic ultrasonography: high-frequency ultra sonography (HFUS) with mini-probes ranging from 12 to 30 MHz, represents an additional method to improve the local staging, through assessment of wall infiltration and lymph node status (overall diagnostic accuracy 70%) [42]; furthermore, HFUS can be adopted to choose the proper resection strategy (endoscopic versus surgical).

12.4.2 Technique

The initial requirement for performing polypectomy safely is a good field of view, maintaining the polyp in a standard-fashion position (between 5 and 11 o'clock) close to the operative channel of the scope, and using scope rotation.

Factors that primarily affect the choice for the proper technique (and accessories) are the *size* and the *aspect* of polyps.

- *Small polyps* (1–10 mm) account for about 90% of all the lesions and the occurrence of cancer is lower than 1%. Polyps <5 mm (diminutive polyps) can be safety removed by forceps in the "cold" or "hot" fashion; for larger lesions, mini-snare cautery should be used to provide complete ablation. Although "hot biopsy" provides more-radical ablation compared to cold mode [43], the risk of colonic wall burning is higher, especially in the cecum and right colon [44,45]; both traction of the polyp during burning, and submucosal injection can make the procedure safer; otherwise, the "cold" technique provides a minimal bleeding risk but recurrence is observed [46].
- *Polyps >10 mm*: *pedunculated* polyps can usually be removed in en-bloc fashion, with single snare resection or by preliminary clip positioning around the stalk, and needle knife incision as well as ESD technique [47]. *Sessile* polyps can be safely removed by assisted submucosal injection, with both piecemeal and en-bloc technique. The proper snare position can be assured by positioning the distal end of the sheath in the middle of the stalk, or at the borders for sessile polyps, and the tip of the snare must be carefully controlled during closure, to avoid entrapment of unaffected tissue.
- *ESG power settings* should be fitted according to the thickness of the tissue, regardless of the aspect of the polyp and can be therefore summarized as follows: for pedunculated polyps >3 mm, and sessile polyps >8 mm, blended current (t_{on} 50 ms, t_{off} 300–500 ms) can be safely applied; in all the other cases the cut (60–200 W) or coagulation current (60–120 W) is preferable [48].

12.4.3 Safe Polypectomy

Outcomes of polypectomy depend upon the risk of complications and the completeness of the resection (radicality), regardless of the morphology, the site, and the size of the lesion, and are strongly influenced by the following factors.

- *The choice of the proper accessories* (hot or cold biopsy forceps; small or large, flexible or stiff snares) according to the different resection technique (hot versus cold method, en-bloc versus piecemeal) influences complete removal. Treatment of the base and scar can be additionally performed whenever there is an increased likelihood for residue, due to piecemeal resection (sessile, large base, or flat lesions); thermal ablative methods such as bipolar or hot forceps, BICAP probe, or APC can substantially reduce the risk for recurrence by up to 50% [49], although intensive follow-up is recommended (4–12 week intervals).

- In order to minimize adverse effects of polypectomy, *trained and skilled assistants* should support any operative step and assess the functioning and the proper use of the equipment. The ensnaring technique, according to the aspects of the polyps (degree of the snare resistance, handle snare marking at the point of closure, evaluation of the volume of the lesion), is the factor that mainly influences the process, to avoid bleeding risks after cheese-wiring. Predisposing factors are considered the snare stiffness (hexagonal, monofilament, reinforced braided), the small size of the lesion, and the modality of snare closing (2:1 ratio of snare/handle movement). During the ensnaring time, normal adjacent mucosa or submucosa can be entrapped into the wire, exposing tissue to the risk of deep thermal injury: it is therefore recommended to check the polyp position in the snare, with eventual reopening and repositioning of the device.
- *Techniques for preventing complications* are essential for polyps greater than 10 mm. For large, floppy-stalked polyps, both detachable nylon loops (Fig. 12.1a) and clips (prior to resection) can be positioned [50]; however, clips can also be attached subsequently to avoid thermal injuries due to contact with the snare [40]. Endoloop usually persist for 4–7 days, whereas clips move away in a different time, ranging from 1–2 (Triclip® by Cook and HX-5L® by Olympus) to 4–5 weeks (Resolution Clip® by Boston Scientific).
- Submucosal injection (SI) is a well-standardized technique for insulating the muscular plane from the layers above, leading to reduction of the risk for thermal injuries [51,52], and ameliorating polyp capture. Several solutions can be applied, such 50% dextrose, glycerol, and 0.5% sodium hyaluronate, providing significantly longer-lasting effect (from 5 to 23 min) compared to normal saline solution with or without epinephrine (2–3 min) [53].
- The lifting effect can be maintained by subsequent injections or, recently, by a new pressure-controlled injective system without needle (Hydrojet®-Erbe), able to dissect the different tissues types through precisely adjusted water pressure (right or left colon, rectum) [54]. Moreover, the overall volume of the solution injected can differ greatly, according to the size and the polyp extension, from 3–4 mL up to 50 mL [55]. SI is furthermore proposed when base coagulation after polypectomy is required, or when residue is found at follow-up. Injection can be unsuccessful (no lifting sign) due to deep neoplastic infiltration, inaccurate needle placement through the wall, or fibrotic scar after previous piecemeal resection.

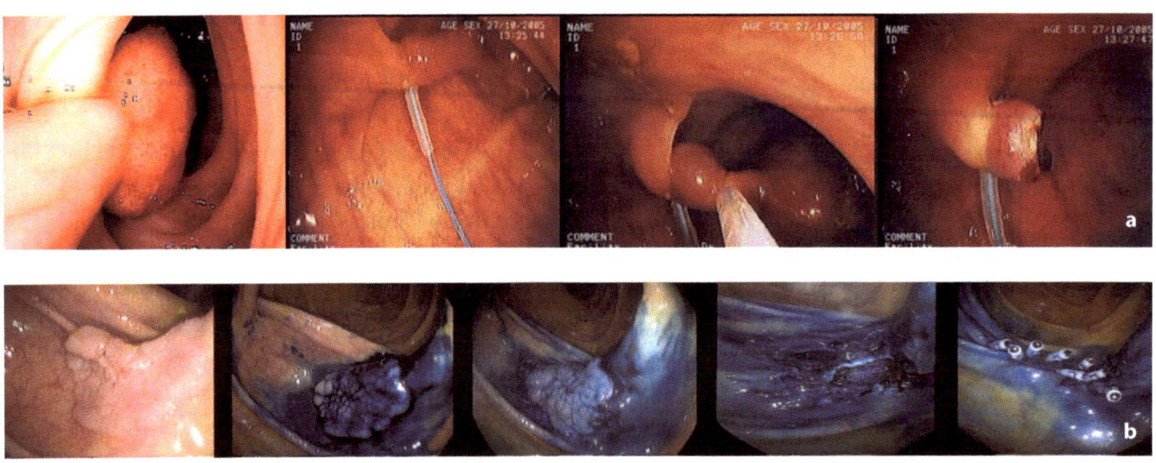

Fig. 12.1a,b Large sigmoid-stalked polyp: detachable nylon loop prior to snare resection (**a**). Flat right colon polyp: indigo carmine submucosal injection, circular mucosal incision by snare tip and resection (**b**)

12.4.4 Difficult Polypectomy

Factors concerning a "difficult" polypectomy are related to the endoscopist's skill, resection technique, polypoid elements (position, localization, aspect, and size), and patient characteristics (comorbidities, use of antiplatelets or anticoaugulants).

- Sessile polyps, larger than 20 mm, extended for more than one-third of the circumference, over two consecutive folds, "clamshell"-like, or in a flexure site (overall prevalence from 0.8% to 5.2%), can be challenging even for skilled endoscopists [56], and a "second look" procedure (with prior endoscopic marking) should be attempted by experts to complete resection; this reduces the need for surgery in up to 50% [57–59].
- Polyps >10 mm should be removed after cecal evaluation; any additional cancer or large polyp found might change the strategy [60].
- For hidden polyps (between folds or diverticula), or when fluids are abundant, the ability to change patient position can lead to better detection [61]; similarly, manual abdominal compression, SI in the proximal part of the lesion, and scope retroflexion (easy in the rectum and right colon, difficult in the left colon) may improve identification [62]. Additional methods to increase the polyp detection rate are plastic cap attachment [63], use of different stiffness [64] or double-channel instruments [65], or the "dual-endoscope technique" [66].
- Sessile polyps larger than 20 mm located between two folds can be resected in one session by the use of stiffer snares (asymmetric or rotating mini) [67], which can provide resection of the polyp proximally; moreover, circular mucosal incision by the tip of the snare can make the ensnaring safer [68] (Fig. 12.1b).
- Polyps larger than 35 mm, LSTs, or suspected for malignancy should be resected with EMR and ESD techniques.
- Surgery should be considered when the polyp is too large in size and extension (>30% in the colon, >60% in the rectum), intra-appendicular, or inside the ileocecal valve; when perforation occurs after polypectomy; when endoscopic removal is incomplete and the follow-up will not be adopted [69].

12.4.5 Radical Polypectomy

The completeness of polypectomy influences therapeutic efficacy directly, as more than 25% of cancers detected after a negative colonoscopy are performed with conventional techniques considered to be a result of incomplete removal [70,71].

Factors involved in the radicality can be distinguished into histology non-dependent factors and those directly related to the nature of the lesion.

12.4.5.1 Histology Non-Dependent Factors

The colonoscopist's skill is the main factor influencing the outcomes of polypectomy [16], with significant reduction in morbidity and increase in the rate of complete resection, particularly for large sessile polyps [72,73].

The polypectomy technique ("piecemeal" or "en-bloc") influences the success rate significantly, as confirmed by the overall failure rate after piecemeal resections ranging from 14% to 55% [57,74,75], with residue and recurrence of large sessile polyps in 50% of cases [76].

12.4.5.2 Histology-Dependent Factors

The main controversy about efficacy of polypectomy concerns the histological definition of malignant polyps and the differentiation in low- and high-risk lesions. Malignant polyps can be defined when cancer cells penetrate through the muscularis mucosae, and negative predictive factors are involvement of resection margins (<2 mm), detection of poorly differentiated cells, and vascular and lymphatic invasion [77,78]. At least one of these elements provides the high-risk definition, and indicates surgery regardless [79], in order to treat any neoplastic residues and to assess lymph node status. Moreover, since more accurate histological specimen evaluation is achievable by endoscopic resection techniques, such as EMR and ESD, predictive factors for lymph node metastases are considered to be the depth and width of submucosal invasion and the presence of "tumor budding" [80].

12.5 Colonic Endoscopic Mucosal Resection and Endoscopic Submucosal Dissection

For large sessile (>2 cm), flat, LSTs, or polyps suspected of malignancy, EMR and ESD techniques have been adopted, since they have the ability to achieve "en-bloc", margin-free resection [81] in most cases, with a significant improvement in the extent of removal and the accuracy of histological analysis (90%) [82–84]; EMR and ESD completely eliminate the affected mucosa by the resection of superficial layers (with "organ-sparing" modality) through the middle or deeper part of the submucosa. In order to perform EMR and ESD, endoscopists should be confident with the "lifting sign" after SI, which is predictive of the absence of muscularis propria infiltration (positive predictive value for invasive cancer 83%), suggesting a lesion confined to the superficial layers (sensitivity 100% and specificity 99%) (Fig 12.2a) [53,85].

12.5.1 EMR Technique

According to the modality, EMR techniques can be classified into "suction" (cap-assisted endoscopic mucosal resection (EMR-C), and endoscopic mucosal resection with ligation (EMR-L)) and "non-suction" (inject and cut, inject-lift and cut). In EMR-C the lesion is preliminarily suctioned through a plastic cap distally attached to the scope, and then resected with asymmetric cautery snare [86]; despite the wide application in the upper gastrointestinal tract, the adoption of this technique for colonic lesions is still limited due to the high risk of perforation compared with "non suction" tecniques (Fig 12.2a); in any case, the modified Soehendra technique, with monofilament stiff snare can be adopted [87].

Non-suction resection can be performed in several ways, with both standard or dual-channel scope, with snares (monofilament or stiffers are preferable) and forceps, with pre-cut technique or overtube attachment; the lesion (smaller than 20 mm), preliminarily lifted, can be resected in "en-bloc" fashion with cautery snare following some *technical tips*:

- avoid marginal SI
- preliminary incision of the non-affected mucosa
- colonic deflation while ensnaring.

Although EMR can be compared to standard polypectomy, experience, skill, and prudence must be reserved especially for sessile lesions in the right colon, due to the thinner wall; elevated injection volume and complete lifting are therefore needed; conversely, hypertonic solutions might lead to parietal injuries and prolong the healing process after EMR [88]. For very large lesions, preliminary chromoendoscopy to define the margins is indicated, as well as a circumferential incision with a needle-knife prior to submucosal injection.

The main limitations of EMR for colorectal polyps are the following:

- polyps greater than 2 cm (oncological resection

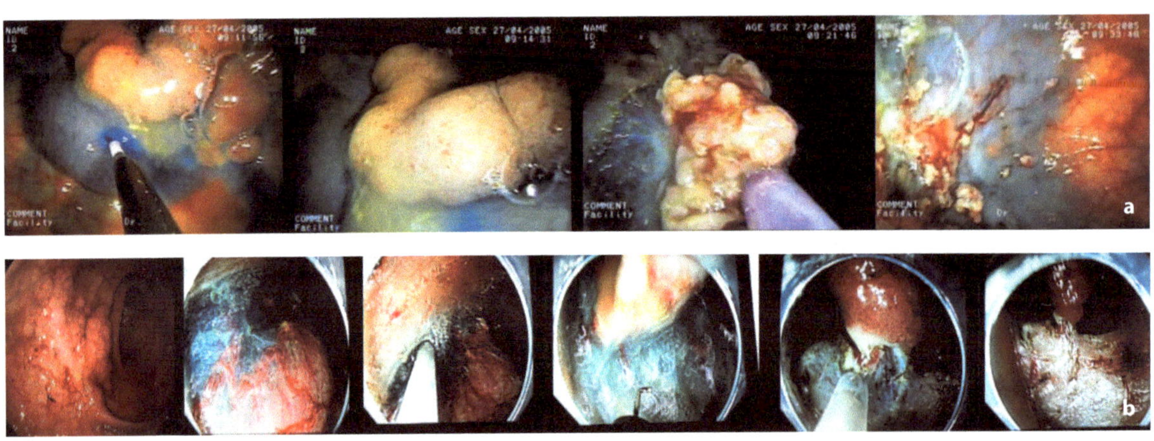

Fig. 12.2a,b Lift-and-cut EMR for large rectal sessile polyp **a**. Standardized ESD technique for rectal LST **b**

cannot always be warranted by piecemeal resection)
- features suspected for malignancy: type V pit pattern, ulcerated and irregular depressed surface
- Sm2–3 invasion detected by HFUS
- fold convergence and incomplete lifting; multiple folds involvement (>2); extension >1/3 of the luminal circumference.

12.5.2 ESD Technique

Conventional polypectomy techniques, as well as EMR, expose patients to the risk of residue, especially for LSTs, large sessile, elevated type, and superficial-type neoplasia; moreover, the maximum size of the lesion for "en-bloc" resection is about 2 cm (due to snare size).

ESD represents a promising, new technique for the "en-bloc" removal of large and extended lesions that are not suitable for EMR, through the recognition of the dissection plane between submucosa and muscularis propria, with reduction of the risk of parietal burning and perforation [89,90]. ESD can be conducted in surgical-like fashion, but only by using specifically designed instruments, including knives and supporting devices (injection needles, plastic caps, metallic clips, and coagulation grasps).

12.5.2.1 Knives

- The IT Knife® is the first knife introduced and the most widely adopted, consisting of a ceramic insulated spherical tip at the distal end of a needle knife.
- The Hook Knife® with a distal L-shape hooks the mucosa and submucosa prior to cutting, achieving control of the depth of resection; the rotary function makes it possible to alter the resection plane vertically or horizontally, and the back of the distal hook can be applied for marking the tissue, reducing the risk of perforation.
- The Flex Knife® is a covered-sheath device, with a distal looped wire end, and is flexible, which provides different cutting planes (vertical, horizontal, or oblique).
- The Triangle Tip Knife® has a distal triangular non-insulated tip, whereby no cutting rotation is required. It can be used from marking and pre-cutting to incision and dissection, and to stop mild bleeding as well.
- The IT Knife 2® combines an insulated triangular and distal ceramic tip.

12.5.2.2 Hemostatic Forceps

Several models, of different shape and cup sizes are currently available:
- Coagrasper®: rotating with anti-slip mechanism applied on the cups which allows grasping of the bleeding points securely for faster and more reliable hemostasis.
- HotClaw® is a rotating device, with claws at the end of the grasp to anchor the tissue more strongly, resulting in minimum submucosal injury.
- HotBite® is useful for marking and incision prior to IT Knife dissection; it is suitable as a needle knife even on flat mucosa.

12.5.2.3 Procedure

The procedure (coupled with ERBE® ICC 200/VIO ESG system) is performed in a well-standardized fashion (Fig. 12.2b):
- a plastic transparent cap is attached to the distal end of a therapeutic gastroscope (3.8 mm channel)
- preliminary cautery marking (forced coagulation 40 W/swift coagulation 60 W) on the non-affected mucosa (2 mm away from the lesion)
- multiple liftings and injections (indigo carmine or methylene blue) to dissociate the muscularis propria (white appearance) from the submucosal layer (blue appearance)
- diathermic circular incision around the lesion is created (endo-cut effect 3, 120 W/swift coagulation effect 4, 40 W)
- specific devices must be used
- neoplastic mucosal dissection from the surrounding colonic tissue is made along the interspaces between the submucosa and muscularis propria (forced coagulation 60 W/swift coagulation 40 W)
- step-by-step and submucosal vessel coagulation with hemostatic forceps (soft coagulation 80 W).

Current limitations of colorectal ESD are:
- difficulties in maintaining the proper scope position
- slimness of the colorectal wall and folds compared to the upper gastrointestinal tract
- luminal angulations and peristaltic colonic movements
- high morbidity rate when perforation occurs.

12.5.2.4 Results

Bleeding is the most frequent complication of ESD, ranging from 1% up to 45%, generally occurring immediately or within 24 hours, while in 13.9% of patients it can be delayed [85,91]; intra-operative hemorrhage cannot be considered as a complication, but rather as a procedural event, usually managed consensually. Perforation rates have shown to be higher (4–10%), compared with the EMR technique (0.3–0.5%), although small defects can be successfully closed by clip positioning.

Local recurrence (both marginal or basal) after ESD reported in the literature ranges from 0 to 46%, due to the different skill of endoscopists, heterogeneity of the lesions (in size and histological type), and the different resection techniques adopted [91–94].

An intensive surveillance protocol must be scheduled after ESD (within 6–12 months), in order to detect any recurrence early, and to re-treat residues with additional ablative methods if necessary; however, the success rate of this strategy is still debatable, with eradication rates ranging from 40% to 50% (49); Conio et al have shown no significant differences in recurrence rate by the addition of APC [82].

Controversies about EMR and ESD for both polyps and early neoplastic lesions (T1m-N0) concern the difficulty and the time-consuming nature of the procedures (25.8 ± 25.9 min and 84.0 ± 54.6 min respectively), the complexity of the resection technique, the need for skilled and trained operators, and the accurate selection of patients and lesions.

EMR and ESD can therefore be considered as innovative techniques, complementary to standard polypectomy, and able to increase the indications for therapeutic endoscopic options.

12.6 Check List

12.6.1 Pre-Polypectomy

A skilled and trained team (endoscopist-assistant-pathologist) is fundamental for achieving the best results. The following are also important:

- optimal colonic cleansing
- bleeding risk should be assessed before the procedure through family history and co-medications
- snare handle marking prior to resection
- polyp position maintained in safe mode and with complete visualization
- for polyps >1 cm, protective methods should be adopted to avoid complications
- immediate bleeding should be managed by available techniques (thermal, injection, or mechanical) at the same time
- permanent marking should be adopted for polyps that are suspected of malignancy, widely extended, or difficult to detect.

12.6.2 Post-Polypectomy

- The examination report should contain details of the technique applied and the equipment used for the resection, the extent of colonic visualization (cecum, distal ileum), and the characteristics of the lesion removed.
- Patients must be informed about probable immediate symptoms, such as pain (common but transient after air discharge); furthermore, complication complaints must be illustrated.
- After operative procedures, patients who are not looked after by family members, or who live a long distance from the hospital must be observed for 24–48 h.
- Therapeutic regimens should be re-assumed by agreement with the patients.
- High-risk patients should refer to the endoscopic unit immediately in case of bleeding, in order to be managed and reduce the need for blood transfusion and surgery.
- If pain persists, after perforation has been excluded (by x-ray or CT scan), patients should be observed for 24–48 h.

References

1. Winawer SJ, Stewart ET, Zauber AG, et al (2000) A comparison of colonoscopy and double-contrast barium enema for surveillance after polypectomy. N Engl J Med 342:1766–1772.
2. Lambert R (1999) European Panel on the Appropriateness of Gastrointestinal Endoscopy: colonoscopy (special section). Endoscopy 31:627–683.
3. A consensus document on preparation before colonoscopy: Prepared by a Task Force From The American Society of Colon and Rectal Surgeons (ASCRS), the American Society for Gastrointestinal Endoscopy (ASGE), and the Society of

American Gastrointestinal and Endoscopic Surgeons (SAGES). (2006) Gastrointest Endosc 63:896–909.
4. Cohen SM, Wexner SD, Binderow SR, et al (1994) Prospective, randomized, endoscopic-blinded trial comparing precolo-noscopy cleansing methods. Dis Colon Rectum 37:689–696.
5. Froenlich F, Fried M, Schnegg JF, et al (1991) Palatability of a new solution compared with standard polyethylene glycol solution for gastrointestinal lavage. Gastrointest Endosc 37:325–328.
6. Rosch T, Classen M (1987) Fractional cleansing of the large bowel with GoLytely for colonoscopic preparation: a controlled trial. Endoscopy 19:198–200.
7. Henderson JM, Barnett JL, Turgeon DK, et al (1995) Single-day, divided dose oral sodium phosphate laxative versus intestinal la-vage as preparation for colonoscopy: efficacy and patient tolerance. Gastrointest Endosc 42:238–243.
8. Kolts BE, Lyles WE, Achem SR, et al (1993) A comparison of the effectiveness and patient tolerance of oral sodium phosphate, castor oil, and standard electrolyte lavage for colonoscopy or sigmoidoscopy preparation. Am J Gastroenterol 88:1218–1223.
9. Vanner SJ, MacDonald PH, Paterson WG, et al (1990) A randomized prospective trial comparing oral sodium phosphate with standard polyethylene glycol-based lavage solution (GoLytely) in the preparation of patients for colonoscopy. Am J Gastroenterol 85:422–427.
10. Huynh T, Vanner S, Paterson W (1995) Safety profile of 5-h oral sodium phosphate regimen for colonoscopy cleansing: lack of clinically significant hypocalce-mia or hypovolemia. Am J Gastroenterol 90:104–107.
11. ASGE guideline on the management of anticoagulation and antiplatelet therapy for endoscopic procedures (2002) Gastrointest Endosc 55:775–779.
12. ASGE guidelines for antibiotic prophylaxis for GI endoscopy (2003) Gastrointest Endosc:58:475–482.
13. Froenlich F, Schwizer W, Thorens J, et al (1995) Conscious sedation for gastroscopy: patient tolerance and cardiorespiratory parameters. Gastroenterology 108:697–704.
14. Waring JP, Baron TH, Hirota WK, et al (2003) Guidelines for conscious sedation and monitoring during gastrointestinal endoscopy. Gastrointest Endosc 58:317–322.
15. Rex DK, Bond JH, Winawer S, et al (2002) Quality in the technical performance of colonoscopy: recommendations in the U.S. multisocietary task force on colorectal cancer. Am J Gastroenterol 97:1296–1308.
16. ASGE. Quality and outcomes assessment in gastrointestinal endoscopy (2000) Gastrointest Endosc 52:827–830.
17. Hurlstone DP, Fujii T, Lobo AJ (2002) Early detection of colorectal cancer using high magnification chromoendoscopy. Br J Surg 89:272–282.
18. Kudo S, Rubio CA, Teixeria CR, et al (2001) Pit pattern in colorectal neoplasia: endoscopic magnifying view. Endoscopy 33:367–373.
19. Hurlstone DP (2002) High resolution magnification chromoendoscopy:common problems encoured in "pit pattern" interpretation and correct classification of flat colorectal lesions. Am J Gastroenterol 97:1069–1070.
20. Gono K, Obi T, Yamaguchi M, et al (2004) Appearance of enhanced tissue features in narrow band endoscopic imaging. J Biomed Opt 9:568–577.
21. Pohl J, Nguyen-Tat M, Petch O, et al (2008) Computed virtual chromoendoscopy for classification of small colorectal lesions: a prospective comparative study. Am J Gastroenterol 103:562–569.
22. Hirata M, Oka S, Kaneko I, et al (2007) Magnifying endoscopy with narrow band imaging for diagnosis of colorectal cancer. Gastrointest Endosc 65:988–995.
23. Tiscendorf WJ, Wasmuth HE, Koch A, et al (2007) Value of magnifying chromoendoscopy and narrow band imaging (NBI) in classifying colorectal polyps: a prospective controlled study. Endoscopy 39:1031–1114.
24. The Paris endoscopic classification of superficial neoplastic lesions. (2003) Gastrointest Endosc 58(suppl):S3–S27.
25. Okamoto T, Tanaka S, Haruma K, et al (1996) Clinicopathological evaluation on colorectal laterally spreading tumor (LST) (in Japanese with English abstract). Nippon Geka Gakkai Zasshi 93:83–89.
26. Kudo S, Kashida H, Nakajima T, et al (1997) Endoscopic diagnosis and treatment of early colorectal cancer. World J Surg 21:694–701.
27. Kato S, Fuji T, Koba I, et al (2001) Assessment of colorectal lesions using magnifying colonoscopy and mucosal dye spraying: can significant lesions be distinguished? Endoscopy 33:306–310.
28. ASGE (2003) Electrosurgical generators Gastroint Endosc 58:656–660.
29. Faigel D, Eisen G, Baron T, et al (2003) Tissue sampling and analysis. Gastrointest Endosc 57:811–816.
30. Tucker RD, Platz CE, Sievert CE, et al (1990) In vivo evaluation of monopolar versus bipolar electrosurgical polypectomy snares. Am J Gastroenterol 85:1386–1390.
31. Yang R, Mabansag R, Laine L (2003) Rotatable polypectomy snares: a randomized, prospective comparison with standard snares. Gastrointest Endosc 57:T1480.
32. Miller K, Waye JD (2001) Polyp retrieval after colonoscopic polypectomy: use of the Roth retrieval net. Gastrointest Endosc 54:505–507.
33. Nelson DB, Bosco JJ, Curtis WD, et al (1999) Endoscopic retrieval devices. Gastrointest Endosc 50:932–934.
34. Nelson DB, Bosco JJ, Curtis WD, et al (1999) ASGE technology status report: injection needles. Gastrointest Endosc 50:928–931.
35. Nelson DB, Block D, Bosco JJ, et al (2000) ASGE technology status evaluation report: endoscopic mucosal resection. Gastrointest Endosc 52:860–863.
36. McArthur CS, Roayaie S, Waye JD (1999) Safety of preoperation endoscopic tattoo with india ink for identification of colonic lesions Surg Endosc 13:397–400.
37. ASGE (2001) Endoscopic hemostatic devices Gastrointest Endosc 54:833–840.
38. Binmoeller KF, Thonke F, Soehendra N (1993) Endoscopic hemoclip treatment for gastrointestinal bleeding. Endoscopy 25:167–170.
39. Chung IK, Ham JS, Kim HS (1999) Comparison of the hemostatic efficacy of the endoscopic hemoclip method with hypertonic saline-epinephrine injection and a combination of the two for the management of bleeding peptic ulcers. Gastrointest Endosc 491:13–18.
40. Matsushita M, Hajiro K, Takawuwa H, et al (1998) Ineffective use of a detachable snare for colonoscopic polypectomy of large polyps. Gastrointest Endosc 47:496–499.
41. Watson JP, Bennett MK, Griffin SM, et al (2000) The tissue

effect of argon plasma coagulation on esophageal and gastric mucosa. Gastrointest Endosc 52:342–345.
42. Larghi A, Lightdale CJ, Memeo L, et al (2005) EUS followed by EMR for staging of high-grade dysplasia and early cancer in Barrett's esophagus. Gastrointest Endosc 62:16–23.
43. Ellis K, Schiel M, Marquis S, et al (1997) Efficacy of hot biopsy forceps, cold or cautery micro-snare techniques in the removal of diminutive colonic polyps. Gastrointest Endosc 45:AB107.
44. Peluso F, Coldrier F (1991) Follow-up of hot biopsy forceps treatment of diminutive colonic polyps. Gastrointest Endosc 37:604–606.
45. Wadas DD, Sanowski RA (1988) Complications of the hot biopsy forceps technique. Gastrointest Endosc 34:32–37.
46. Woods A, Sanowski RA, Wadas DD (1989) Eradication of diminutive polyps: a prospective evaluation of bipolar coagulation versus conventional biopsy removal. Gastrointest Endosc 35:536–540.
47. Cipolletta L, Bianco MA, Rotondano G, et al (1999) Endoclip-assisted resection of large pedunculated colon polyps. Gastrointest Endosc 50:405–406.
48. Van Gossum A, Cozzoli A, Adler M, et al (1992) Colonoscopic snare polypectomy: analysis of 1485 resections comparing two types of current. Gastrointest Endosc 38:472–475.
49. Zlatanic J, Waye JD, Kim PS, et al (1999) Large sessile colonic adenomas: use of argon plasma coagulator to supplement piecemeal snare polypectomy. Gastrointest Endosc 49:731–735.
50. Uno Y, Satoh K, Tuji K, et al (1999) Endoscopic ligation by means of clip and detachable snare for management of colonoscopic post-polypectomy hemorrhage. Gastrointest Endosc 49:113–115.
51. Walsh RM, Ackroyd FW, Shelito PC (1992) Endoscopic resection of large sessile colorectal polyps. Gastrointest Endosc 38:303–309.
52. Tsuga K, Harama K, Fujimura J, et al (1998) Evaluation of the colorectal wall in normal subjects and patients with ulcerative colitis using an ultrasonic catheter probe. Gastrointest Endosc 48:477–484.
53. Conio M, Rajan E, Sorbi D, et al (2002) Comparative performance in the porcine esophagus of different solutions used for submucosal injection. Gastrointest Endosc 56:513–516.
54. Kaehler GF, Collet PH, Moritz GS, et al (2007) Waterjet for mucosal elevation in the GI tract – First clinical experiences. Gastrointest Endosc 65:T1546.
55. Kanamori T, Itoh M, Yokoyama Y, et al (1996) Injection-incision-assisted snare resection of large sessile colorectal polyps. Gastrointest Endosc 43:189–193.
56. Fukami N, Lee J (2006) Endoscopic treatment of large sessile and flat colorectal lesions. Curr Opin Gastroenterol 22:54–59.
57. Church JC (2003) Experience in the endoscopic management of large colonic polyps. ANZ J Surg 73:988–995.
58. Voloyiannis Lipof T, Bartus C, et al (2005) Preoperative colonoscopy decreases the need for laparoscopic management of colonic polyps. Dis Colon Rectum 48:1076–1080.
59. Voloyiannis T, Snyder MJ, Bailey R, et al (2008) Management of the difficult colon polyp referred for resection: resect or rescope? Dis Colon Rectum 51:292–295.
60. Waye JD (1991) Endoscopic treatment of adenomas. World J Surg; 15:14–19.
61. East J, Suzuki N, Arebi N, et al (2007) Position changes improve visibility during colonoscope withdrawal: a randomized, blinded, crossover trial. Gastrointest Endosc 65:263–269.
62. Pishvaian AC, Al-Kawas FH (2006) Retroflexion in the colon: a useful and safe technique in the evaluation and resection of sessile polyps during colonoscopy. Am J Gastroenterol 101:1479–1483.
63. Kondo S, Yamaji Y, Watabe H, et al (2007) A randomized controlled trial evaluating the usefulness of a transparent hood attached to the tip of the colonoscope. Am J Gastroenterol 102:75–81.
64. Brooker JC, Saunders BP, Shah SG, et al (2000) A new variable stiffness colonoscope makes colonoscopy easier randomised controlled trial. Gut 46:801–805.
65. Valentine JF (1998) Double-channel endoscopic polypectomy technique for the removal of large peduncolated polyps. Gastrointest Endosc 48:314–316.
66. Ng AJ, Kortsen MA (2002) The difficult polipectomy: description of a new dual-endoscope technique. Gastrointest Endosc 55:430–432.
67. McAfee JH, Katon RM (1994) Tiny snares prove safe and effective for removal of diminutive colorectal polyps. Gastrointest Endosc 40:301–303.
68. Kanamori T, Itoh M, Yokoyama Y, et al (1996) Injection-incision-assisted snare resection of large sessile colorectal polyps. Gastrointest Endosc 43:189–195.
69. Rex D (2006) Difficult colonic polyps: detection and removal. ASGE annual postgraduate course 2006. Gastrointestinal endoscopy live course syllabus. ASGE, Oak Brook, IL, pp 105–110.
70. Pabby A, Schoen RE, Weissfeld JL, et al (2005) Analysis of colorectal cancer occurrence during surveillance colonoscopy in the dietary Polyp Prevention Trial. Gastrointest Endosc 61:385–391
71. Farrar WD, Sawhney MS, Nelson DB, et al (2006) Colorectal cancers found after a complete colonoscopy. Clin Gastroenterol Hepatol 4:1259–1264.
72. Brooker JC, Saunders BP, Shah SG, et al (2002) Endoscopic resection of large sessile colonic polyps by specialist and non-specialist endoscopists Br J Surg 89:1020–1024.
73. Dell'Abate P, Iosca A, Galimberti A, et al (2001) Endoscopic treatment of colorectal benign-appearing lesions 3 cm or larger: techniques and outcome. Dis Colon Rectum 44:112–118.
74. Doniec JM, lohnert MS, Schienwind B, et al (2003) Endoscopic removal of large colorectal polyps: prevention of unnecessary surgery? Dis Colon Rectum 46:340–348.
75. Higaki S, Hashimoto S, Harada K, et al (2003) Long-term follow up of large flat colorectal tumors resected endoscopically. Endoscopy 35:845–849.
76. Repici A, Tricerri R (2004) Endoscopic polypectomy: techniques, complications and follow-up. Tech Coloproctol 8:S283–290.
77. Netzer P, Forster C, Biral R, et al (1998) Risk factor assessment of endoscopically removed malignant colorectal polyps. Gut 43:669–674.
78. Nivatvongs S, Rojanasakul A, Reiman HM, et al (1991) The risk of lymph node metastasis in colorectal polyps with invasive adenocarcinoma. Dis Colon Rectum 34:323–328.

79. Hassan C, Zullo A, Risio M, et al (2005) Histologic risk factor and clinical outcome in colorectal malignant polyp: a pooled-data analysis. Dis Colon Rectum 48:1588–1596.
80. Sakuragi M, Togashi K, Konishi F, et al (2001) Predicitve factors for lymph node metastasis in T1 stage colorectal carcinomas. Dis Colon Retum 46:1626–1632.
81. Kudo S (1993) Endoscopic mucosal resection of flat and depressed types of early colorectal cancer. Endoscopy 25:455–461.
82. Conio M, Repici A, Demarquay JF, et al (2004) EMR of large sessile colorectal polyps. Gastrointest Endosc 60:234–241.
83. Das A (2006) Endoscopic submucosal dissection: cure in one piece. Endoscopy 38:1044–1046.
84. Kodashima S, Fujishiro M, Yahagi N, et al (2006) Endoscopic submucosal dissection using flexknife. J Clin Gastroenetrol 40:378–384.
85. Iishi H, Tatsuta, Iseki K, et al (2000) Endoscopic piecemeal resection with submucosal saline injection of large sessile colorectal polyps. Gastrointest Endosc 51:697–700.
86. Inoue H, Takeshita K, Hori H, et al (1993) Endoscopic mucosal resection with cap-fitted panendsocope for esophagus, stomach, and colon mucosal lesions. Gastrointest Endosc 39:58–62.
87. Yoshikane H, Hidano H, Sakakibara A, et al (2001) Efficacy of distal attachment in Endoscopic resection of colorectal polyps situated behind semilunar Folds. Endoscopy 33:440–442.
88. Fujishiro M, Yahagi N, Kukushima M, et al (2005) Tissue damage of different submucosal injection solutions for EMR. Gastrointest Endosc 62:933–942.
89. Yamamoto H, Koiwai H, Yube T, et al (1999) A successful single-step endoscopic resection of a 40 millimeter flat-elevated tumor in the rectum: endoscopic mucosal resection using sodium hyaluronate. Gastrointest Endosc 50:701–704.
90. Larghi A, Waxman I (2007) State of the art on endoscopic mucosal resection and endoscopic submucosal dissection Gastrointest Endosc Clin North Am 17:441–469.
91. Tanaka S, Haruma K, Oka S, et al (2001) Clinicopathological features and endoscopic treatment of superficial spreading colorectal neoplasms lager than 20 mm. Gastrointest Endosc 54:62–66.
92. Ahmad NA, Kochman ML, Long WB, et al (2001) Efficacy, safety, and clinical outcomes of endoscopic mucosal resection: a study of 101 cases. Gastrointest Endosc 55:390–396.
93. Bergman U, Beger HG (2003) Endoscopic mucosal resection for advanced non-polypoid colorectal adenoma and early stage carcinoma. Surg Endosc 17:75–79.
94. Hurlstone DP, Sanders DS, Cross SS, et al (2004) Colonoscopic resection of lateral spreading tumors: a prospective analysis of endoscopic mucosal resection. Gut 53:1334–1339.

Management and Treatment of Complications in Diagnostic and Therapeutic Lower Gastrointestinal Tract Endoscopy

13

Giampaolo Angelini and Laura Bernardoni

Abstract Safe and efficient performance of endoscopy depends on the availability not only of expert endoscopists, but also of properly trained personnel and adequate facilities and equipment. Complications are inherent to all phases of the endoscopic procedure (preparation, sedation and analgesia, diagnostic or, more frequently, therapeutic endoscopy, and reprocessing post-procedure), immediately or within 30 days of the procedure. The most important complications are perforation and intestinal bleeding, which can be lethal if not properly diagnosed. Frequently, surgical exploration is still the treatment of choice, but new endoscopic techniques, such as clipping, argon plasma coaguation, endoloop or injection of drug, can allow patients to be treated conservatively.

Keywords Bleeding • Clip • Colonoscopy • Complication • Diagnostic colonoscopy • Endoscopic polypectomy • Lower gastrointestinal endoscopy • Management • Perforation • Therapeutic colonoscopy

13.1 Introduction

In the past, complications related to endoscopic procedures were thought to be associated with poor patient care. It is now well known that complications are inherent to all phases of the endoscopic procedure, including patient preparation, sedation and analgesia, and instrument-related maneuvers. Today, the physician endoscopist has the responsibility of understanding the associated risks of gastrointestinal endoscopy and needs to know all the measures available to minimize their occurrence. Additionally, when complications occur, the endoscopist should properly diagnose and treat them as soon as possible. The safe and efficient performance of endoscopy depends on the availability not only of expert endoscopists, but also of properly trained personnel, and adequate facilities and equipment. In this way, the mortality and morbidity of complications related to lower gastrointestinal endoscopy may be reduced. Colonoscopy-related mortality is extremely low, near zero [1–3]. The percentage of complications is also very low and these are mainly represented by acute events, recorded during the examination, but procedure-related complications can occur within 30 days of colonoscopy. In studies of diagnostic colonoscopies, the rates for bleeding have varied from 0.02% to 0.03%, and for perforations from 0.04% to 0.6%. The corresponding figures for therapeutic colonoscopy are 0.31% to 2.7% [4].

G. Angelini (✉)
Unit of Digestive Endoscopy, Institute of Gastroenterology, University of Verona, Verona, Italy

G.G. Delaini, T. Skřička, G. Colucci (eds.), *Intestinal Polyps and Polyposis*,

13.2 Complications of Bowel Preparation

Colonic lavage is essential to clear the lumen of mucus, fecal matter, combustible gases [5–7], and blood. Inadequate colonic lavage reduces mucosal visibility and can increase the risk for perforation and failure to diagnose disease [8,9]. Oral purgatives are a standard regimen for colonic cleansing [10,11].

Polyethylene glycol (PEG) lavages are non-digestible and non-absorbable solutions that are iso-osmolar with plasma so they are relatively safe with no significant changes in fluid and electrolyte balance [12]. Minor side-effects may include nausea, vomiting, and abdominal discomfort [13]. Gastrointestinal tolerance can be improved with administration of a prokinetic agent 30 min before ingestion of lavages [14,15]. Other reported complications are aspiration with hypoxemia or adult respiratory distress syndrome (ARDS) [16], Mallory–Weiss tears, anal irritation, angioedema, systemic allergic reactions, cardiac arrhythmias, and death.

Oral sodium phosphate solutions are osmotic, buffered, saline laxatives that offer a small-volume alternative for bowel preparation. They have the same efficacy and are well tolerated [17,18] in comparison to PEG solutions [19,20], but have been associated with severe electrolyte disturbances and so should be used with caution in patients with underlying renal insufficiency, congestive heart failure, liver cirrhosis with ascites, and advanced age [21–28]. Active colitis and aphthous ulcerations have been reported [11].

Sennoside preparations have been associated with a mononuclear infiltrative colitis, making interpretation of colon biopsies difficult [29].

In patients with chronic constipation or neuromuscular disorders, a clear liquid diet for 24–48 h before examination, with addition of magnesium citrate, may improve the quality of bowel cleansing [30]. In patients with suspected obstruction or toxic megacolon, bowel lavage should be carried out very cautiously because of the risk of perforation.

13.3 Complications of Sedation and Analgesia

Although colonoscopy is generally well tolerated, some patients regard the procedure as unpleasant and painful. Lower body mass index (BMI), history of hysterectomy, diarrhea, first-time colonoscopy, and anxiety level are independent factors related to painful colonoscopy; older age, lower BMI, and history of hysterectomy are independent factors related to difficult colonoscopy [31–33].

Intravenous benzodiazepines, alone or in combination with narcotic analgesic, are the preferred regimens for short-term conscious sedation [34]. Midazolam and diazepam are the most commonly used and have proven to be relatively safe and effective. Midazolam has a more rapid onset of action, shorter duration of effects, and more significant degree of anterograde amnesia.

13.3.1 Cardiopulmonary Complications

Cardiopulmonary complications (respiratory depression with hypoxemia, with potential risk of myocardial infarction or cardiac arrhythmias) may occur in 0.01–2% of patients [35]. To prevent potential complications, the endoscopist should identify high-risk patients by prior anamnestic evaluation (pregnancy, obesity, drug or alcohol abuser, uncooperative, old age), and administer intravenous benzodiazepines in small, incremental doses [36]. Careful titration of sedatives and analgesics to achieve patient comfort, and reassessment of drug effect between boluses reduce the occurrence of over-sedation and subsequent risk of cardiopulmonary complications [37].

Opiate preparations are commonly used with benzodiazepines to prolong the sedative effect and provide short-term analgesia. The combined use of benzodiazepines and opiates causes a synergistic effect with increased risk of respiratory depression, severe hypoxia, and potential myocardial ischemia. Administering opiates prior to benzodiazepines leads to a 50–70% of reduction in total drug dose [38]. For prolonged procedures, such as endoscopic retrograde cholangiopancreatography (ERCP), and in children [39], a short-acting general anesthetic agent (e.g. propofol) is necessary to induce deeper sedation and more-potent analgesia [40–42].

Treatment of cardiopulmonary complications includes discontinuation of procedure, hemodynamic support, and administration of reversal agents (flumazenil for benzodiazepines, naloxone for opioid drug) even if naloxone has been associated with serious side-effects (tachyarrhythmias and withdrawal syndrome that can be confused with other endoscopic complications) and should not be used routinely after conscious sedation [43].

13.3.2 Paradoxical Reactions

Paradoxal reactions, such as combativeness, agitation, disorientation, and tachycardia, are rare events due to administration of sedative medications and can be confused with insufficient sedation. Management includes the use of pulse oximetry aids, and administering the reversal agent flumazenil [44].

It is generally accepted that rapid detection of changes in cardiopulmonary parameters may prompt early patient assessment during the endoscopic procedure and decrease the risk of progression to serious complications. Guidelines of the American Society of Anesthesiologists (1996) recommend electronic monitoring before sedation and during and after the endoscopic procedure [45]. Continuous pulse oximetry monitoring is recommended for all the patients submitted to colonoscopy, irrespective of whether sedation is used or not. Continuous oxygen supplementation is recommended in patients with baseline-limited cardiopulmonary reserve [46,47].

13.4 Complications of Diagnostic Colonoscopy

The demand for colonoscopy has dramatically increased in recent years, for the important role of diagnosis in colorectal cancer screening, surveillance of prior cancer or polypectomy, evaluation of hematochezia, changes in bowel habits, etc. Colonoscopy continues to be an invasive procedure with infrequent, but potentially severe, complications as a result of preparation and premedication, mechanical trauma, anatomical variation of the colon, and operator inexperience. The complication rate in diagnostic colonoscopy is 0.1–0.3%, although reports of these data are very different in the literature [4,48].

Perforation is the most important complication, with a frequency of 0.03–0.9% [49] and a mortality rate of 0.2% in diagnostic colonoscopy. The incidence of perforation increases when polypectomy or other therapeutic interventions are performed. Major mechanisms leading to perforation are [50]:

- excessive mechanical pressure of the colonscope on the colonic wall and too many manipulations during torsion and straightening of the instrument
- excessive air insufflation
- poor visibility due to inadequate bowel preparation.

These events may result in seromuscular stripping and serosal tears without the development of free perforation. Procedure-related colonic perforations are most commonly localized in the rectosigmoid tract (74%) and occur predominantly in females [48,51,52]. Risk factors for perforation are: the presence of inflammatory activity, prior pelvic radiotherapy, the presence of diverticula, and prior abdominal surgery [53]. For some authors, the use of anesthetic does not increase the risk of perforation and is not accompanied by severe complications, but the extent of sedation and analgesia is a controversial subject [50].

The definitive management remains controversial [54]. Sometimes, immediate operative intervention is not mandatory. It seems that non-operative management could be undertaken in highly selected patients who do not exhibit signs of peritoneal contamination or abdominal sepsis [49]. In patients with obvious perforation, who develop abdominal pain, fever, cardiac arrhythmias, or hemodynamic instability, immediate surgical exploration is the management of choice. In patients with few symptoms or with abdominal symptoms but without evidence of free air, a conservative treatment can be proposed. Endoscopic clip application is recommended for iatrogenic perforations (Fig. 13.1); endoclips create successful mucosal and submucosal apposition, while apposition of the muscularis propria and serosa is not possible because of the superficial bite of the clips. Endoluminal repair of colonic perforations with clips, and further conservative treatment with no oral intake, and intravenous broad-spectrum antibiotics, results in a shorter length of hospitalization and lower morbidity [55]. An alternative method to treat colonic perforation may be endoscopic suturing using an EndoCinch endoscope-suturing device, but experience is, at present, very limited [49].

Hemorrhage is a rare complication of diagnostic colonoscopy (risk 0.02%). Some authors report that post-biopsy bleeding is too low to be deemed noteworthy apart from the rarity of such a complication [4,48]. Minor mucosal or submucosal bleeding may occur as a result of manipulations within the colon, but these events are usually self-limited. In very rare cases,

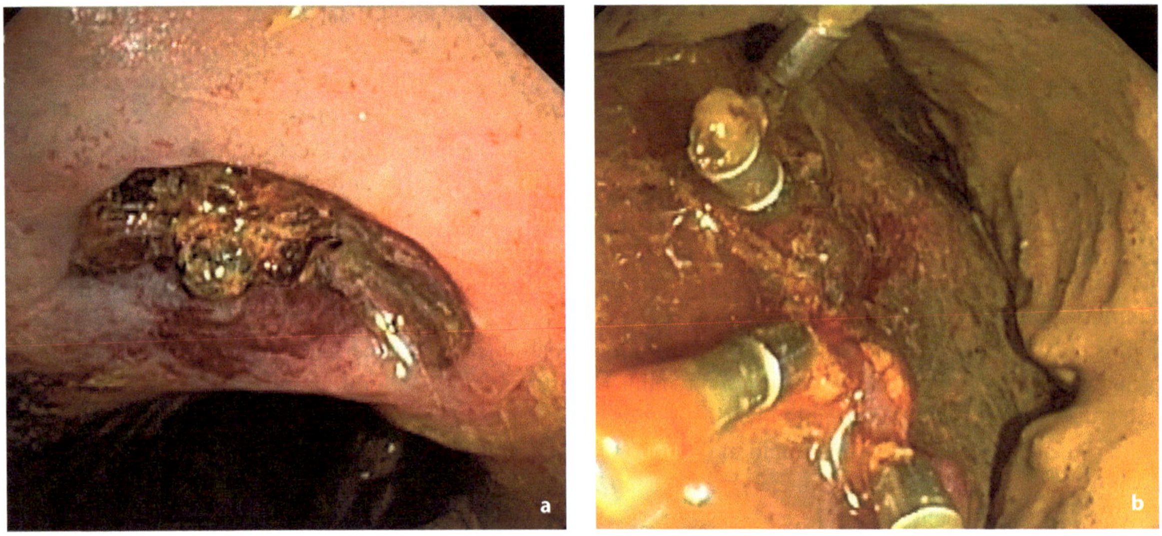

Fig. 13.1a,b Repair of iatrogen perforation with clips; **a**, perforation of rectum after enema in an old patient; **b**, close of iatrogenic perforation with clips

excessive manipulation and force can cause hemoperitoneum. Both hot and cold biopsies have been previously associated with clinically significant bleeding, mainly in patients with underlying inflammatory disorders or changes in hemocoagulation parameters [56].

Cardiorespiratory changes are frequently observed during routine lower endoscopy, performed with or without sedation or analgesia. Vasovagal reactions characterized by bradycardia and hypotension may occur as a result of abdominal pain for looping of the colonoscope with mesenteric stretching, excessive air insufflation, and premedication. Although minor cardiorespiratory complications occur frequently, severe events are rare.

Abdominal visceral injury is a rare complication due to excessive traction on the splenocolic ligament, blind passage along the splenic flexure, or direct trauma induced by the colonoscope loop. Risk factors for splenic injury are serosal adhesion for prior surgery and inflammatory bowel disease. Rarely, liver laceration may occur [57–65].

Acute diverticulitis and appendicitis, caused by microscopic perforation of the colon, have been reported as complications of diagnostic colonoscopy [51,66,67].

We can consider failure to diagnose disease a complication of colonoscopy. The missing of lesions has been directly demonstrated in some studies in which patients had consecutive back-to-back colonoscopies [68]. Depending on the type of study and the population studied, the colorectal cancer diagnosis rate, 3–5 years after a "clean colonoscopy", ranges from 0.5% to 5% [69]. Multiple factors can limit the exploration of colon mucosa during colonoscopy, such as poor bowel cleansing, tortuous colon, colonic spasm, characteristics of polyps (diameter, flat pattern), and the number of polyps in the patient. The "miss rate" is significantly higher for flat polyps, whatever the size of the lesions, and in the proximal colon [70]. Even when the examination is performed by an experienced endoscopist, it is generally accepted that 5–10% of mucosa may not be visualized. Experimental methods [71] to obtain a better view include wide-angle optics [72], cap or hood-fitted colonoscopy [73], and the third-eye retroscope [74]. It is very important to take enough time to examine the colon; in 2002, the US Multi-Society Task Force recommended that the withdrawal time in normal colonoscopies in patients with intact colon should average at least 6–10 min [71], as was recently also underlined by Simmons [75].

Other rare complication of diagnostic colonoscopy are postcolonoscopy syndrome [76] (abdominal pain and moderate distension) due to overinsufflation of air [77, 78], with radiographic evidence of intramural air or free air in the peritoneum. Conservative management is indicated.

13.5 Complications of Therapeutic Colonoscopy

Removal of polyps through biopsy with a snare or forceps increases the risk of serious complication nearly nine-fold compared with colonoscopy without biopsy. Post-polypectomy bleeding is the most common complication. Most bleeding is self-limited, not requiring surgery or transfusions [51]. Post-polypectomy bleeding occurs with an overall frequency of 0.3% to 6.1% in studies where endoscopic polypectomies have been exclusively evaluated [79]. Hemorrhage can occur during the procedure, within 24 h, or delayed more than 24 h after polypectomy. The first and second type occur more frequently and can generally be resolved with endoscopic therapy (submucosal injection of epinephrine, hemoclip or endoloop placement, electrocoagulation, or argon plasma coagulation (APC) used alone or in combination) (Figs. 13.2 and 13.3). Patients with the highest risk of bleeding are those undergoing snare polypectomy of a large polyp (with diameter of 20 mm or greater), with sessile morphology, or pedunculated with large stalks, and those of advanced age [4,79]. Other risk factors for post-polypectomy bleeding are the endoscopist's experience, the use of pure cutting current, and proximal location [80]. Delayed bleeding may be related not only to direct damage to vessels perfusing the resected polyp, but also to disturbance in tissue healing or vascular regeneration [81], and is more frequent in patients with underlying coagulation disorders or undergoing anticoagulant therapy, particularly when anticoagulant therapy has been reinstituted. Low-dose aspirin does not increase the risk of post-polypectomy bleeding; only warfarin usage is associated with an increase bleeding rate. Definitive data on the risk of polypectomy in patients taking newer antiplatelet drugs (such as clopidogrel) are not yet available, and caution is advised [80]. Although the current American Society of Gastroenterology (ASGE) guidelines [82] suggest that aspirin and not steroidal anti inflammatory drugs (NSAID) do not need to be discontinued prior to polypectomy, it would be helpful to stop these drugs electively for 7 days before any planned resection of large polyps or when multiple polypectomies have to be performed, and to keep off these drugs for 2 weeks after these procedures [80]. Restarting anticoagulation within 72 h after a polypectomy is associated with a five-fold increased risk of post-polypectomy bleeding [83]. The decision to reverse anticoagulation, thereby risking thromboembolic complications, must be carefully weighed against the increased risk of bleeding when maintaining anticoagulation [84]. In Fig. 13.4, a proposed approach for patients on anticoagulation undergoing an endoscopic procedure is described. Table 13.1 reports the ASGE

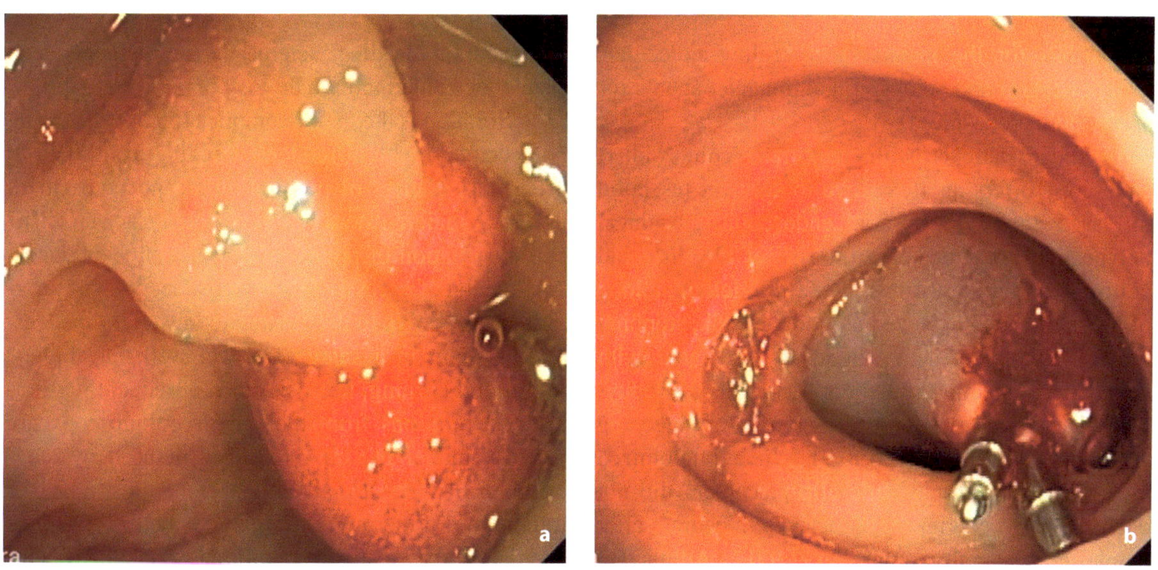

Fig. 13.2a,b Hemostatic technique with injection of epinephrine and placement of two clips; **a**, large peduncolated polyp in the sigma; **b**, bleeding post-polipectomy stop with injection of epinephrine at the foundation of polyp and position of two clips

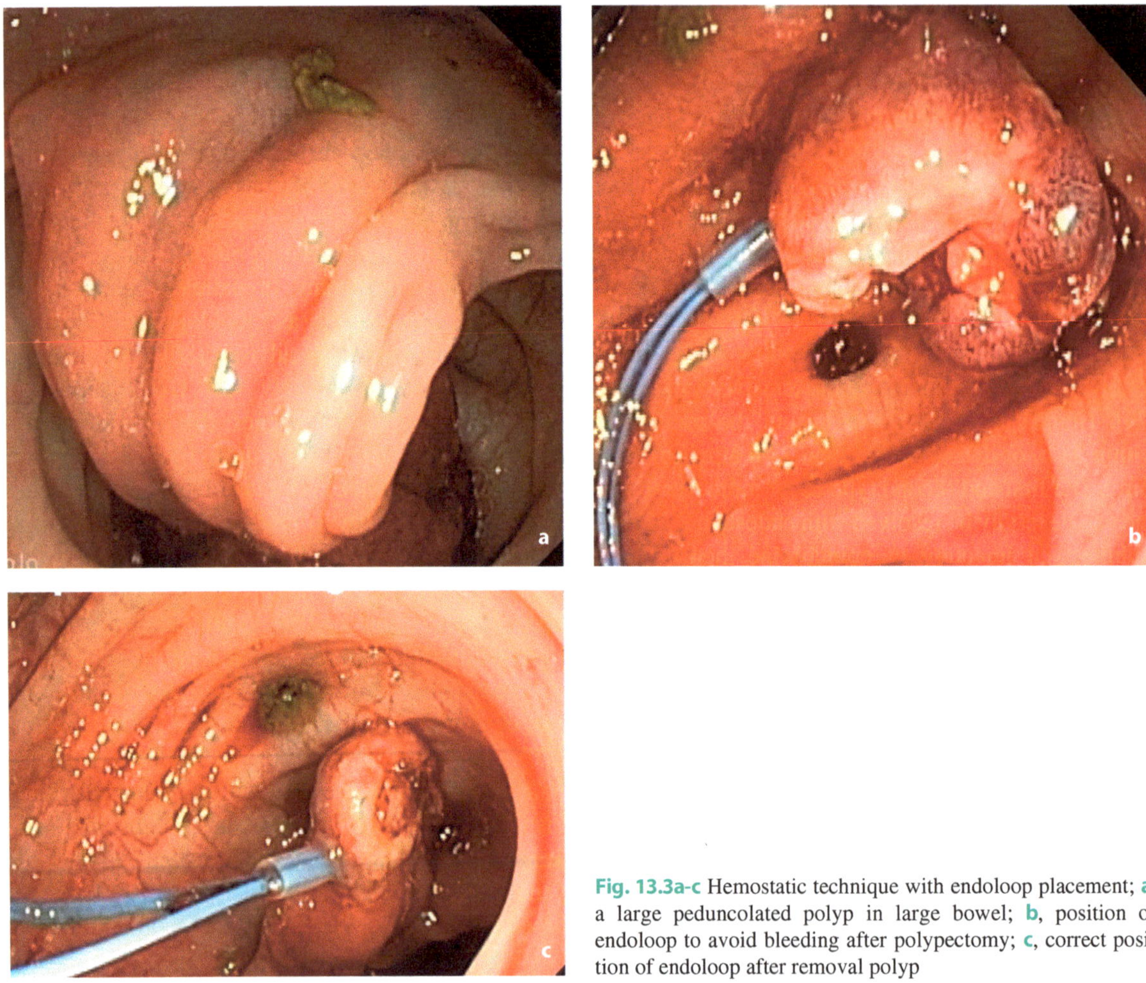

Fig. 13.3a-c Hemostatic technique with endoloop placement; a, a large peduncolated polyp in large bowel; b, position of endoloop to avoid bleeding after polypectomy; c, correct position of endoloop after removal polyp

indications for the management of low-molecular-weight heparin and non-aspirin antiplatelet agents for endoscopic procedures [85].

In older patients with important comorbidity, endoscopy offers an alternative to surgical resection for large polyps [86]; serial endoscopic piecemeal polypectomy may decrease the risk of bleeding and perforation in these settings [87,88]. Endoloop, a loop of nylon thread is tightened around the stalk, stopping blood flow in the stalk, alone or associated with epinephrine injection [89]; this is used as a prophylactic measure to prevent post-polypectomy bleeding for pedunculated polyps [90,91]. In a randomized controlled trial, prophylactic clip placement did not decrease the occurrence of delayed bleeding after colonoscopic polypectomy [92,93]. Adjunctive thermal modalities such as APC, heater and gold probe, and hot biopsy forceps are used to stop post-polypectomy bleeding. Sclerosants are no longer used since they could increase the risk of perforation [80].

Perforation is the major complication of endoscopic polypectomy, with a frequency ranging from 0.3% to 3.1% [94]. Causes of perforation are poor visualization for remaining stool or blood, colonic spasm, difficult localization of polyp, and operator inexperience. Application of excessive electric current or close proximity of the polyp to the contralateral wall, or inadvertent catching in a snare of bowel wall may induce transmural burning, necrosis, and subsequent perforation. Perforations from therapeutic colonoscopy are usually smaller in size than those induced by diagnostic maneuvers, and abdominal contamination is at a minimum. Furthermore, if colonic lavage is adequate and the patient's general condition is good, these types of perforation could respond satisfactorily to conservative treatment. Management includes hospitalization

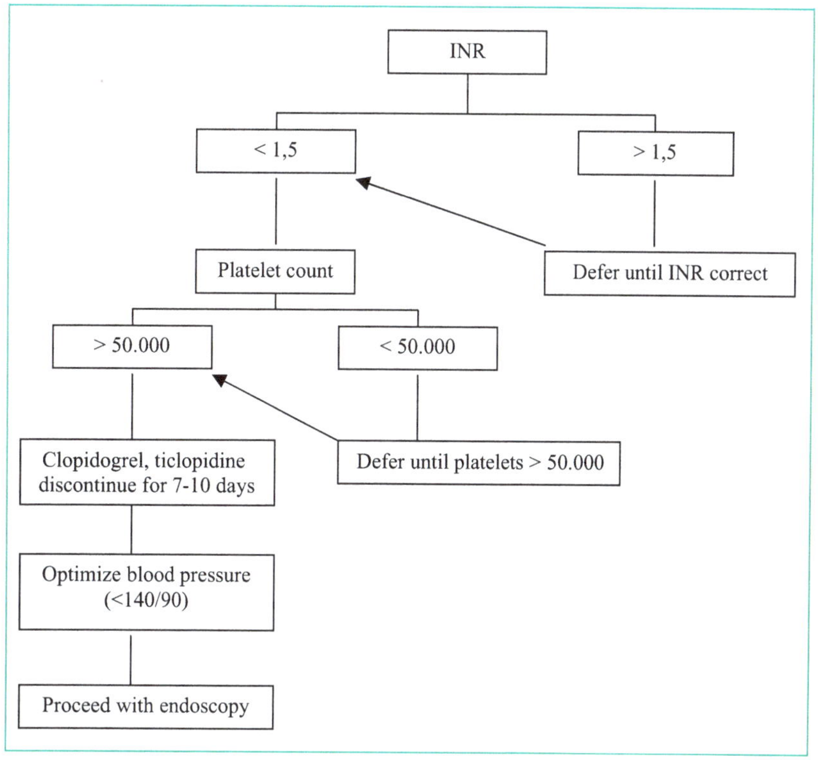

Fig. 13.4 Proposed approach for patients on anticoagulation undergoing an endoscopic procedure. *INR*, international normalized ratio. Reproduced from [84], with permission from Elsevier

Table 13.1 American Society of Gastroenterology guidelines for management of low-molecular-weight heparin and non-aspirin antiplatelet agents for endoscopic procedures. Reproduced from [85], with permission from Elsevier

Management of LMWH in patients undergoing endoscopic procedures	
Procedure	Recommendation
High	Consider discontinuation at least 8 h before procedure
Low	No change in therapy
Reinstitution of LMWH should be individualized	
Management of antiplatelet medication (clopidogrel or ticlopidine) in patients undergoing endoscopic procedures	
Procedure	Recommendation
High	Consider discontinuation 7-10 h before procedure
Low	No change in therapy
Patient on combination therapy (e.g., dopidogrel and aspirin) may be at an additional increased risk of bleeding.	
For acute GI hemorrehage in the patient on clopidogrel or ticlopidine, the decision to transfuse platelets should be individualized, usually weighing the risk of an acute cardiovascular even against the risk of continued bleeding.	
Reinstitution of clopidogel or ticlopidine should be individualized	
Procedure risk	
High-risk procedures	Low-risk procedures
Polypectomy	Diagnostic
Biliary sphinterotomy	EGD ± biopsy
Pneumatic or bougie dilation	Flexible sphincterotomy ± biopsy
PEG placement	Colonoscopy ± biopsy
EUS-guided FNA	ERCP without endoscopic sphincterotomy
Laser ablation and coagulation	Biliary/pancreatic stent without endoscopic sphincterotomy
Treatment of varices	EUS without FNA
	Enteroscopy

EGD, esophagogastroduodenoscopy; *ERCP*, endoscopic retrograde cholangiopancreatography; *EUS*, endoscopic ultrasound scan; *FNA*, fine needle aspiration; *LMHW*, low-molecular-weight heparin; *PEG*, percutaneous endoscopic gastrostomy

for observation, and supportive measure with radiological abdominal examination; surgical exploration is indicated when the patient does not improve or worsens with conservative therapy [49,50,52,55,95]. Perforations occurring during diagnostic colonoscopy are diagnosed sooner and patients undergo operation earlier compared with perforations arising during/after therapeutic colonoscopy. This may be at least partly explained by the different pathophysiological mechanisms of the two types of perforation, and by the different length of time before the perforation becomes clinically evident [4,96].

Other specialized endoscopic techniques like APC or laser therapy, stent placement, and dilatation have a major risk of complication, mainly bleeding and perforation (4.1% complication for APC, 7.8% for dilatation, 14% for stent placement) [94].

13.6 Prevention of Infection

Infectious complications as a result of endoscopy involve the following situations:

- transfer of pathogens related to contaminated equipment
- transient bacteremia associated with biopsy or therapeutic procedures
- aspiration pneumonia for regurgitation and inhalation of stomach contents
- perforation.

In particular, transient bacteremia may cause infectious endocarditis; past gastrointestinal guidelines recommended prophylaxis in patients with cardiovascular disease who undergo a high-risk endoscopic procedure. The new guidelines of the American Heart Association in 2007 state that administration of prophylactic antibiotics solely to prevent endocarditis is not recommended for patients who undergo genitourinary (GU) or gastrointestinal (GI) procedures, because no published data demonstrate a conclusive link between these procedures and development of infectious endocarditis, or that the administration of antimicrobial prophylaxis prevents infectious endocarditis in association with procedures performed on the GI or GU tract. For patients with a prosthetic cardiac valve, previous infectious endocarditis, congenital heart disease, or cardiac transplantation recipients who develop cardiac valvulopathy, or who have an established GI or GU tract infection or sepsis, it may be reasonable that the antibiotic regimen includes an agent active against enterococci [97].

References

1. Rathgaber SW, Wick TM (2006) Colonoscopy completion and complication rates in a community gastroenterology practice. Gastrointest Endosc 64:556–562.
2. Nelson DB (2002) Procedural success and complications of large-scale screening colonoscopy. Gastrointest Endosc 55:307–314.
3. Tulchinsky H Madhala-Givon O, Wasserberg N, et al (2006) Incidence and management of colonoscopic perforations: 8 years' experience. World J Gastroenterol 12:4211–4213.
4. Dafnis G (2001) Complications of diagnostic and therapeutic colonoscopy within a defined population in Sweden. Gastrointest Endosc 54:302–309.
5. Ladas SD, Karamanolis G, Ben-Soussan E (2007) Colonic gas explosion during therapeutic colonoscopy with electrocautery. World J Gastroenterol 13:5295–5298.
6. Hofstad B (2007) [Explosion in the rectum.] Tidsskr Nor Laegeforen 127:1789–1790.
7. Josemanders DF, Spillenaar Bilgen EJ, van Sorge AA, et al (2006) Colonic explosion during endoscopic polypectomy: avoidable complication or bad luck? Endoscopy 38:943–944.
8. Bond JH (2007) Should the quality of preparation impact postcolonoscopy follow-up recommendations? Am J Gastroenterol 102:2686–2687.
9. Froehlich F, Wietlisbach V, Gonvers JJ, et al (2005) Impact of colonic cleansing on quality and diagnostic yield of colonoscopy: the European Panel of Appropriateness of Gastrointestinal Endoscopy European multicenter study. Gastrointest Endosc 61:378–384.
10. Shawki S, Wexner SD (2008) Oral colorectal cleansing preparations in adults. Drugs 68:417–437.
11. Belsey J, Epstein O, Heresbac D (2007) Systematic review: oral bowel preparation for colonoscopy. Aliment Pharmacol Ther 25:373–384.
12. Rothfuss KS, Bode JC, Stange EF, Parlesak A (2006) Urinary excretion of polyethylene glycol 3350 during colonoscopy preparation. Z Gastroenterol 44:167–172.
13. Dykes C, Cash BD (2008) Key safety issues of bowel preparations for colonoscopy and importance of adequate hydration. Gastroenterol Nurs 31:30–35; quiz 36–37.
14. Mishima Y, Amano Y, Okita K, et al (2008) Efficacy of prokinetic agents in improving bowel preparation for colonoscopy. Digestion 77:166–172.
15. Katsinelos P, Pilpilidis I, Paroutoglou G, et al (2005) The administration of cisapride as an adjuvant to PEG-electrolyte solution for colonic cleansing: a double-blind randomized study. Hepatogastroenterology 52:441–443.
16. de Graaf P, Slagt C, de Graaf JL, Loffeld RJ (2006) Fatal aspiration of polyethylene glycol solution. Neth J Med 64:196–198.
17. Kastenberg D, Barish C, Burack H, et al (2007) Tolerability and patient acceptance of sodium phosphate tablets compared with 4-L PEG solution in colon cleansing:

combined results of 2 identically designed, randomized, controlled, parallel group, multicenter phase 3 trials. J Clin Gastroenterol 41:54–61.
18. Rapier R, Houston C (2006) A prospective study to assess the efficacy and patient tolerance of three bowel preparations for colonoscopy. Gastroenterol Nurs 29:305–308.
19. Tan JJ, Tjandra JJ (2006) Which is the optimal bowel preparation for colonoscopy – a meta-analysis. Colorectal Dis 8:247–258.
20. Schanz S, Kruis W, Mickisch O, et al (2008) Bowel preparation for colonoscopy with sodium phosphate solution versus polyethylene glycol-based lavage: a multicenter trial. Diagn Ther Endosc 2008:713521.
21. Wechsler A, Schneider R, Sapojnikov M, et al (2006) Bowel cleansing in patients with chronic renal failure—an often overlooked hazard. Nephrol Dial Transplant 21:1133–1134.
22. Rodriguez-Alcalde D, Marín-Gabriel JC, Rodríguez-Muñoz S, et al (2008) [Tolerability, safety, and efficacy of sodium phosphate preparation for colonoscopy: the role of age.] Rev Esp Enferm Dig 100:17–23.
23. Niemeijer ND, Rijk MC, van Guldener C (2008) Symptomatic hypocalcemia after sodium phosphate preparation in an adult with asymptomatic hypoparathyroidism. Eur J Gastroenterol Hepatol 20:356–358.
24. Mehta BP, Shmerling RH, Moss AC (2008) Pseudogout after polyethylene glycol bowel cleansing. J Clin Gastroenterol Jun 3 [Epub ahead of print].
25. Khurana A, McLean L, Atkinson S, Foulks CJ (2008) The effect of oral sodium phosphate drug products on renal function in adults undergoing bowel endoscopy. Arch Intern Med 168:593–597.
26. Sunada K, Yano T, Arashiro M, Miyata T, et al (2008) [Endoscopic therapy using double balloon endoscopy.] Nippon Rinsho 66:1268–1276.
27. Gonlusen G, Akgun H, Ertan A, et al (2006) Renal failure and nephrocalcinosis associated with oral sodium phosphate bowel cleansing: clinical patterns and renal biopsy findings. Arch Pathol Lab Med 130:101–106.
28. Gumurdulu Y, Serin E, Ozer B, et al (2004) Age as a predictor of hyperphosphatemia after oral phosphosoda administration for colon preparation. J Gastroenterol Hepatol 19:68–72.
29. Radaelli F, Meucci G, Imperiali G, et al (2005) High-dose senna compared with conventional PEG-ES lavage as bowel preparation for elective colonoscopy: a prospective, randomized, investigator-blinded trial. Am J Gastroenterol 100:2674–2680.
30. Hookey LC, Vanner S (2007) A review of current issues underlying colon cleansing before colonoscopy. Can J Gastroenterol 21:105–111.
31. Chung YW, Han DS, Yoo KS, Park CK (2007) Patient factors predictive of pain and difficulty during sedation-free colonoscopy: a prospective study in Korea. Dig Liver Dis 39:872–876.
32. Bafandeh Y, Khoshbaten M, Eftekhar Sadat AT, Farhang S (2008) Clinical predictors of colorectal polyps and carcinoma in a low prevalence region: results of a colonoscopy based study. World J Gastroenterol 14:1534–1538.
33. Boustiere C (2008) [Complications of routine digestive endoscopy.] Rev Prat 58:701–705.
34. Radaelli F , Meucci G, Sgroi G, Minoli G; Italian Association of Hospital Gastroenterologists (AIGO) (2008) Technical performance of colonoscopy: the key role of sedation/analgesia and other quality indicators. Am J Gastroenterol 103:1122–1130.
35. Sharma VK, Nguyen CC, Crowell MD, et al (2007) A national study of cardiopulmonary unplanned events after GI endoscopy. Gastrointest Endosc 66:27–34.
36. Padmanabhan U, Leslie K (2008) Australian anaesthetists' practice of sedation for gastrointestinal endoscopy in adult patients. Anaesth Intensive Care 36:436–441.
37. McQuaid KR, Laine L (2008) A systematic review and meta-analysis of randomized, controlled trials of moderate sedation for routine endoscopic procedures. Gastrointest Endosc 67:910–923.
38. Waring JP, Baron TH, Hirota WK, et al (2003) Guidelines for conscious sedation and monitoring during gastrointestinal endoscopy. Gastrointest Endosc 58: 317-322.
39. Lee KK, Anderson MA, Baron TH, et al; ASGE Standards of Practice Commitee (2008) Modifications in endoscopic practice for pediatric patients. Gastrointest Endosc 67:1–9.
40. Mandel JE, Tanner JW, Lichtenstein GR, et al (2008) A randomized, controlled, double-blind trial of patient-controlled sedation with propofol/remifentanil versus midazolam/fentanyl for colonoscopy. Anesth Analg 106:434–439.
41. Weinstock LB, Cohen AM, Volotsky GR (2007) How deep should "deep sedation" be? Am J Gastroenterol 102:906–907; author reply 907–908.
42. Sipe BW, Scheidler M, Baluyut A, Wright B (2007) A prospective safety study of a low-dose propofol sedation protocol for colonoscopy. Clin Gastroenterol Hepatol 5:563–566.
43. Grant C. Ludbroock G, Hampson EA, et al (2008) Adverse physiological events under anaesthesia and sedation: a pilot audit of electronic patient records. Anaesth Intensive Care 36:222–229.
44. Leung FW (2008) Unsedated colonoscopy for paradoxical agitation: an unusual practice for an uncommon complication in US veterans. Am J Gastroenterol 103:1578–1579.
45. [No authors listed] (1996) Practice guidelines for sedation and analgesia by non-anesthesiologist. A report by the American Society of Anesthesiologists Task Force on sedation and analgesia by non-anesthesiologists. Anesthesiology 84:459-471.
46. Rex DK (2006) Review article: moderate sedation for endoscopy: sedation regimens for non-anaesthesiologists. Aliment Pharmacol Ther 24:163–171.
47. Froehlich F, Ludbrook G, Hampson EA, et al (2006) Current sedation and monitoring practice for colonoscopy: an International Observational Study (EPAGE). Endoscopy 38:461–469.
48. Paspatis GA (2008) Complications of colonoscopy in a large public county hospital in Greece A 10-year study. Dig Liver Dis Apr 14. [Epub ahead of print].
49. Magdeburg R, Collet P, Post S, Kaehler G (2008) Endoclipping of iatrogenic colonic perforation to avoid surgery. Surg Endosc 22:1500–1504.
50. Garcia Martinez MT, Ruano Poblador A, Galán Raposo L, et al (2007) [Perforation after colonoscopy: our 16-year experience.] Rev Esp Enferm Dig 99:588–592.
51. Levin TR, Zhao W, Conell C, et al (2006) Complications of

colonoscopy in an integrated health care delivery system. Ann Intern Med 145:880–886.
52. Luning TH, Keemers-Gels ME, Barendregt WB, et al (2007) Colonoscopic perforations: a review of 30,366 patients. Surg Endosc 21:994–997.
53. Cobb W (2004) Colonoscopic perforations: incidence, management and outcomes. Am Surg 70:750–757.
54. Ker TS, Wasserberg N, Beart RW Jr (2004) Colonoscopic perforation and bleeding of the colon can be treated safely without surgery. Am Surg 70:922–924.
55. Iqbal CW, Chun YS, Farley DR (2005) Colonoscopic perforations: a retrospective review. J Gastrointest Surg 9:1229–1235; discussion 1236.
56. Zuccaro G (2008) Epidemiology of lower gastrointestinal bleeding. Best Pract Res Clin Gastroenterol 22:225–232.
57. Parker WT, Edwards MA, Bittner JG 4th, Mellinger JD (2008) Splenic hemorrhage: an unexpected complication after colonoscopy. Am Surg 74:450–452.
58. Petersen CR, Adamsen S, Gocht-Jensen P, et al (2008) Splenic injury after colonoscopy. Endoscopy 40:76–79.
59. Lalor PF, Mann BD (2007) Splenic rupture after colonoscopy. JSLS 11(1):151-156.
60. Di Lecce F, Viganò P, Pilati S, et al (2007) Splenic rupture after colonoscopy. A case report and review of the literature. Chir Ital 59:755–757.
61. Cappellani A, Di Vita M, Zanghì A, et al (2008) Splenic rupture after colonoscopy: report of a case and review of literature. World J Emerg Surg 3:8.
62. Schilling D, Kirr H, Mairhofer C, Rumstadt B (2008) [Splenic rupture after colonoscopy.] Dtsch Med Wochenschr 133:833–835.
63. Naini MA, Masoompour SM (2005) Splenic rupture as a complication of colonoscopy. Indian J Gastroenterol 24:264–265.
64. Shah P (2007) Splenic rupture as complication of colonoscopy. Indian J Gastroenterol 26:150.
65. Holubar S, Dwivedi A, Eisdorfer J, et al (2007) Splenic rupture: an unusual complication of colonoscopy. Am Surg 73:393–396.
66. Chae HS, Jeon SY, Nam WS, et al (2007) Acute appendicitis caused by colonoscopy. Korean J Intern Med 22:308–311.
67. Horimatsu T, Fu KI, Sano Y, et al (2007) Acute appendicitis as a rare complication after endoscopic mucosal resection. Dig Dis Sci 52:1741–1744.
68. Heresbach D (2008) Miss rate of colorectal neoplastic polyps: a prospective multicenter study of back-to-back video colonoscopies. Endoscopy 40:284–290.
69. Kiesslich R, von Bergh M, Hahn M, et al (2001) Chromoendoscopy with indigocarmine improves the detection of adenomatous and nonadenomatous lesions in the colon. Endoscopy 33:1001–1006.
70. Heresbach D (2008) Miss rate colorectal neoplastic polyps: a prospective multicenter study of back-to-back video colonoscopies. Endoscopy 40:284–290.
71. Douglas KRM (2006) Maximizing detection of adenomas and cancer during colonoscopy. Am J Gastroenterol 101:2866–2877.
72. Fatima H (2008) Cecal insertion and withdrowal times with wide-angle versus standard colonoscopy: a randomized controlled trial. Clin Gastroenterol Hepatol 6:109–114.
73. Matsushita M (1998) Efficacy of total colonoscopy with a transparent cap in comparison with colonoscopy without the cap. Endoscopy 30:444–447.
74. Triadafilopoulos G (2006) A novel retrograde-viewing auxiliary imaging device improve the detection of simulated polyps in anatomical model of the colon. Gastrointest Endosc 63:AB103.
75. Simmons DT (2006) Impact of endoscopist withdrawal speed on polyp yield: implications for optimal colonoscopy withdrawal time. Aliment Pharmacol Ther 24:965–971.
76. Lee YC, Wang HP, Chiu HM, et al (2006) Factors determining post-colonoscopy abdominal pain: prospective study of screening colonoscopy in 1000 subjects. J Gastroenterol Hepatol 21:1575–1580.
77. Leung FW (2008) Methods of reducing discomfort during colonoscopy. Dig Dis Sci 53:1462–1467.
78. Saito Y, Uraoka T, Matsuda T, et al (2007) A pilot study to assess the safety and efficacy of carbon dioxide insufflation during colorectal endoscopic submucosal dissection with the patient under conscious sedation. Gastrointest Endosc 65:537–542.
79. Consolo P (2008) Efficacy, risk factors and complication of endoscopic polypectomy: 10 year experience at a single center. World J Gastroenterol 21:2364–2369.
80. Fraser CSB (2004) Preventing postpolypectomy bleeding: obligatory and optional steps. Endoscopy 36:898–900.
81. Watabe H, Yamaji Y, Okamoto M, et al (2006) Risk assessment for delayed hemorrhagic complication of colonic polypectomy: polyp-related factors and patient-related factors. Gastrointest Endosc 64:73–78.
82. Eisen G (2002) Guidelines on the management of anticoagulation and antiplatelet therapy for endoscopic procedures. Gastrointest Endosc 55:775–779.
83. Sawhney MS, Salfiti N, Nelson DB, et al (2008) Risk factors for severe delayed postpolypectomy bleeding. Endoscopy 40:115–119.
84. Harewood GC (2006) Recommendations for endoscopy in the patient on chronic anticoagulation: apply with care! Gastrointest Endosc 64:79–81.
85. Zuckerman MJ, Hirota WK, Adler DG, et al (2005) ASGE guideline: the management of low-molecular-weight heparin and non aspirin antiplatelet agents for endoscopic procedure. Gastrointest Endosc 61:189–194.
86. Katsinelos P, Chatzimavroudis G, Papaziogas B, et al (2008) Endoclipping-assisted resection of large colorectal polyps. Surg Laparosc Endosc Percutan Tech 18:19–23.
87. Zerey M, Paton BL, khan PD, et al (2007) Colonoscopy in the very elderly: a review of 157 cases. Surg Endosc 21:1806–1809.
88. Tsutsumi S, Fukushima H, Osaky K, et al (2007) Feasibility of colonoscopy in patients 80 years of age and older. Hepatogastroenterology 54:1959–1961.
89. Paspatis GA, Paraskeva K, Theodoroupolou A, et al (2006) A prospective, randomized comparison of adrenaline injection in combination with detachable snare versus adrenaline injection alone in the prevention of postpolypectomy bleeding in large colonic polyps. Am J Gastroenterol 101:2805; quiz 2913.
90. Kaltenbach T, Milkes D, Friedland S, et al (2008) Safe endoscopic treatment of large colonic lipomas using endoscopic looping technique. Dig Liver Dis Apr 21 Epub ahead of print.
91. Di Giorgio P, de Luca L, Calcagno G, et al (2004)

Detachable snare versus epinephrine injection in the prevention of postpolypectomy bleeding: a randomized and controlled study. Endoscopy 36:860–863.
92. Kazuhiko Shioji, Yutaka Suzuki, Masaaki Kobayashi, et al (2003) Prophylactic clip application does not decrease delayed bleeding after colonoscopic polypectomy. Gastrointest Endosc 57:691–694.
93. Harewood GC (2007) Prophylactic clip application after colonic polypectomy. Gastrointest Endosc 65:183; author reply 183.
94. Adams TL, Benjamin SB (2002) Complications of gastrointestinal endoscopy. In: Di Marino AJ, Benjamin SB (eds) Gastrointestinal Disease-An endoscopic approach. Second Edition, Slack incorporated. p. 65.
95. Avgerinos DV, Llaguna OH, Lo AY, et al (2008) Evolving management of colonoscopic perforations. J Gastrointest Surg 12(10): 1783-1789
96. Taku K, Sano Y, Fu KI, et al (2007) Iatrogenic perforation associated with therapeutic colonoscopy: a multicenter study in Japan. J Gastroenterol Hepatol 22:1409–1414.
97. American Heart Association (2007) AHA guideline. Prevention of infective endocarditis. Circulation 10:1736–1754.

Follow-up After Endoscopic Polypectomy

14

Walter Piubello, Fabrizio Bonfante, Irene Zagni and Morena Tebaldi

Abstract Follow-up after endoscopic polypectomy should be led by the number, size, and histopatologic characteristics of the polyps at the baseline colonoscopy. In a patient with large sessile polyps and incomplete excision, a follow-up colonoscopy at 2–6 months is suggested. After curative resection of invasive carcinoma or malignant polyps, intense colonoscopy follow-up at 1 year is mandatory. Patients with high-grade dysplasia adenoma, villous histology, polyps larger than 10 mm, or more than three polyps have an intermediate risk of advanced neoplasia, and surveillance at 3 years is recommended. Patients with 1–2 small tubular adenomas should undergo surveillance colonoscopy at 5–10 years; in patients with negative or sporadic hyperplastic polyps at baseline colonoscopy, surveillance is now suggested after 10 years.

Keywords Advanced adenoma • Colonoscopic quality • Colorectal polyps • Hyperplastic polyps • Malignant polyps • New guidelines after polypectomy • Surveillance colonoscopy

14.1 Introduction

In this cost-conscious era, all economically advanced countries are having difficulty coping with the consequences of advancing technology that can increase the overall welfare cost.

It is clear that the cost of polypectomy surveillance is now a large part of endoscopic practice, draining resources for screening and diagnosis. In addition, some surveys have shown that a large proportion of endoscopists and primary care physicians are conducting surveillance examinations at shorter intervals than recommended in guidelines [1–3].

This suggests an overuse of surveillance colonoscopy, which constitutes approximately one-quarter of the procedures performed in the USA [4]. Factors influencing how often surveillance colonoscopy should be carried out after polypectomy are patients' fear of harboring undetected cancer, and physicians' fear of missing an opportunity to prevent cancer. Moreover, Van Rijn [5] observed that the first colonoscopy missed polyps in about 22% of patients (2.1% >1 cm; 26% <6 mm).

Finally, the evidence provides little guidance on optimal surveillance frequency [6].

14.2 Diagnostic Tests After Polypectomy

Even if it is not an infallible "gold standard", colonoscopy should today be considered the best way

W. Piubello (✉)
Department of Internal Medicine and Digestive Endoscopic Unit, Desenzano Hospital, Desenzano del Garda (BS), Italy

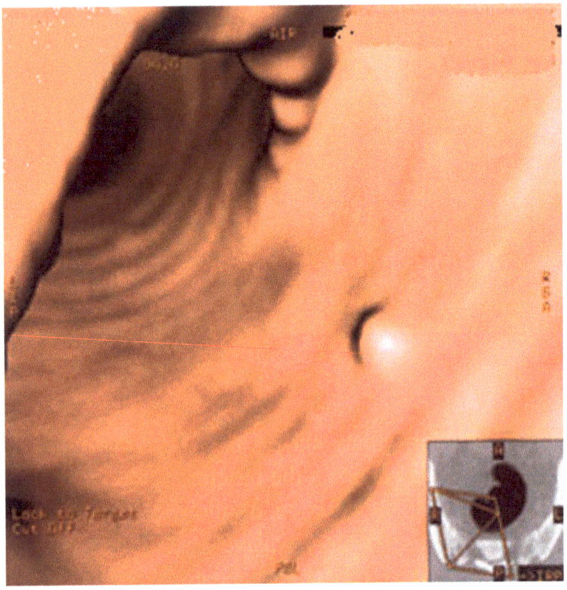

Fig. 14.1 Virtual coloscopy: small polyp

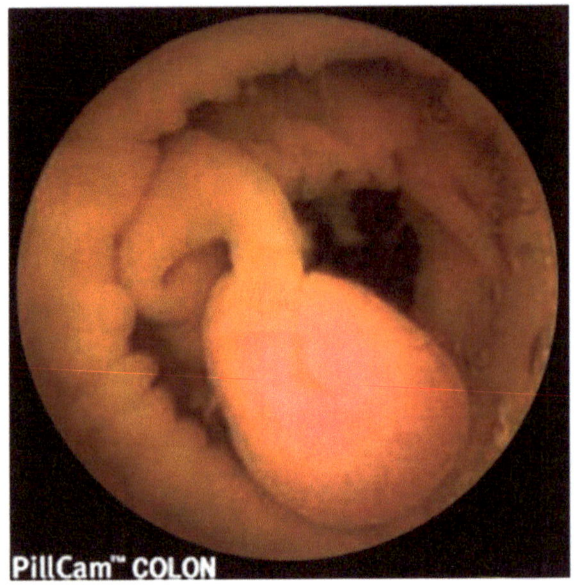

Fig. 14.2 PillCam colon capsule: adenoma

to follow up patients after polypectomy. High quality depends mainly on complete visualization of the entire colonic mucosa, the ability to detect and remove polyps safely, and timely and appropriate management of adverse events. Individuals should be informed about the limitations of colonoscopy, including the very important fact that it may miss some cancers and significant adenomas. Controlled studies have, in fact, shown that the colonoscopy "miss rate" for large adenomas is 6–12%, and for cancer it is about 5% [7,8]. In addition, during the colonoscopy there is a risk, albeit small, of complication such as perforation or hemorrhage (following polypectomy), with subsequent hospitalization and, in rare circumstances, more-serious harm. Sedation is usually used to minimize discomfort during the examination, and thus a companion is required to provide transportation after the examination.

Because of its low sensitivity for adenomas (48–73%) [9,10], air-contrast barium enema should not be proposed in follow-up patients after polypectomy, even though it is a relatively safe procedure with a lower perforation rate when compared with colonoscopy (1 in 25,000 versus 1 in 1000–2000) [11].

Virtual colonoscopy is a minimally invasive imaging examination of the entire colon and rectum taking approximatively 10 min, with no sedation or recovery time. However, this technique has not yet been validated in a multicenter trial (Fig. 14.1) [12].

A new modality for colonic evaluation, PillCam colon capsule endoscopy (Fig. 14.2) has much improved results when compared with conventional colonoscopy, with a current "miss rate" of 25–40% for smaller polyps [13].

14.3 Colonoscopy Follow-up Over the Years

In the last few years the concept of colonoscopic surveillance after polypectomy has changed.

Since adenomatous polyps are the most common neoplastic findings discovered when colonoscopy is performed, in the 1970s it was common practice for colonoscopy patients to have annual follow-up surveillance examinations to detect additional new adenomas and/or missed synchronous adenomas. The National Polyps Study report in 1993 and The Funnen Adenoma Follow-up Study in 1995 showed clearly, in a randomized design, that the first post-polypectomy examination could be deferred for 3 years [14,15]. In 2003 these guidelines were updated, and colonoscopy was recommended as the only follow-up examination. Stratification at baseline into low risk (1–2 small adenomas

<10 mm) and higher risk (>3 small adenomas or advanced adenomas) for subsequent adenomas was suggested [4].

14.4 Factors Influencing Colonoscopic Quality

The most recent guidelines for surveillance intervals after polypectomy need to clarify the key questions, i.e. whether baseline colonoscopic findings can predict recurrent advanced neoplasia. It has been shown that the first screening colonoscopy and polypectomy produces the greatest effects on reducing the incidence of colorectal cancer in patients with adenomatous polyps [16].

The two main factors influencing high quality at baseline colonoscopy are excellent patient preparation [17] and adequate withdrawal time. When the withdrawal time is less than 6 min, the polyp detection rate decreases from 28% to 11.8%, and the advanced neoplasia detection rate from 6.4% to 2.6% [18].

14.5 Advanced Adenoma

The current definition of advanced adenoma includes adenomas with high-grade dysplasia (Fig. 14.3), villous histopathology (Fig. 14.4), size >10 mm, or >3 tubular adenomas at index colonoscopy.

The relative risk of incidence of advanced adenomas at 3-year colonoscopy is 2.52 (95% confidence interval (CI) 1.07–5.97) for patients with ≥3 adenomas at index colonoscopy.

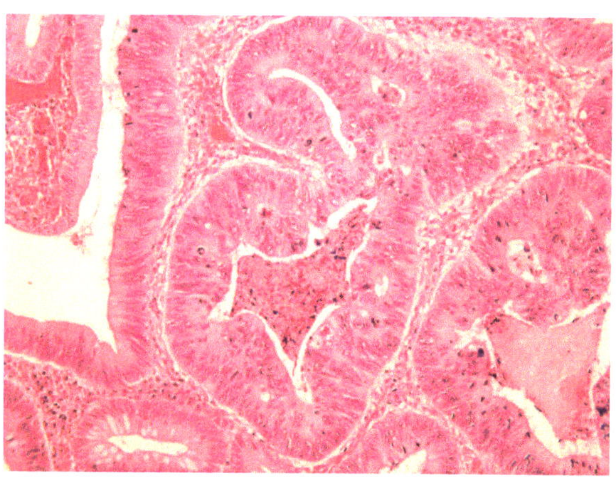

Fig. 14.3 Tubular adenoma with high-grade dysplasia

The relative risk for patients with high-grade dysplasia is 1.84 (95% CI 1.06–3.19) [19].

14.6 Malignant Colorectal Polyps

Invasive carcinoma may be found in approximately 2–4% of colonic polyps removed endoscopically (Fig. 14.5). Polypectomy may be curative in selected, superficially invasive colon cancer [20].

A malignant polyp is defined as one containing invasive carcinoma penetrating trough the muscularis mucosa into the submucosa. In this case close collabo-

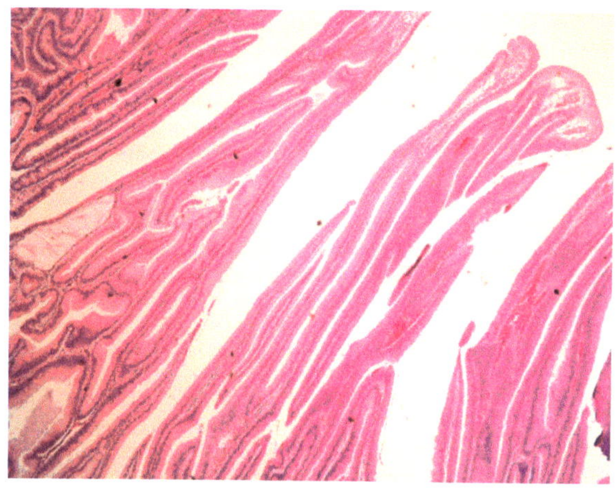

Fig. 14.4 Villous adenoma

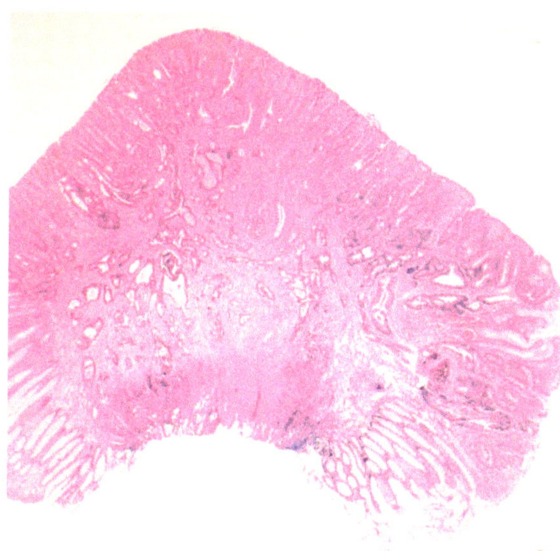

Fig. 14.5 Tubular adenoma with invasive carcinoma

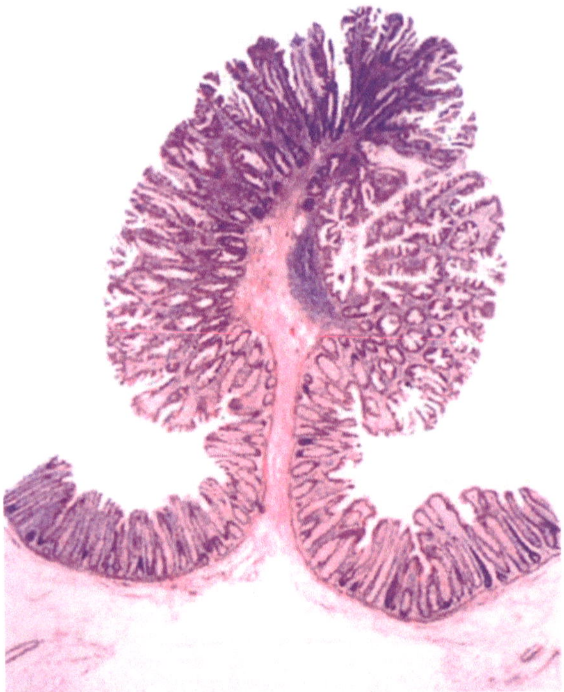

Fig. 14.6 Hyperplastic polyp

ration between pathologist and surgeon is necessary. In all cases of large polyps in younger patients (<50 years), surgical treatment should be considered because the risk of local lymph node metastasis is very high [21].

14.7 Hyperplastic Polyps

Although there is some controversy regarding the clinical significance of hyperplastic polyps (Fig. 14.6), they do not appear to increase the risk of proximal neoplasia or proximal advanced neoplasia in asymptomatic individuals undergoing screening [22]. However, patients with hyperplasic polyposis syndrome may have an increased risk of colorectal cancer [23].

14.8 The New Guidelines for Colonoscopic Follow-up After Polypectomy

After the detection and removal of colonic polyps, surveillance using colonoscopy can detect new polyps, even if it is evident that most patients with adenomas will not benefit from surveillance. It is known that 30–50% of adults will develop colon adenomas during their lifetime, but only 6% will develop colorectal cancer [4].

Recently, two major articles were published about this problem. Lieberman and colleagues [24] reported the 5.5-year risk for advanced neoplasia in a cohort of 1171 veterans who underwent screening colonoscopy. The risk was 2.4% when no baseline adenomas were detected, 4.6% in those with one or two small tubular (<10 mm) adenomas, 11.9% in those with three or more small tubular (<10 mm) adenomas, 15.5% in those with one large tubular adenoma (>10 mm), and 16–17% in those with villous adenomas or high-grade dysplasia.

Laiyemo and colleagues [25] used data from the Polyps Prevention Trial to asses the clinical utility of the recommended surveillance intervals. Among the entire cohort of 1905 patients, the overall risk for an advanced adenoma of 4 years was 6%. Among the 715 patients with baseline high-risk adenomas (1 advanced adenoma or >3 adenomas of any size), the 4-year risk for an advanced adenoma was 9%. The remaining 1190 patients had a 5% risk. The 4-year risk for advanced adenomas was higher for people with an advanced adenoma at baseline (9% versus 5%), a villous polyp (12% versus 5%), a large adenoma (8% versus 6%), or a high-grade dysplastic polyp (10% versus 6% low-grade dysplasia). All the differences were statistically significant. Moreover, this study shows two interesting predictors of advanced adenoma. Firstly, patients with proximal adenomas had a higher risk than those with distal adenoma (9% versus 5%). Patients with two non-advanced adenomas were at higher risk (9%) if at least one was proximal, compared with a single adenoma anywhere, or two distal adenomas. Secondly, patients with three or more non-advanced adenomas had the same 6% risk of advanced adenoma as the entire cohort.

Although the advanced adenoma is the driver of a polyp surveillance policy, far too little is known about its natural history. In a longitudinal study in 226 patients with large polyps, followed by serial barium enemas, the rate of progression to cancer was about 1% per year [26].

In the study of Laiyemo and colleagues [25], a villous polyp at baseline was the only independent endoscopic predictor of recurrence, and villous polyps are more likely to express mutations of both ki-ras and p53, the two mutation involved in the adenoma-to-carcinoma sequence [27]. This suggests that we will need to consider other predictors than the baseline colonoscopy to identify a large subgroup with a less than 1–2% probability of recurrence of advanced adenoma [5].

Another major problem of follow-up after polypectomy is the compliance of the patients. In a study conducted by Siddiqui and colleagues [28], the compliance rate was significantly associated with previous colonoscopy, previous polyps (size, numbers), and asymptomatic status. Independent predictors for follow-up colonoscopy at multivariate analysis were statin use, first-degree relatives with colon cancer, and compliance with outpatient clinic follow-up.

It is known, moreover, that the compliance with follow-up colonoscopy may affect the estimates of reduction in colorectal cancer incidence as demonstrated by the National Polyp Study. In this study the compliance was 80%, although the real compliance rate for follow-up colonoscopy in the general population is unknown [29].

Table 14.1 summarizes the guidelines for follow-up after polypectomy [30,31]. In patients with sessile adenomas removed piecemeal (Fig. 14.7), subsequent surveillance needs to be individualized based on the endoscopist's judgement. Completeness of removal should be based on both endoscopist and pathologist assessement. In patients with more than ten adenomas (Fig. 14.8) on a single examination, the possibility of an underlying familial syndrome should be considered, and colonoscopy can be proposed for first-degree relatives. In patients with between three and ten adenomas, or one adenoma >10 mm (Fig. 14.9), or any adenoma with villous features (Fig. 14.4), or high-grade dysplasia (Fig. 14.3), all adenomas must be completely removed. If the follow-up colonoscopy is normal, the interval for subsequent examination should be 5 years.

Table 14.1 Follow-up after colonoscopy polypectomy

Baseline colonoscopy	Follow-up
Large sessile polyp with incomplete excision	2–6 months
Malignant polyp (or colon cancer)	1 year curative (then 3 years and 5 years if results normal)
Advanced neoplasia (high-grade dysplasia, villous histology, size >10 mm), or 3–10 tubular adenomas	3 years
Small tubular adenoma (≤2, <10 mm)	5–10 years
Hyperplastic polyp (sporadic) or negative surveillance polyp	10 years

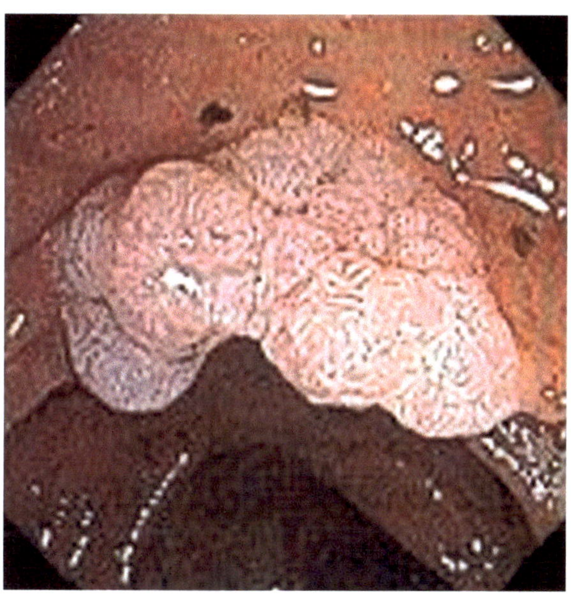

Fig. 14.7 Sessile adenoma

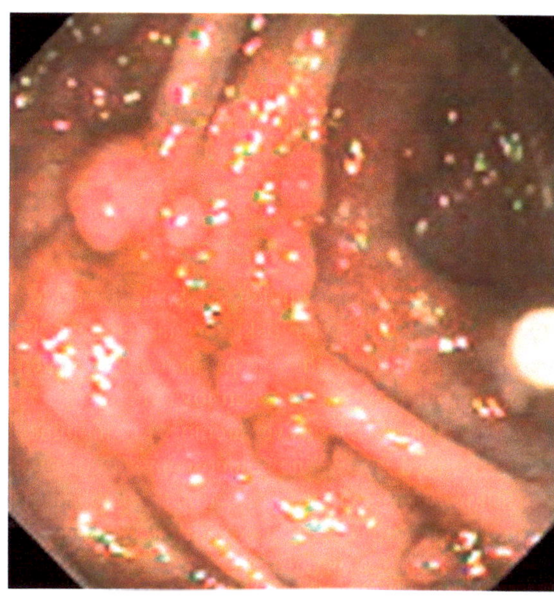

Fig. 14.8 Several polyps (familial polyposis syndrome)

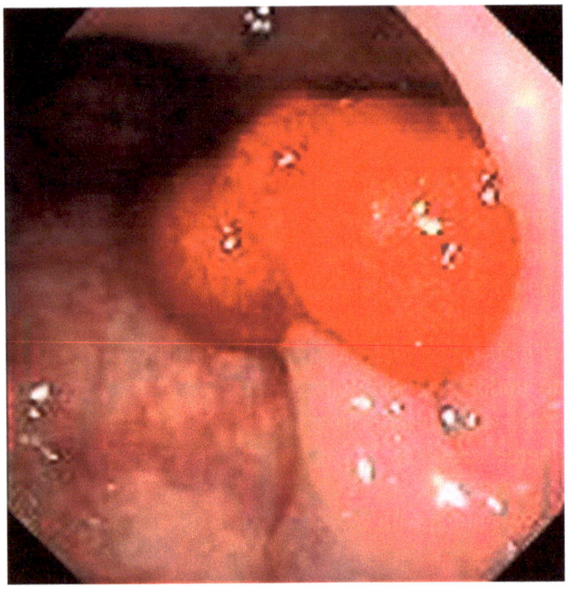

Fig. 14.9 Pedunculated adenoma

Finally, in the patients with one or two small tubular adenomas with low-grade dysplasia, the precise time interval for follow-up should be based on other clinical factors (such as prior colonoscopy findings, family history, preferences of the patient, and judgment of the physician). Moreover, colonoscopy in patients undergoing curative resection for colon or rectal cancer at 1 year is suggested in addition to perioperative colonoscopy for synchronous tumors. If the examination after 1 year is normal, then the interval before the next examination should be 3 years. If that colonoscopy is normal, then the interval before the subsequent examination should be 5 years.

14.9 Conclusion

The aim of colonoscopy guidelines should be to identify individuals at higher risk, and target them for surveillance, similar to the risk-prediction recommendations for other disease (e.g. cardiovascular).

Patients who have invasive carcinoma, or malignant polyps detected at baseline colonoscopy, need intensive follow-up after curative resection to establish that all important lesions were discovered at baseline and completely removed, and that they do not develop new significant pathology after a short time.

The new guidelines suggest that repeating colonoscopy is recommended within 1 year if cancer is detected at baseline, even if a complete examination was performed at the time of resection. Patients with adenomas with high-grade dysplasia, villous histology, size >10 mm, or >3 tubular adenomas have an intermediate risk of advanced neoplasia during surveillance. In these patients, surveillance at 3 years is suggested if the initial examination was complete and adequate. Patients with one or two small adenomas (<10 mm) have a lower risk of advanced neoplasia within 5 years. Surveillance is now recommended at 5–10 years in this group.

In patients with negative or sporadic hyperplasic polyps at baseline colonoscopy, surveillance is now recommended at 10 years.

Finally, because it is very likely that important lesions are missed or incompletely removed at the baseline colonoscopy and then discovered during surveillance, this may represent an important limitation of optical colonoscopy as it is currently performed.

Future advances in optics or the use of chromoendoscopy may enhance the ability to identify neoplasia on the first examination, although it is yet to be proven that this improves clinical outcomes.

References

1. Mysliwiec PA, Brown ML, Klabunde CN, Ransohoff DF (2004) Are physicians doing too much colonoscopy? A national survey of colorectal surveillance after polypectomy. Ann Intern Med 141:264–271.
2. Boolchand V, Olds G, Singh J, et al (2006) Colorectal screening after polypectomy: a national survey study of primary care physicians. Ann Intern Med 145:654–659.
3. Klabunde CN, Frame PS, Meadow A, et al (2003) A national survey of primary care physicians' colorectal cancer screening recommendations and practices. Prev Med 36:352–362.
4. Winawer S, Fletcher R, Rex D, et al; Gastrointestinal Consortium Panel (2003) Colorectal cancer screening and surveillance: clinical guidelines and rationale – update based on new evidence. Gastroenterology 124:544–560.
5. Van Riyn JC, Reitsma JB, Stoker J, et al (2006) Polyp miss rate determined by tandem colonoscopy: a systematic review. Am J Gastroenterol 101:343–350.
6. Imperiale TF, Sox HC (2008) Guidelines for surveillance intervals after polypectomy: coping with the evidence. Ann Intern Med 148:477–479.
7. Pickhardt PJ, Nugent PA, Mysliewiec PA (2004) Location of adenomas missed by optical colonoscopy. Ann Intern Med 141:352–359.
8. Bressler B, Paszat LF, Vinden C (2004) Colonoscopic miss

rates for right-sided colon cancer: a population-based analysis. Gatroenterology 127:452–456.
9. Williams CB, Macrae FA, Bartram CI (1982) A prospective study of diagnostic methods in adenoma follow-up. Endoscopy 14:74–78.
10. Winawer SJ, Stewart ET, Zauber AG (2000) A comparison of colonoscopy and double contrast barium enema for surveillance after polypectomy. National Polyps Study Work Group. N Engl J Med 352:1766–1772.
11. Blakeborough A, Sheridan MB, Chapman AH (1997) Complications of barium enema examinations: a survey of UK consultant radiologists 1992 to 1994. Clin Radiol 52:142–148.
12. Iannacone R, Laghi A, Catalano C (2004) Computed tomography colography without cathartic preparation for the detection of colorectal polyps. Gastroenterology 127:1300–1311.
13. Eliakim R, Fireman Z, Gralnek IM, et al (2006) Evolution of the PillCam colon capsule in the detection of colonic pathology: results of the first multicenter, prospective comparative study. Endoscopy 38:963–970.
14. Winawer SJ, Zauber AG, O'Brien MJ, et al (1993) Randomized comparison of surveillance intervals after colonoscopic removal of newly diagnosed adenomatous polyps. The National Polyp Study Workgroup. N Engl J Med 328:901–906.
15. Jørgensen OD, Kronborg O, Fenger C (1995) A randomized surveillance study of patients with pedunculated and small sessile tubular and tubulovillous adenomas. The Funen Adenoma Follow-up Study. Scand J Gastroenterol 30:686–692.
16. Winawer SJ, Zauber AG, Fletcher RH, et al; US Multi-Society Task Force on Colorectal Cancer; American Cancer Society (2006) Guidelines for colonoscopy surveillance after polypectomy: a consensus update by the US Multi-Society Task Force on Colorectal Cancer and the American Cancer Society. Gastroenterology 130:1872–1885.
17. Ben-Horin S, Bar-Meir S, Avidan B (2007) The impact of colon cleanliness assessment on endoscopists' recommendations for follow-up colonoscopy. Am J Gastroenterol 102:2680–2685.
18. Barclay RL, Vicari JJ, Doughty AS, et al (2006) Colonoscopic withdrawal times and adenoma detection during screening colonoscopy. N Engl J Med 355:2533–2541.
19. Saini SD, Kim HM, Schoenfeld P (2006) Incidence of advanced adenomas at surveillance colonoscopy in patients with a personal history of colon adenomas: a meta-analysis and systematic review. Gastrointest Endosc 64:614–626.
20. Morson BC, Whiteway JE, et al (1984) Histopathology and prognosis of malignant colorectal polyps treated by endoscopic polypectomy. Gut 25:437–444.
21. Hassan C, Zullo A, Winn S, et al (2007) The colorectal malignant polyp: scoping a dilemma. Dig Liver Dis 39:92–100.
22. Lin OS, Gerson LB, Soon MS, et al (2005) Risk of proximal colon neoplasia with distal hyperplastic polyps: a meta-analysis. Arch Intern Med 165:382–390.
23. Hyman NH, Anderson P, Blasyk H (2004) Hyperplastic polyposis and the risk of colorectal cancer. Dis Colon Rectum 47:2101–1204.
24. Lieberman DA, Weiss DG, Harford WV, et al (2007) Five-year colon surveillance after screening colonoscopy. Gastroenterology 133:1077–1085.
25. Laiyemo AO, Murphy G, Albert PS, et al (2008) A Post-polypectomy colonoscopy surveillance guidelines: predictive accuracy for advanced adenoma at 4 years. Ann Intern Med 148:419–426.
26. Stryker SJ, Wolff BG, Culp CE, et al (1987) Natural history of untreated colonic polyps. Gastroenterology 93:1009–1013.
27. Barry FL, Baron JA, Grau MV, et al (2006) K-ras mutation in incident sporadic colorectal adenomas. Cancer 106:1036–1040.
28. Siddiqui AA, Patel A, Huerta S (2006) Determinants of compliance with colonoscopy in patients with adenomatous colon polyps in a veteran population. Aliment Pharmacol Ther 24:1623–1630.
29. Colquhoun P, Chen HC, Kim JI, et al (2004) High compliance rates observed for follow up colonoscopy post polypectomy are achievable outside of clinical trials: efficacy of polypectomy is not reduced by low compliance for follow up. Colorectal Dis 6:158–161.
30. Davila RE, Rajan E, Baron TH, et al; Standards of Practice Committee American Society for Gastrointestinal Endoscopy (2006) ASGE guideline: colorectal cancer screening and surveillance Gastrointest Endosc 63:546–557.
31. Levin B, Lieberman DA, McFarland B, et al (2008) Screening and surveillance for the early detection of colorectal cancer and adenomatous polyps, 2008: a joint guideline from the American Cancer Society, the US Multi-Society Task Force on colorectal cancer and The American College of Radiology.Gastroenterology 134:1570–1595.

Surgical Options for Familial Adenomatous Polyposis

15

Gian Gaetano Delaini, Chiara Zugni, Tania Magro, Filippo Nifosì, Maurizio Mainente and Gianluca Colucci

Abstract Familial adenomatous polyposis (FAP) is caused by a mutation of *APC* gene. The use of prophylactic surgery has significantly changed patients' destiny, avoiding the development of colorectal cancer. It is now clear that the different patterns of FAP are based on different *APC* mutations. For this reason, it is important to offer a tailored approach, where the type of operation and the timing is based on aggressiveness of the disease and the risk of other manifestations such as desmoid tumors. Furthermore, it is also important to take into account the wishes and expectations of these patients. For example, we are increasingly coming across asymptomatic young women with FAP who want to have children and who are worried about the consequences of surgery for pregnancy. The decision process is not always easy, especially since there is often not a unique right answer. The final choice should often be taken by the patients, guided but not forced by the surgeon.

Keywords Extracolonic manifestation • Genetic • Ileorectal anastomosis • Pouch • Proctocolectomy

15.1 Introduction

Familial adenomatous polyposis (FAP) is an autosomal dominant disease characterized by the development of hundreds of adenomatous polyps distributed throughout the colon and rectum, and various extracolonic lesions such as upper gastrointestinal tract polyps (gastric, duodenal, periampullary), desmoids, and retinal lesions [1,2]. FAP is based on mutations in the *APC* gene (on chromosome 5), which encodes a protein involved in the regulation of intracellular signal transmission, and whose alteration causes a constant activation and transcription of growth factors [3]. The number of colorectal polyps varies from a minimum of 100 (necessary for diagnosis) up to more than 7000, with the onset of disease in the second and third decade of life (although cases of onset in children or in 60-year-old individuals are reported) [3]. This pattern of disease implies that people affected by FAP are at 100% risk of developing colorectal cancer within the third and fourth decade of life [1]. Therefore nowadays, the optimal management of this disease consists of a "prophylactic colectomy" to be performed in asymptomatic carriers in order to avoid the development of cancer.

The ideal intervention for such an aim should have the following characteristics: preserving fecal continence, sparing the pelvic innervation to the sexual organs, and eliminating the risk of cancer, while minimizing postoperative morbidity and mortality [4].

G.G. Delaini (✉)
Department of Surgery and Gastroenterology, University of Verona, Verona, Italy

15.2 Surgical Options

There are three main surgical options in people affected by FAP: proctocolectomy with ileal pouch-anal anastomosis (the so called restorative proctocolectomy) (IPAA), total colectomy with ileorectal anastomosis (IRA), and proctocolectomy with definitive terminal ileostomy [5].

Restorative proctocolectomy with ileal pouch anal anastomosis was introduced at the beginning of the 1980s, and today is considered as the gold standard procedure for FAP.

Since every epithelial cell in the colorectal mucosa is affected by the *APC* mutation, the aim of surgery should be the complete removal of the tissue at risk for cancer [6]. IPAA fulfils this requirement as it removes all the diseased colorectal mucosa, and at the same time provides an acceptable quality of life by avoiding ileostomy [7,8].

The concept of restoring bowel continuity after proctocolectomy in order to avoid a permanent ileostomy and retain fecal continence is quite old. Since the beginning of the 20th century, many authors have tried to achieve such an aim through different types of intervention, but often without acceptable functional results. Between the end of the 1970s and the beginning of the 1980s, several authors published quite large series of patients affected by ulcerative colitis or FAP and undergoing restorative proctocolectomy with good functional results [9–12]. The idea of "building" a reservoir in order to collect and dehydrate the stool and to reduce the defecation frequency was introduced in 1950 by Champeau who first performed an ileal reservoir [10,11,13–15]. Since then, different shapes of ileal reservoirs have been considered: J, S, W, U, and H shape. Nowadays the most common type of reservoir combining good functional results, a relatively easily performed technique, and a low rate of long-term complications is the J-shaped pouch, which was first described by Utsonomiya [6,16].

The first cases of restorative proctocolectomy consisted of a hand-sewn ileal pouch-anal anastomosis performed with a perineal approach at the level of the dentate line. Before performing the suture, an endoanal mucosectomy was carried out in order to completely eliminate the pathological epithelium; this approach theoretically reduces to a minimum the risk of adenoma recurrence. At the same time, it is quite a complex technique, requiring a long procedure and an experienced surgeon. In addition, some authors experienced that the use of anal retractors, which is necessary to perform the endoanal mucosectomy and the hand-sewn anastomosis, increases the rate of fecal incontinence, especially night soiling [17,18].

Double-stapled ileal pouch anal anastomosis does not require previous mucosectomy, and appears to be easier to perform from the technical point of view. That is why this technique, which has been introduced more recently, has widely diffused to all centers and is chosen by the majority of surgeons [19,20]. Nowadays a matter of discussion is whether to choose the hand-sewn anastomosis or double-stapled technique. The former requires surgical experience and an extended surgical time but grants an excellent result from the pathological point of view, while the latter technique is simpler, faster, and executable by most surgeons in a reasonably short time, but is not "complete" in pathological epithelium removal as a 1.5–2 cm cuff of rectal mucosa remains in place [21–24]. Van Duijvendijk et al [24] compared two series of patients who underwent IPAA with mucosectomy and hand-sewn anastomosis versus a double-stapled anastomosis without mucosectomy respectively; according to their observations 31% of patients of the second group developed adenomas in the site of anastomosis within 7 years versus 10% of those in the first group.

15.3 Ileostomy

The IPAA procedure originally also included a temporary loop ileostomy in order to prevent septic complications linked to the anastomosis leakage [6,16,23,25–28]. Another advantage of ileostomy is the diversion of fecal transit in the immediate postoperative period when a transient incontinence can be present, due to both sphincter distress and the liquid consistency of stool. Metcalf et al, in 1986, first tried to omit the ileostomy in a group of selected patients and reported good results in term of complications after one-time IPAA [29]. These results were confirmed by other studies, which reported almost the same rate of septic complications in patients undergoing a one-step procedure (without diverting ileostomy) or a two-step procedure [30–34]; apart from this study, only one prospective randomized study showing these results has been performed, by Grobler et al [35], so that at present the construction of a temporary ileostomy is usually performed except for selected patients who are considered at low risk of anastomotic complications. The selection of patients who can avoid ileostomy is based on their general health status (including age, co-morbidity, surgical history, and nutrition status) and on surgical technical aspects (how adequately vascularized and without tension the anastomosis is at the end of the intervention) [30,31,33,34,36–38]. In particular, this latter aspect is a basic one: lack of tension on the anastomosis is the main element to consider for avoiding ileostomy [30]. Mesentery-lengthening techniques have been suggested by many authors in order to reduce the tension on the pouch-anal anastomosis [39–41]. In patients undergoing IPAA for FAP, the ileostomy could be theoretically avoided in most cases, since these are usually young and otherwise healthy patients with a good performance status and without inflammatory complications in the site of intervention [42,43].

IPAA should be performed in every patient affected by FAP, particularly those with more than 20 rectal polyps or more than 1000 colonic polyps, those with an adenocarcinoma anywhere in the colon or rectum, and those with large rectal (more than 3 cm) or severe dysplastic rectal adenomas [5,7,43–45].

Subtotal colectomy with ileorectal anastomosis is a simple and less complicated procedure with a rapid postoperative recovery and good functional results [5].

Before the early 1980s, when the IPAA procedure was not yet introduced and the choice was between subtotal colectomy with IRA and *proctocolectomy with definitive ileostomy*, the IRA option was often performed in order to avoid a mutilating operation with a permanent ileostomy resulting in a low quality of life in young people [2]. Since restorative proctocolectomy has been introduced and has widely spread to many centers, the indications for IRA have become more limited. IRA is a reasonable option in patients with fewer than 20 rectal adenomas or fewer than 1000 colonic adenomas, in attenuated FAP [43,45], and in young women before pregnancy [6,46]. According to some authors, it grants a better postoperative quality of life than proctocolectomy with IPAA, but on the other hand it does not completely eliminate the risk for rectal cancer since it does not completely remove the pathological epithelium.

When a rectal stump is left in place (IRA), a strict follow-up is mandatory. The risk of dying from rectal cancer after IRA is 12.5% by age 65 years; when compared with IRA, IPAA would increase life expectancy by 1.8 years [1]. The patients must be informed in detail about the importance of follow-up after IRA, and only compliant individuals can be candidates for the intervention. Vasen et al showed that 75% of patients with previous IRA who developed a cancer on the rectal stump had a negative rectoscopy within 12 months before the diagnosis [1]. Nugent and Phillips suggest a follow-up with rectoscopy and biopsies every 4–6 months [44,47–50]. If a high-grade dysplastic adenoma or a carcinoma is found at endoscopic follow-up, or a large number of polyps without the possibility of complete endoscopic removal is found, conversion to IPAA is necessary. On the other hand, if a low rectal cancer is found at follow-up after IRA, the abdominoperineal resection must be performed with sphincter excision.

Conversion from IRA to IPAA is not always possible because of technical problems: if desmoids are present or the mesentery is short, it may not be possible to perform the procedure.

As the risk for developing desmoids increases at every intervention the patient undergoes, the probability of finding a desmoid tumor is higher when proctectomy and ileal pouch-anal anastomosis is performed as the second step of surgical treatment in a patient who has previously undergone IRA. Mutation within codons 1445–1580 is connected to a higher risk of the development of desmoids after surgery [51], and this must be considered when there is a choice between IRA and IPAA.

Patients who may be eligible for IRA as the first intervention must be informed not only about the strict follow-up they must undergo, but also about the fact that IPAA as a second-step intervention can be more difficult to perform from a technical point of view. Aside from this, it can be attributed to a higher rate of complications, in particular short-term complications such as wound infection.

Nowadays proctocolectomy with definitive ileostomy is rarely performed as the procedure of choice. It was the definitive surgical treatment before the introduction of IPAA, but is now only reserved for when cancer in the lower third of the rectum is found at the first intervention, whenever a rectal cancer arises after the IPAA procedure, or when IPAA demolition is required [5]. Definitive ileostomy is also necessary whenever the anatomical conditions do not permit use of the ileum to build a pouch, for example due to shortness of the mesentery.

It is difficult for patients to accept this procedure because they are usually young and asymptomatic [5,52].

A valid alternative to conventional definitive ileostomy in selected patients is the continent ileostomy, or Kock continent ileostomy, which offers the comfort of avoiding an external appliance but can be attributed to a high rate of re-operation [53].

15.4 Genetics and Familial Adenomatous Polyposis

Surgical options for FAP should be based not only on clinical criteria but also on the genetic aspects [54]. In fact, the kind of *APC* mutation determines the synthesis of different types of altered protein associated with different patterns of disease; this fact should be considered in making a decision about the kind of intervention to perform [44]. FAP severity is defined according to the number of polyps in the colon: if fewer than 1000 polyps are found, FAP is called mild or attenuated FAP (aFAP), if more than 1000 polyps are present at colonoscopy, the disease is considered severe [55].

Many authors [6,43,50,56,57] have found that *APC* germline mutations between codons 1250 and 1465 are associated with severe disease; in particular, Wu et al [55] observed that mutations in codons 1309 and 1328 in exon 15G always cause a severe polyposis phenotype with thousands of adenomas. *APC* mutations occurring in the 5' end of the gene (in particular exons 3 and 4) are associated with the attenuated form of FAP, which is characterized by fewer polyps and a later onset of cancer [6,7,58,59]. In attenuated FAP, a higher rate of right-sided polyps rather than rectal polyps has been observed [60]. Total colectomy with ileorectal anastomosis appears to be the appropriate treatment for people affected by the attenuated phenotype of disease [8,58,60–62]. To date, a matter of discussion is whether to perform a colectomy with IRA or a restorative proctocolectomy, in patients carrying a mutation related to a potentially severe genotype with high risk of rectal cancer development, but without important rectal involvement at the time of surgery [2].

There is no doubt that people with severe rectal involvement (more than 20 adenomas) or colonic involvement (more than 1000 adenomas), those with a severe dysplastic rectal polyp, or a cancer anywhere in the large bowel or a large (more than 3 cm) rectal adenoma, should have a primary restorative proctocolectomy [43,45,63,64].

15.5 Fertility, Pregnancy, and Delivery

Patients with FAP are usually young and childless at the time of operation [65,66]. The majority of reports about fertility after IPAA are related to women affected by ulcerative colitis; studies about fertility and FAP are still rare and inconclusive [66,67]. Olsen et al report a decrease in fertility rate (54%) after IPAA [65]. Gorgun et al studied 300 women, 206 of whom were attempting to conceive, and observed a 38% failure rate before the operation, rising to 56% after the intervention [68]. The reason for the decrease in fertility rate could be related to the effect of the surgical procedure on pelvic anatomy. According to this hypothesis, Oresland et al used a postoperative hysterosalpingography and discovered unilateral (43%) and bilateral (10%) occlusion of the fallopian tubes, and adhesion of the tubes to the pelvis (48%) [69]. The desire for future pregnancy and the postoperative decrease of fertility rate are two basic aspects to consider before surgery. Patients should be informed in detail about this problem, and the best timing of surgical procedure should be discussed and decided together with the patient.

All published studies have observed that restorative proctocolectomy with IPAA is compatible with a safe pregnancy and normal vaginal delivery [70–76].

Many authors have noted an increase in stool frequency during the third trimester of pregnancy, persisting for about three months after delivery [70–74,77,78].

With regard to vaginal delivery, the risk for sphincteric injury should be considered, particularly if an episiotomy is performed [73].

Vaginal delivery should be avoided in patients with a scarred and non-compliant perineum; otherwise, the method of delivery should be dictated by obstetric considerations [74,79,80].

15.6 Urinary Dysfunction

Many authors suggest that urinary and sexual dysfunction are technique dependent. There are two different techniques for proctectomy: the "close rectal wall" dissection and the "total mesorectal excision" (TME). The first intervention consists of a dissection as close as possible to the rectum, while the second one is carried out in the anatomical plane between the mesorectum and the presacral fascia.

The St-Antoine's group has used the first approach and reported a 1.7% rate of transient postoperative dysuria and urinary retention, 0.6% transient impotence, and 0% retrograde ejaculation [16]. Kartheuser et al have also achieved good functional results by using the "close rectal wall technique" [81]; so it seems that urinary and sexual problems can be reduced to a minimum by choosing the first technique of dissection [6,16,81].

TME dissection must be used in cases of rectal cancer or high-grade dysplasia; in all these cases the rate of urinary and sexual disturbance is higher by 10% or more [75,82,83].

Colwell and Grey found a rate of 0.5–1.5% for erectile dysfunction and 3–4% for ejaculatory dysfunction in males who had undergone a TME dissection; this was probably due to denervation of the pelvic plexus [84].

Regarding women, the same authors noted a postoperative dyspareunia rate of 3–22%.

The cause of this dysfunction is injury to the autonomic nerves and also the removal of the rectum, which changes the pelvic anatomy [75,84,85]. In fact, Metcalf et al found that dyspareunia is worse in patients with proctectomy and terminal ileostomy than in patients with an ileal pouch filling the space posterior to the vagina [85]. In addition, sexual relations are also inhibited by the fear of stool leakage [84].

Cornish et al, in a systematic review, evidenced many different causes for sexual dysfunction after ileal pouch anal anastomosis: dyspareunia, vaginal dryness, pain interfering with sexual pleasure, and fear of stool leakage.

According to many studies [16,86–91], the frequency of these problems should decrease as surgical experience and standardisation of the IPAA procedure improve.

15.7 Extracolonic Manifestation

FAP is characterized by an increased risk of developing desmoid tumors and small bowel tumors. In particular, neoplastic lesions can arise in the stomach, in the duodenum, in the ampullary and biliary area, and in the ileum, with particular emphasis on the ileal pouch itself and the ileostomy (when a permanent one has been performed) [3].

Almost 50% of people affected by FAP develop gastric polyps: about 40% of patients show fundic gland polyps (FGPs), while adenomas are less common with a 5% incidence [92,93]. FGPs are the most common sporadic gastric lesions, whose development seems to be related to the use of proton pump inhibitors (PPIs) and prolonged exposure to bilirubin [92,94]. FGPs in patients affected by FAP have the same histopathological appearance as sporadic FGPs, but a higher rate of dysplasia of every grade (up to 40% of FGPs in FAP patients show dysplastic aspects versus 3% among sporadic ones) [95–97]. Gastric adenomas usually develop on atrophic gastritis epithelium and may also be found in microscopic specimens of macroscopically normal mucosa [95,98]. Although gastric polyps are quite common in FAP patients, the incidence of severe dysplasia is very low (less than 6% in different series) [95,99], and cases of gastric adenocarcinoma arising on polyps are rare.

Duodenal and periampullary dysplasia is very common among patients affected by FAP. A study by Spigelman et al [99] on 102 FAP patients undergoing endoscopic surveillance reported a 92% incidence of duodenal dysplasia, and a 74% incidence of periampullary dysplastic biopsies. Of these cases, only 5% develop malignancies [100,101]. As the progression of duodenal polyps is not yet clear, a scoring system has been proposed by Spigelman to classify the extent of duodenal involvement and the subsequent management in terms of surveillance and/or treatment. This score cannot be applied to ampullary disease.

According to Kashiwagi and Spigelman, endoscopic surveillance on the upper gastrointestinal tract should be started at age 25 years. The endoscopic removal of polyps can be performed with Nd-YAG laser, argon plasma coagulation, and photodynamic therapy [102]. Pharmacological therapy can be performed with sulindac and celecoxib, which have shown a regression of duodenal and periampullary polyps [103–105].

Surgical procedures must be considered as the last option in patients with Spigelman stage IV. Open polypectomy through duodenotomy can be considered as the less-invasive surgical procedure. Pylorus-preserving pancreaticoduodencectomy and Whipple's procedures are usually reserved to deal with invasive neoplasms or whenever a definitive treatment for duodenal, periampullary, and ampullary disease is pursued. These surgical procedures have a high rate of complications, so the indication must be based on an accurate clinical evaluation.

Ampullary adenomas are quite common in FAP patients, with an incidence from 28% to 50% in different series [106,107]. Ampullary lesions can be symptomatic, with bile or pancreatic obstruction features, and can become malignant [106,108,109]. In fact, the risk for malignant transformation is not so high (lower than the rate of duodenal polyp progression to malignancy), and most ampullary adenomas have been found to remain benign over long-term follow-up [106,109,110]. Ampullary lesions can be treated by endoscopic ampullectomy when benign; malignant forms must be treated with surgical ampullectomy when severe dysplasia or focal areas of carcinoma develop, and the ampullectomy can be performed with a free-from-disease section margin or with pancreaticoduodenectomy when an infiltrating form is found.

Jejunal and ileal polyps in FAP patients are quite rare, and only found in exceptional cases where cancer arising from them is reported. Endoscopic removal is possible whenever these polyps are found with ileoscopy after restorative proctocolectomy.

Individuals with proctocolectomy and definitive ileostomy for FAP can develop inflammatory pseudopolyps on the mucosa of their stoma. Adenomas can also infrequently occur at this site, yet rarely become malignant [111,112].

Endoscopic surveillance of the ileal pouch in FAP patients who have undergone restorative proctocolectomy is mandatory. First we must consider the risk of developing adenomas and cancer in the epithelium of the rectal "cuff", which remains after ileal pouch-anal double-stapled anastomosis [24,113–115]. A rate of 28–31% dysplastic polyps has been estimated in the residual mucosa. After a hand-sewn anastomosis with previous mucosectomy, the rate of dysplasia is lower (10–14%) but not absent, and islets of rectal mucosa can remain and give rise to adenomatous transition. This is why every FAP patient must undergo endoscopic surveillance after IPAA, whatever technique of anastamosis was performed [24,113].

Secondly, the ileal pouch mucosa itself can develop adenomas; the estimated incidence of pouch polyps is about 60% [116,117], much higher than the incidence of ileal polyps, with a risk for dysplasia increasing with pouch age. This suggests that the intervention can act as a trigger towards the development of adenomas on the pouch epithelium, and the changing mucosal environment when a pouch is constructed with terminal ileum can itself act as a stimulus for polyps arising [116,118–120]. Endoscopic polypectomy can be performed when a polyp arises in the pouch. Pouch excision is necessary when severe dysplasia is found.

15.8 Conclusion

The treatment of patients with FAP is more complex than a simple surgical decision. Many factors need to be taken into account. The disease can have many different manifestations, the implications of which can guide us in making the right choices. Over the years, the increasing knowledge gained about the influence of genetics on the progression and extracolonic manifestation of FAP have given insight into different surgical options available, and have also have changed the choices available to surgeons. Of the techniques now available, we also have an increased knowledge of the functional outcomes of these techniques. For example, it is now more acceptable to perform an IRA as a first option, which can be converted into an IPAA, if needed, with good functional results.

The successes of the outcomes are highly dependent on the patient's level of participation, and on their compliance with the treatment program chosen. Patient compliance has become an increasingly important factor in the decision-making process, especially when less aggressive surgical choices are made with an increased need for a strict follow-up program. From this we can conclude that the decision should be made by a strict interaction between the surgeon and the patient, considering not only the technical factors involved in treatment but also the patient's compliance and desires. It is evident that more studies and data are required to consolidate existing knowledge on this topic.

References

1. Vasen HFA, van Duijvendijk P, Buskens E, et al (2001) Decision analysis in the surgical treatment of patients with

familial adenomatous polyposis: a Dutch-Scandinavian collaborative study including 659 patients. Gut 49:231–235.
2. Contessini-Avesani E, Botti F, Carrara A, et al (2006) Genetic mutations in FAP and conventional or laparoscopic surgical approach. In: Delaini GG (ed) Inflammatory bowel disease and familial adenomatous polyposis. Clinical management and patients' quality of life. Springer, Milan Berlin Heidederlberg, New York, pp 329–344.
3. Will OCC, Man RF, Phillips RKS, Tomlinson IP, Clark SK (2008) Familial adenomatous polyposis and the small bowel: a loco-regional review and current management strategies. Pathol Res Pract 204:449–458.
4. Ambrose WL Jr, Dozois RR, Pemberton JH, et al (1992) Familial adenomatous polyposis: results following ileal pouch-anal anastomosis and ileorectostomy. Dis Colon Rectum 35:12–15.
5. Jagelman DG (1991) Choice of operation in familial adenomatous polyposis. World J Surg 15:47–49.
6. Kartheuser A (1997) Surgery, genetics and experimental models. PhD thesis, Université Catholique de Louvain, Brussels.
7. Kartheuser A, Stangherlin P, Brandt D, Remue C, Sempoux C (2006) Restorative proctocolectomy and ileal pouch-anal anastomosis for familial adenomatous polyposis revisited. Fam Cancer 5:241–260.
8. Möslein G, Pistorius S, Saeger HD, Schackert HK (2003) Preventive surgery for colon cancer in familial adenomatous polyposis and hereditary nonpolyposis colorectal cancer syndrome. Langenbecks Arch Surg 388:9–16.
9. Martin LW, LeCoultre C, Schubert WK (1977) Total colectomy and mucosal proctectomy with preservation of continence in ulcerative colitis. Ann Surg 186:477.
10. Fonkalsrud EW, Ament ME (1978) Endorectal mucosal resection without proctectomy as an adjunct to abdominoperineal resection for non-malignant conditions: clinical experience with five patients. Ann Surg 188:245.
11. Parks AG, Nicholls RJ, Belliveau P (1980) Proctocolectomy with ileal reservoir and anal anastomosis. Br J Surg 67:533.
12. Utsonomiya J, Iwama T, Imajo M, et al (1980) Total colectomy, mucosal proctectomy, and ileoanal anastomosis. Dis Colon Rectum 23:459.
13. Parc YR, Radice E, Dozois RR (1999) Surgery for ulcerative colitis: historical perspective. A century of surgical innovations and refinements. Dis Colon Rectum 42:299–306.
14. Valiente MA, Bacon HE (1955) Construction of a pouch by using a "pantaloon" technique for pull through of ileum following total colectomy. Am J Surg 90:742–750.
15. Dozois RR (1989) Technique of ileal pouch-anal anastomosis. Perspect Colon Rectal Surg 2:85–94.
16. Kartheuser A, Parc R, Penna C, et al (1996) Ileal pouch-anal anastomosis as the first choice operation in patients with familial adenomatous polyposis. A ten year experience. Surgery 119:615–623.
17. Remzi F, Ooi B, Preen M, Bast J, et al (2002) Prospective evaluation of functional outcome and quality of life in over 2000 patients undergoing mucosectomy handsewn (MHS) versus stapled (STP) ileal pouch-anal anastomosis (IPAA). American Society of Colon and Rectal Surgeons: Annual Meeting 2002, poster P57
18. Tuckson WB, Lavery LC, Fazio VW, et al (1991) Manometric and functional comparison of ileal pouch-anal anastomosis with and without anal manipulation. Am J Surg 161:90–96.
19. Ooi BS, Remzi FH, Gramlich T, et al (2003) Anal transitional zone cancer after restorative proctocolectomy and ileoanal anastomosis in familial adenomatous polyposis. Dis Colon Rectum 46:1418–1423.
20. Deen KI, Williams JG, Grant EA, et al (1995) Randomized trial to determine the optimum level of pouch-anal anastomosis in stapled restorative proctocolectomy. Dis Colon Rectum 38:133–138.
21. Duff SE, O'Dwyer ST, Hulten L, et al (2002) Displasia in the ileo-anal pouch. Colorectal Dis 4:420–429.
22. Thompson-Fawcett MW, Mortensen NJ (1996) Anal transitional zone and columnar cuff in restorative proctocolectomy. Br J Surg 83;1947–1955.
23. Wexner SD, Cera SM (2005) Laparoscopic surgery for ulcerative colitis. Surg Clin N Am 85:35–47.
24. Van Duijvendijk P, Vasen HF, Bertario L, et al (1999) Cumulative risk of developing polyps or malignancy at the ileal pouch-anal anasrtomosis in patients with familial adenomatous polyposis. J Gastrointestinal Surg 3:325–330.
25. Dozois RR, Kelly KA, Welling DR, et al (1989) Ileal pouch-anal anastomosis: comparison of results in familial adenomatous polyposis and chronic ulcerative colitis. Ann Surg 210:268–273.
26. Becker JM, Stucchi AF (2004) Proctocolectomy with ileoanal anastomosis. J Gastrointestinal Surg 8:376–386.
27. Nicholls RJ, Bartolo DC, Mortensen N (1993) Restorative proctocolectomy. Blackwell Scientific Publications, Oxford, p 166.
28. Gullberg K, Liljeqvist L (2001) Stapled ileoanal pouches without loop ileostomy: a prospective study in 86 patients. Int J Colorectal Dis 16:221–227.
29. Metcalf AM, Dozois RR, Beart RW, et al (1986) Temporary ileostomy for ileal pouch-anal anastomosis: function and complications. Dis Colon Rectum 29:30–33.
30. Kartheuser A, Brandt D, Detry R, et al (2003) Ileal pouch-anal anstomosis: avoiding ileostomy by Riolan's arcade preservation. Colorectal Dis 5:42.
31. Heuschen UA, Hinz U, Allemeyer EH, et al (2002) Risk factors for ileoanal J pouch-related septic complications in ulcerative colitis and familial adenomatous polyposis. Ann Surg 235:207–216.
32. Sugerman HJ, Newsome HH (1994) Stapled ileoanal anastomosis without a temporary ileostomy. Am J Surg 167:58–66.
33. Galandiuk S, Wolff BG, Dozois RR, Beart RW (1991) Ileal pouch-anal anastomosis without ileostomy. Dis Colon Rectum 34:870–873.
34. Gorfine SR, Gelernt IM, Bauer JJ, et al (1995) Restorative proctocolectomywithout diverting ileostomy. Dis Colon Rectum 38:18–94.
35. Grobler SP, Hosie KB, Keighley MR (1992) Randomized trial of loop ileostomy in restorative proctocolectomy. Br J Surg 79:903–906.
36. Cohen Z, MacLeod RS, Stephen W, et al (1992) Continuing evolution of the pelvic pouch procedure. Ann Surg 216:506–511.
37. Onaitis MW, Mantyh C (2003) Ileal pouch-anal anastomosis for ulcerative colitis and familial adenomatous polyposis. Historical development and current status. Ann Surg 65:S42–48.

38. Fazio VW, Ziv Y, Church JM, et al (1995) Ileal pouch-anal anastomosis complications and functions in 1005 patients. Ann Surg 222:120–127.
39. Smith L, Friend WG, Medwell SJ (1984) The superior mesenteric artery: the critical factor in the pouch pull-through procedure. Dis Colon Rectum 27:741–744.
40. Martel P, Blanc P, Bothereau H, et al (2002) Comparative anatomicalstudy of division of the ileocolic pedicle or the superior mesenteric pedicle for mesenteric lengthening. Br J Surg 89:775–778.
41. Goes RN, Coy CS, Amaral CA, Fagundes JJ, Medeiros RR (1995) Superior mesenteric artery syndrome as a complication of ileal pouch-anal anastomosis. Dis Colon Rectum 38:543–544.
42. Gignoux BM, Parc R, Tiret E (2002) Ileal pouch-anal anastomosis without covering ileostomy. Gastroenterol Clin Biol 26:671–674.
43. Church J, Simmang C (2003) Standards task force of the American Society of Colon and Rectal Surgeons. Practice parameters for treatment of patients with dominantly colorectal cancer (familial adenomatous polyposis and Hereditary nonpolyposis colorectal cancer). Dis Colon Rectum 46:1001–1012.
44. Ziv Y, Church JM, Oakley JR, McGannon E, Fazio VW (1995) Surgery for the teenager with familial adenomatous polyposis: ileorectal anastomosis or restorative proctocolectomy? Int J Colorectal Dis 10:6–9.
45. Church J, Burke C, MacGannon E, et al (2003) Risk of rectal cancer in patients after colectomy and ileorectal anastomosis for familial adenomatous polyposis. Dis Colon Rectum 46:1175–1181.
46. Bülow S, Bülow C, Vasen H, et al (2008) Colectomy and ileorectal anastomosis is still an option for selected patients with familial adenomatous polyposis. Dis Colon Rectum 51:1318–1323.
47. Bertario L, Sala P, Radice P, et al (2001) Ileorectal anastomosis in patients with familial adenomatous polyposis. Gastroenterology 121:502–503.
48. Bülow C, Vasen H, Jarvinen H, et al (2000) Ileorectal anastomosis is appropriate for a subset of patients with familial adenomatous polyposis. Gastroenterology 119:1454–1460.
49. Vasen HF, van der Luijt RB, Slors JF, et al (1996) Molecular genetic tests as a guide to surgical management of familial adenomatous polyposis. Lancet 348;433–435.
50. Nugent K, Phillips R (1992) Rectal cancer risk in older patients with familial adenomatous polyposis and an ileorectal anastomosis: a cause of concern. Br J Surg 79:1204–1206.
51. Friedl W, Caspari R, Sengteller M, et al (2001) Can *APC* mutation analysis contribute to therapeutic decision in familial adenomatous polyposis? Experience from 680 FAP families. Gut 48:515–521.
52. Aziz O, Athanasiou T, Fazio VW, et al (2006) Meta-analysis of observational studies of ileorectal versus ileal pouch-anal anastomosis for familial adenomatous polyposis. Br J Surg 93:407–417.
53. Wasmuth HH, Svinsås M, Tranø G, et al (2006) Surgical load and long-term outcome for patients with Kock continent ileostomy. Colorectal Dis 9:713–717.
54. Boardman LA (2002) Heritable colorectal cancer syndromes: recognition and preventive management. Gastroenterol Clin North Am 31:1107–1131.
55. Wu JS, Paul P, MacGannon E, Church J (1998) *APC* genotype, polyp number, and the surgical options in familial adenomatous polyposis. Ann Surg 227:57–62.
56. Giardiello F, Krush A, Petersen G, et al (1994) Phenotypic variability of familial adenomatous polyposis in 11 unrelated families with identical *APC* gene mutation. Gastroenterology 106:1542–1547.
57. Bertario L, Russo A, Sala P, et al (2004) *APC* genotype is not a prognostic factor in familial adenomatous polyposis patients with colorectal cancer. Dis Colon Rectum 47:1162–1169.
58. Hernegger G, Harvey G, Guillem J (2002) Attenuated familial adenomatous polyposis an evolving an poorly understood entity. Dis Colon Rectum 45:127–136.
59. Spirio L, Olschwang S, Groden J, et al (1993) Alleles of the *APC* gene: an attenuated form of familial polyposis. Cell 75:951–957.
60. Soravia C, Berk T, Madlensky L (1998) Genotype-phenotype correlation in attenuated adenomatous polyposis coli. Am J Hum Genet 62:1290–1301.
61. Nieuwenhuis M, Mathus-Vliegen L, Slors F, et al (2007) Genotype-phenotype correlations as a guide in the management of familial adenomatous polyposis. Clin Gastroenterol Hepatol 374–378.
62. Valanzano R, Ficari F, Curia M, et al (2007) Balance between endoscopic and genetic information in the choice of ileorectal anastomosis for familial adenomatous polyposis. J Surg Oncol 95:28–33.
63. Church J, Burke C, MacGannon E, et al (2001) Predicting polyposis severity by proctoscopy: how reliable is it? Dis Colon Rectum 44:1249–1254.
64. Madden M, Neale K, Nicholls R, et al (1991) Comparison of morbidity and function after colectomy with ileorectal anastomosis or restorative proctocolectomy for familial adenomatous polyposis. Br J Surg 78:789–792.
65. Olsen KO, Juul S, Bülow S, et al (2003) Female fecundity before and after operation for familial adenomatous polyposis. Br J Surg 90:890.
66. Gallagher MC, Sturt NJH, Phillipq RKS (2003) Female fecundity before and after operation for familial adenomatous polyposis. Br J Surg 90:227–231.
67. Johansen C, Bitsch M, Bülow S (1990) Fertility and pregnancy in women with familial adenomatous polyposis. Int J Colorectal Dis 5:203–206
68. Gorgun E, Remzi FH, Goldberg JM, et al (2004) Fertility is reduced after restorative proctocolectomy with ileal pouch anal anastomosis: a study of 300 patients. Surgery 136:795–803.
69. Oresland T, Palmblad S, Ellstrom M, et al (1994) Gynaecological and sexual function related to anatomical changes in the female pelvis after restorative proctocolectomy. Int J Colorectal Dis 9:77–81.
70. Hahnloser D, Pemberton JH, Wolff BG, et al (2004) Pregnancy and delivery before and after ileal pouch-anal anastomosis for inflammatory bowel disease: immediate and long-term consequences and outcomes. Dis Colon Rectum 47:1127–1135.
71. Metcalf A, Dozois RR, Beart RW, Wolff BG (1985) Pregnancy following ileal pouch.anal anastomosis. Dis Colon Rectum 28:859–861.
72. Nelson H, Dozois RR, Kelly KA, et al (1989) The effect of pregnancy and delivery on the ileal-pouch-anal anastomosis

functions. Dis Colon Rectum 32:384–388.
73. Wax JR, Pinette MG, Cartin A, Blackstone J (2003) Female reproductive health after ileal pouch anal anastomosis for ulcerative colitis. Obstet Gynecol Surv 58:270–274.
74. Ravid A, Richard CS, Spencer LM, et al (2002) Pregnancy, delivery and pouch function after ileal pouch-anal anastomosis for ulcerative colitis. Dis Colon Rectum 45:1283–1288.
75. Dozois RR, Nelson H, Metcalf AM (1993) Fonction sexuelle après anastomose iléo-anale. Ann Chir 47:1009–1013.
76. Farouk R, Pemberton GH, Wolff BG, et al (2000) Functional outcomes after ileal pouch-anal anastomosis for chronic ulcerative colitis. Ann Surg 231:919–926.
77. Scott HJ, McLeod RS, Blair J, et al (1996) Ileal pouch-anal anastomosis: pregnancy, delivery and pouch function. Int J Colorect Dis 11:84–87.
78. Cornish JA, Tan E, Teare J, et al (2007) The effect of restorative proctocolectomy on sexual function, urinary function, fertility, pregnancy and delivery: a systematic review. Dis Colon Rectum 50:1128–1138.
79. Juhasz ES, Fozard B, Dosoiz RR, Ilstrup DM, Nelson H (1995) Ileal pouch-anal anastomosis function following childbirth. An extended evaluation. Dis Colon Rectum 38:159–165.
80. Remzi FH, Gorgun E, Bust J, et al (2005) Vaginal delivery after ileal pouch-anal anastomosis: a word of caution. Dis Colon Rectum 48:1691–1699.
81. Karthauser A, Charre L, Kayser J, et al (2003) Laparoscopic restorative proctocolectomy without ileostomy. Colorectal Dis 5(suppl 2):170.
82. Penna C, Tiret E, Daude F, Parc R (1994) Results of ileal pouch-anal anastomosis in familial adenomatous polyposis complicated by rectal carcinoma. Dis Colon Rectum 37:157–160.
83. Panis Y, Bonhomme N, Hautefeuille P, Valleur P (1996) Ileal pouch-anal anastomosis with mesorectal excision for rectal cancer complicating familial adenomatous polyposis. Eur J Surg 162:817–821.
84. Colwell JC, Grey M (2001) What functional outcomes and complications should be taught to the patient with ulcerative colitis or familial adenomatous polyposis who undergoes ileal pouch anal anastomosis. J Wound Ostotomy Continence Nurs 28:184–189.
85. Metcalf AM, Dozois RR, Kelly KA (1986) Sexual function in women after proctocolectomy. Ann Surg 204:624–627.
86. Tekkis PP, Fazio VW, Lavery IC, et al (2005) Evaluation of the learning curve in ileal pouch-anal anastomosis surgery. Ann Surg 241:262–268.
87. Ziv Y, Fazio VW, Church JM, et al (1996) Stapled ileal pouch-anal anastomoses are safer than hand-sewn anastomoses in patients with ulcerative colitis. Am J Surg 171:320–323.
88. Sagar PM, Lewis W, Holdsworth PJ, Johnston D (1992) One-stage restorative proctocolectomy without temporary defunctioning ileostomy. Dis Colon Rectum 35:582–588.
89. Fazio VW, O'Riordain MG, Lavery IC, et al (1999) Long-term functional outcome and quality of life after stapled restorative proctocolectomy. Ann Surg 230:575–586.
90. Meagher AP, Farouk R, Dozois RR, et al (1998) J ileal pouch-anal anastomosis for chronic ulcerative colitis: complications and long-term outcome in 1310 patients. Br J Surg 85:800–803.
91. Van Duijvendijk P, Slors F, Taat CW, et al (2000) What is the benefit or preoperative sperm preservation for patients who undergo restorative proctocolectomy for benign disease. Dis Colon Rectum 43:838–842.
92. Iida M, Yao T, Watanabe H, Itoh H, Iwashita A (1984) Fundic gland polyposis in patients without familial adenomatosis coli: its incidence and clinical features. Gastroenterology 876:1437–1442.
93. Domizio P, Talbot IC, Spigelman AD, Williams CB, Phillips RK (1990) Upper gastrointestinal pathology in familial adenomatous polyposis: results from a prospective study of 102 patients. J Clin Pathol 43:738–743.
94. Mabrut JY, Romagnoli R, Collard JM, et al (2006) Familial adenomatous polyposis predispodes to pathologic exposure of the stomach to bilirubin. Surgery 140:818–823.
95. Bertoni G, Sassatelli R, Nigrisoli E, et al (1996) Dysplastic changes in gastric fundic gland polyps of patients with familial adenomatous polyposis. Ital J Gastroenterol Hepatol 8:1201–1206.
96. Wu TT, Kornacki S, Rashid A, Yardley JH, Hamilton SR (1998) Dysplasia and dysregulation of proliferation in foveolare and surface epithelia of fundic gland polyps from patients with familial adenomatous polyposis. Am J Surg Pathol 22:293–298.
97. Abraham SC, Nobukawa B, Giardiello FM, Hamilton SR, Wu TT (2000) Fundic gland polyps in familial adenomatous polyposis: neoplasms with frequent somatic adenomatous polyposis coli gene alterations. Am J Pathol 157:747–754.
98. Nakamura S, Matsumoto T, Kobori Y, Iida M (2002) Impact of *Helicobacter pylori* infection and mucosal atrophy on gastric lesions in patients with familial adenomatous polyposis. Gut 51:485–489.
99. Spigelman AD, Willaims CB, Talbot IC, Domizio P, Phillips RK (1989) Upper gastrointestinal cancer in patients with familial adenomatous polyposis. Lancet 2:783–785.
100. Gallagher MC, Phillips RK, Bülow S (2006) Surveillance and management of upper gastrointestinal disease in familial adenomatous polyposis. Fam Cancer 5:263–273.
101. Wallace MH, Phillips RK (1998) Upper gastrointestinal disease in patients with familial adenomatous polyposis. Br J Surg 85:742–750.
102. Kashiwagi H, Spigelman AD (2000) Gastroduodenal lesions in familial adenomatous polyposis. Surg Today 30:675–682.
103. Nugent KP, Farmer KC, Spigelman AD, Williams CB, Phillips RK (1993) Randomized controlled trial of the effect of sulindac on duodenal and rectal polyposis and cell proliferation in patients with familial adenomatous polyposis. Br J Surg 80:1618–1619.
104. Richard CS, Berk T, Bapat BV, et al (1997) Sulindac for periampullary polyps in FAP patients. Int J Colorectal Dis 12:14–18.
105. Phillips RK, Wallace MH, Lynch PM, et al (2002) A randomised, double blind, placebo controlled study of celecoxib, a selective cyclooxygenase 2 inhibitor on duodenal polyposis in familial adenomatous polyposis. Gut 5:857–860.
106. Bertoni G, Sassatelli R, Nigrisoli E, et al (1996) High prevalence of adenomas and microadenomas of the duodenal papilla and periampullary region in patients with

familial adenomatous polyposis. Eur J Gastroenterol Hepatol 8:1201–1206.

107. Church JM, McGannon E, Hull-Boiner S, et al (1992) Gastroduodenal polyps in patients with familial adenomatous polyposis. Dis Colon Rectum 35:1170–1173.
108. Murakami Y, Uemura K, Hayashidani Y, Sudo T, Sueda T (2006) Relapsing acute pancreatitis due to ampullary adenoma in a patient with familial adenomatous polyposis. J Gastroenterol 41:798–801.
109. Burke CA, Beck GJ, Church JM, Van Stolk RU (1999) The natural history of untreated duodenal and ampullary adenomas in patients with familial adenomatous polyposis followed in an endoscopic surveillance program. Gastrointest Endosc 49:358–364.
110. Matsumoto T, Iida M, Nakamura S, et al (2000) Natural history of ampullary adenoma in familial adenomatous polyposis:reconfirmation of benign nature during extended surveillance. Am J Gastroenterol 95:1557–1562.
111. Iizuka T, Sawada K, Hayakawa K, et al (2002) Successful local excision of ileostomy adenocarcinoma after colectomy for familial adenomatous polyposis: report of a case. Surg Today 32:638–641.
112. Quah HM, Samad A, Maw A (2005) Ileostomy carcinomas a review: the latent risk after colectomy for ulcerative colitis and familial adenomatous polyposis. Colorectal Dis 7:538–544.
113. Remzi FH, Church JM, Bast J, et al (2001) Mucosectomy vs stapled ileal pouch-anal anastomosis in patients with familial adenomatous polyposis: functional outcome and neoplasia control. Dis Colon Rectum 44;1590–1596.
114. Malassagne B, Penna C, Parc R (1995) Adenomatous polyps in the anal transitional zone after ileal pouch-anal anastomosis for familial adenomatous polyposis: treatment by transanal mucosectomy and ileal pouch advancement. Br J Surg 82:1634.
115. Tulchinsky H, Keidar A, Strul H, et al (2005) Extracolonic manifestations of familial adenomatous polyposis after proctocolectomy. Arch Surg 140:159–163.
116. Parc YR, Olschwang S, Desaint B, et al (2001) Familial adenomatous polyposis: prevalence of adenomas in the ileal pouch after restorative proctocolectomy. Ann Surg 233:360–364.
117. Parc Y, Piquard A, Dozois RR, Parc R, Tiret E (2004) Long-term outcome of familial adenomatous polyposis patients after restorative proctocolectomy. Ann Surg 239:378–382.
118. Polese L, Keighley MR (2003) Adenomas at resection margins do not influence the long-term development of pouch polyps after restorative proctocolectomy for familial adenomatous polyposis. Am J Surg 186:32–34.
119. Thompson-Fawcett MW, Marcus VA, Redston M, Cohen Z, McLeod RS (2001) Adenomatous polyps develop commonly in the ileal pouch of patients with familial adenomatous polyposis. Dis Colon Rectum 44:347–353.
120. Wu JS, McGannon EA, Church JM (1998) Incidence of neoplastic polyps in the ileal pouch of patients with familial adenomatous polyposis after restorative proctocolectomy. Dis Colon Rectum 41:552–556.

Restorative Proctocolectomy with Ileal Pouch-Anal Anastomosis for FAP

16

The Role of Laparoscopy - An Overview

Filippo Nifosì, Maurizio Mainente, Gianluca Colucci and Gian Gaetano Delaini

Abstract The interest in a minimally invasive technique for familial adenomatous polyposis (FAP) has extensively increased in the last 10 years. As shown in many published studies, the laparoscopic approach has already been used in many patients, with outcomes that are at least comparable to open surgery, as far as functional results are concerned. Furthermore, it seems that laparoscopic surgery can offer less tissue trauma, fewer postoperative complications (e.g. incisional hernias), and better cosmetic results. Unfortunately, there is a steep learning curve for these procedures, and the costs are higher than for open surgery. More studies are needed to clarify the roles of these approaches in FAP patients.

Keywords Cosmetic result • Hand-assisted surgery • Ileal pouch • Laparoscopy • Minimally invasive technique • Quality of life

16.1 Introduction

Restorative proctocolectomy (RPC) with ileal pouch-anal anastomosis (IPAA) has evolved as a procedure devised to preserve the normal route of defecation and continence function in patients who need removal of the colon and rectum [1]. RPC with IPAA, completely removing all diseased tissues, offers a very satisfactory functional outcome, good quality of life, and high level of acceptance by patients [2], and has become the standard operative procedure for classic familial adenomatous polyposis (FAP) [3–5]. Based on the initial experiences of Ravitch and Sabinston [6], who documented the use of a straight ileo-anal anastomosis following proctocolecromy, Parks and Nicholls [7] refined the procedure, adding an ileal reservoir anastomosed to the dentate line, and achieved superior functional results after such a modification [1,8]. Since its introduction in clinical practice, the procedure has undergone several technical refinements aiming to simplify the procedure, reduce morbidity, and improve outcome [3,8]. Of the well-known controversies regarding the technique of rectal dissection, the hand-sewn or stapled approach to ileo-anal anastomosis, the use or omission of a diverting loop ileostomy [8], and, recently, the role and possible benefits of laparoscopy, have been added to the debate [9] .

16.2 Laparoscopy and Colorectal Surgery

During the last decade, laparoscopic segmental colonic resections, whether for benign or malignant disease, have proved to be associated with some advantages to patients, such as diminished postoperative pain,

N. Nifosì (✉)
Department of Surgery and Gastroenterology, University of Verona, Verona, Italy

reduced need for narcotic use, faster recovery of intestinal function, and shorter hospital stay [10,11]. Moreover, recent evidence-based experience has shown the safety and feasibility of the laparoscopic approach to colorectal tumors, as well as an improved long-ternm oncologic outcome [11,12]. Following technological innovations and implementation of experience in laparoscopic surgery, minimally invasive approaches have been applied to more complex colorectal procedures including RPC with IPAA [13,14]. For FAP patients, this operation is prophylactic, and a minimally invasive technique has much to offer to a group of young asymptomatic patients in a critical period of their lives [15,16].

16.3 Laparoscopic Restorative Proctocolectomy

Restorative proctocolectomy is a complex undertaking with a steep learning curve, and early experiences with laparoscopic IPAA demonstrated no advantages when compared to standard open operation. Moreover, laparoscopy was associated with excessively long operative times, extended length of hospital stay, and higher level of morbidity [13,14,17]. However, in recent series, an earlier return of bowel activity and shorter postoperative hospitalization favoring laparoscopy has been observed [16,18,19]. In addition, safety issues and surgical morbidity compared well to the open approach [9,20–22]. To date, there are about 70 reports on laparoscopic IPAA, most case series and case-matched studies, including mainly ulcerative colitis (UC) rather than FAP patients, preventing a selective analysis specifically addressed to FAP cases. The operative steps usually involve complete laparoscopic mobilization of the colon, with intracorporeal ligation of the vascular pedicles, and exteriorization of the bowel through a Pfannenstiel incision about 7–8 cm wide [16,23]. Through the minilaparotomy, the steps of proctectomy, pouch fashioning, and IPAA (with or without protective ileostomy) follow as in the open procedure (laparoscopic-assisted IPAA) [16,19,23,24]. With the aim of shortening total operative time and making the intracorporeal dissection easier, particularly the demanding mobilization of the transverse colon, some surgeons advocate a hand-assisted approach by means of an hand-port device (hand-assisted laparoscopic IPAA) [25,26]. In addition, the hand-assisted technique would carry the benefit of a more rapid learning curve because the surgical maneuvers involved are more similar to the traditional ones [26,27]. By contrast, in almost totally laparoscopic operations, rectal dissection and ileal-anal anastomosis construction [9,27] are carried out through only a small peri-umbilical incision of 3–4 cm from which the bowel is delivered. This approach is the preferred choice of those laparoscopic purists who argue that performing a part of the procedure through the Pfannenstiel incision could offset the potential benefits of a truly minimally invasive surgery [9,20,28,29].

16.3.1 Early Postoperative Results

One of the largest reported experiences in laparoscopic RPC comes from the University of Heidelberg [9]. Fifty patients (UC 23, FAP 27) underwent a procedure where all the steps were conducted laparoscopically, except for a peri-umbilical incision of 4 cm for pouch fashioning. Median operative time was 320 minutes; complications occurred in 30% of these patients. Median hospital stay was 12 days, and there was a conversion rate of 8%. Ulcerative colitis was associated with a higher overall rate of complications, and increased body mass index with higher conversion events. The authors emphasize the reduced blood loss (mean 200 ml) and consequent lack of requirement for perioperative blood transfusion as a direct benefit of the minimally invasive approach, allowing the pelvic dissection to be carried out in a more precise and bloodless way [9] due to the magnifying effect of the laparoscope [16]. Agha and colleagues [26] reported somewhat overlapping results for operative time (median 210 minutes), conversion rate (5%), intraoperative blood loss (median 70 ml), complication rate (20%), and length of hospital stay (mean 11 days), when employing a hand-assisted technique on 20 patients (UC 18, FAP 2). A series of 14 patients (UC 13, FAP 1) from Bristol underwent laparoscopic-assisted restorative pouch surgery, with the pelvic phase of the procedure being completed through the Pfannenstiel access. In these patients, the median duration of surgery was 260 minutes, oral feeding was resumed 2 days postoperatively, and the median length of hospital stay was 7 days. It was reported that all patients were fully continent in the long term, and had a median of four daily bowel movements. All rated functional and cosmetic results with a high degree of satisfaction. McNevin and colleagues [23] reported that of 32 laparoscopic-assisted RPCs (27 IPAA, 5 Brooke ileostomy), the surgery took a mean of 197 minutes, the estimated blood loss was in the range of 200 ml, the mean stay in hospital was 4.8 days, and 4.7% of the cases were converted to open surgery. In a small series of 10 patients (UC 6, FAP 4), Lòpez-Rosales and colleagues were able to show further improvement in the operative time (187 minutes), blood loss (mean 46 ml) and days spent in the hospital (mean 3.4), with no conversion, and a 20% complication rate [30].

16.3.2 Comparison of Laparoscopic and Open IPAA

Marcello and colleagues [18] and Heise and colleagues [24] carried out comparative studies on 20 and 65 laparoscopic cases respectively, with equal numbers of patients treated conventionally. The authors agree that the resumption of intestinal function and length of hospitalization favor the laparoscopic group, but at the price of longer operative time. A Mayo Clinic trial [20] also reached the same conclusions. Heise and colleagues [24] highlight that, if considered as a whole, the costs (operation, hospital stay, readmission) no longer penalize laparoscopy. In spite of theoretical premises and favorable results evidenced in some case series and case-matched studies [16,18,21–24, 26), data released from the only randomized trial comparing laparoscopic hand-assisted and open surgery for UC and FAP patients [25] failed to substantiate all the previous findings. Maartense and colleagues [25] comment that laparoscopic IPAA is as safe as open procedure, but there are no clear advantages in terms of morphine requirements and postoperative hospital stay. Moreover, the operation takes one hour longer and costs are about 3000 euros more than for the open procedure. The same surgical group extended the comparison to a homogeneous cohort of totally laparoscopically treated patients; however, they did not show meaningful improvements of the clinical indicators analysed (e.g. morbidity, length of surgery, costs, hospital stay, quality of life) [31]. The kinds of data reported to date in the literature, most from numerically small and non-randomized trials, prevent objectively drawing firm and unequivocal conclusions about the merit of laparoscopic RPC. Tan and Tjandra [32] conducted a meta-analysis showing the superiority of laparoscopy highlighted in a more rapid gastrointestinal recovery and a 20% shorter hospitalization and better cosmesis. The length of surgery may be improved in conjunction with the increasing experience of surgical teams [23,30,32]. A number of operative and postoperative parameters as well as complications related to laparoscopic or open RPC have been scrutinized with a meta-analytical technique by Tilney and colleagues [33]. The minimally invasive procedure takes 86 minutes longer to complete, average intracorporeal blood loss is quantifiable as 84 ml, the overall conversion rate is below 1%, and hospitalization is significantly shortened, but only in high-quality studies and those reporting on more than 30 cases. No difference in short-term adverse events compared to open surgery emerges favoring laparoscopy. The authors comment that the current expected benefits of laparoscopic versus open RPC appear to be limited [33]. Most of the patients included in this meta-analysis were given diverting loop ileostomy, and although the practicability and safety of one-stage laparoscopic IPAA has been proven [19,34,35], the topic is controversial [36]. Tilney and colleagues [33] and Casillas and Delaney [36] indeed outline the need for a clear and formal definition of criteria allowing a safe omission of protective stoma during laparoscopic RPC. For Kartheuser and colleagues [37], FAP patients who are young, not under steroid or immunosuppressive treatment, and are devoid of inflammatory anal canal modifications, are ideal candidates for one-stage minimally invasive IPAA. Therefore, the only reason to adopt a diverting ileostomy could be tension at the anastomotic suture line [38].

16.3.3 Functional Results and Quality of Life

Aside from indications, safety, and surgical results [21], the issues concerning functional outcome and quality of life are of the utmost significance [20,25,33]. When examining the global physical functioning, Dunker and colleagues [19] reach the conclusion that after laparoscopic restorative proctocolectomy, patients have fewer health-related constraints in their performance of demanding physical activities.

However, outcome in terms of intestinal and sexual function and quality of life, measured by means of the SF-36 one year postoperatively, show overlapping results regardless of the type of surgery. A case-matched trial from the Mayo Clinic [20] comparing two groups of 33 patients each, laparoscopically and conventionally operated, proved that the functional results, at one year follow-up, were equivalent, allowing the authors to state that intestinal function and quality-of-life indicators are not compromised, in the long term, by the minimally invasive technique, and that well-selected patients may be offered this alternative approach. Finally, in the meta-analysis by Tilney and colleagues [33] as well, there are no statistically clear differences in any of the measures of pouch function, despite a trend towards fewer nocturnal bowel movements benefiting patients in the laparoscopic group.

16.3.4 Cosmetic Results and Body Image

Abdominal wall preservation and better cosmesis are two undisputed benefits of minimally invasive techniques over open laparotomy [16,19,21,33], which meet the expectations of decreased disability and improved body image pursued by young and asymptomatic patients [37]. New evidence of these benefits is substantiated by the increasing demand of laparoscopic RPC in referral centers with expertise in laparoscopic surgery [39]. Polle and colleagues [39] gave a detailed account of these aspects by means of a body image questionnaire and a cosmetic score specifically devised to cover these topics. The randomized study, with a median follow-up of 3 years, strengthens that body image and cosmetic outcomes definitely favor laparoscopically operated patients over those dealt with using the open procedure; these effects are particularly meaningful in female patients. Otherwise, functional results, surgical morbidity, and quality of life overlap. Although body image and cosmesis are generally considered secondary and unconventional outcomes in the field of gastrointestinal surgery, the significance of these factors is highlighted in a Dutch trial by the fact that the majority of the patients who underwent open surgery, when confronted with the cosmetic outcome of laparoscopic surgery, would have chosen laparoscopy instead [39]. If it is true that accelerated postoperative recovery, reduced blood loss, lower morbidity, and shorter hospitalization are the fundamental advantages of laparoscopy [9,18,30,32), it also must be realized that from a patient perspective these are only temporary short-term gains, whereas, body preservation and abdominal cosmesis may be long-lasting benefits of the minimally invasive approach [9,39].

16.4 Discussion and Conclusion

There are still some incontrovertible data missing regarding the advantages of the minimally invasive technique, which are needed to substantiate a widespread adoption of laparoscopic RPC [33,36,37,40]. Laparoscopic IPAA may represent an alternative over conventional procedures for feasibility (conversion rate 1–8%) and surgical safety [9,16,18,20,24,25], while operative time comes close to that of the open approach as the experience of surgical teams is refined [23,30,32]. A lack of randomized prospective trials prevents the expected benefits of the minimally invasive technique from being clearly revealed in comparison to standard surgery [25,33,36,37]. However, accelerated resumption of intestinal function (of about 1–2 days), reduced intraoperative blood loss (nearly 100 ml), and a 20% shortening of hospitalization have been shown in some comparative studies, favoring laparoscopy [18,22,24,32]. Moreover, pouch global functioning and quality of life are not different in comparison with conventional RPC [19,20,25,33]. Data reported in the literature do not facilitate the choice to determine the most profitable approach [31] between laparo-assisted [16,22], hand-assisted [26,27], or totally laparoscopic [9,29] surgery. The need for a diverting stoma after IPAA also remains a controversial issue [33], and a formal standardization of selective criteria for one-stage laparoscopic IPAA is still lacking [33,36]. To date, the major achievement unanimously ascribable to laparoscopic IPAA is its superior cosmetic outcome [16,19,21,25,33,36]; it could be argued that these results in themselves outweigh the longer operating times and higher costs related to laparoscopy [31]. In any case, Polle and colleagues [39] gave objective proof that open IPAA has a negative impact on body image and cosmesis as compared with minimally invasive IPAA; in addition, cosmetic benefit seems to be a long-lasting and appealing advantage for relatively young, fit, and motivated patients, particularly females. Laparoscopic access for bowel operations significantly reduces the incidence of ventral hernia and postoperative adhesions [41]. This, in turn, could translate into lowering the rate of small bowel obstruction after IPAA [21,41,42], and could positively influence fertility in young women with FAP following IPAA [43], providing a potential source of decreased overall morbidity [41]. Since surgery precedes the growth of desmoid tumors in 68–83% of FAP patients [37], an assumption could be made about whether a minimally invasive technique, resulting in a more gentle and less traumatic manipulation of tissues, could lower the risk of intra-abdominal desmoid development [44]. The various possible benefits of laparoscopic RPC over the conventional open approach [21], although not yet fully proven on evidence-based grounds, constitute a positive trend with a substantial chance that they will be confirmed by subsequent randomized trials [37,45]. To date, laparoscopic IPAA represents an effective and promising alternative to traditional surgery for motivated and accurately selected FAP patients [20,21,32,36].

References

1. Onaitis MW, Mantyh C (2003) Ileal pouch-anal anastomosis for ulcerative colitis and familial adenoumatos polyposis . Historical development and current status. Ann Surg 238:542–548.
2. Seidel SA, Peach SE, Newman M, Sharp KW (1999) Ileal pouch procedures: clinical outcomes and quality of life assessment. Am Surg 1:40–46.
3. Fazio VW, Ziv Y, Church JM, et al (1995) Ileal pouch-anal anastomosis, complications and function in 1005 patients. Ann Surg 222:120–127.
4. Nyam DC, Brillant PT, Dozois RR (1997) Ileal pouch-anal anastomosis for familial adenoumatos polyposis. Ann Surg 226:514–521.
5. Penna CH (2002) Gestione chirurgica della poliposi adenomatosa familiare. J Chir 139:260–267.
6. Ravitch MM, Sabiston DC (1947) Anal ileostomy with preservation of the sphincter: a proposed operation in patients requiring total colectomy for benign lesions. Surg Gyn Obst 84:1095–1099.
7. Parks AG, Nicholls RJ (1978) Proctocolectomy without ileostomy for ulcerative colitis. Br Med J 2:85–88.
8. Radice E, Dozois RR (1999) Techniques of ileoanal anastomosis: historical evolution, refinements and current approaches. Tech Coloproctol 3:27–32.
9. Kienle P, Z'graggen K, Schmidt J, et al (2005) Laparoscopic restorative proctocolectomy. Br J Surg 92:88–93.
10. Braga M, Vignali A, Gianotti L, et al (2002) Laparoscopic versus open colorectal surgery. Ann Surg 236:759–767.
11. Lacy AM, Garcia-Valdecasas JC, Delgado S, et al (2002) Laparoscopy-assisted colectomy versus open colectomy for treatment of non metastatic colon cancer: a randomized trial. Lancet 359:2224–2229.
12. The Clnical Outcomes of Surgical Therapy Study Group (2004) A comparison of laparoscopically assisted and open colectomy for colon cancer N Engl J Med 350:2050–2059.
13. Wexner SD, Johansen OB, Nogueras JJ, Jagelman DG (1992) Laparoscopic total abdominal colectomy. A prospective trial. Dis Colon Rectum 35:651–655.
14. Santoro E, Carlini M, Carboni F, Feroce A (1999) Laparoscopic total proctocolectomy with ileal J pouch-anal anastomosis. Hepatogastroenterology 46:894–899.
15. Milsom JW, Ludwing KA, Church JM, Garcia-Ruiz A (1997) Laparoscopic total abdominal colecomy with ileo rectal anastomosis for familial adenomatous polyposis. Dis Colon Rectum 40:675–678.
16. Gill TS, Karantana A, Rees J, et al (2004) Laparoscopic proctocolectomy with restorative ileal-anal pouch. Colorectal Dis 6:458–461.
17. Reissman P, Salky BA, Pfeifer J, et al (1996) Laparoscopic surgery in the management of inflammatory bowel disease. Am J Surg 171:47–50.
18. Marcello PW, Milsom JW, Wong SK, et al (2000) Laparoscopic restorative proctocolectomy. Case-matched comparative study with open restorative proctocolectomy. Dis Colon Rectum 43:604–608.
19. Dunker MS, Bemelman WA, Slors JFM, et al (2001) Functional out come, quality of life, body image and cosmesis in patients after laparoscopic-assisted and conventional restorative proctocolectomy. A comparative study. Dis Colon Rectum 44:1800–1807.
20. Larson DW, Dozois EJ, Piotrowicz K, et al (2005) Laparoscopic-assisted vs open ileal pouch-anal anastomosis: functional out come in a case-matched series. Dis Colon Rectum 48:1845–1850.
21. Kessler H (2006) Laparoscopic surgery in inflammatory bowel disease: is the future already here? Curr Opin Gastroenterol 22:391–395.
22. Zhang H, Hu S, Zhang G, et al (2007) Laparoscopic versus open proctocolectomy with ileal-pouch-anal-anastomosis. Mini Invasive Ther Allied Technol 16:187–191.
23. McNevin SM, Bax T, MacFarlane M, et al (2006) Outcomes of a laparoscopic approach for total abdominal colectomy and proctocolectomy. Am J Surg 5:673–676.
24. Heise CP, Kennedy G, Foley EF, Harms BA (2008) Laparoscopic restorative proctocolectomy with ileal S-pouch. Dis Colon Rectum July 8:epub ahead of print.
25. Maartense S, Dunker MS, Slors FJ, et al (2004) Hand-assisted laparoscopic versus open restorative proctocolectomy with ileal pouch anastomosis. A randomized trial. Ann Surg 240:984–992.
26. Agha A, Moser C, Iesalnieres I, et al (2008) Combination of hand-assisted and laparoscopic proctocolectomy (HALP): technical aspects, learning curve and early postoperative results. Surg Endosc 22:1547–1552.
27. Boller AM, Larson DW (2007) Laparoscopic restorative proctocolectomy for ulcerative colitis. J Gastrointest Surg 11:3–7.
28. Hasegawa S, Nomura A, Kawamura J, et al (2007) Laparoscopic restorative total proctocolectomy with mucosal resection. Dis Colon Rectum 50:1152–1156.
29. Alves A, Panis Y (2005) Anastomoses iléoanales avec

réservoir par voie laparoscopique. Ann Chir 130:421–425.
30. Lòpez-Rosales F, Gonzàles-Contreras Q, Muro LJ, et al (2007) Laparoscopic total proctocolectomy with ileal pouch anal-anastomosis for ulcerative colitis and familial adenomaous polyposis: initial experience in Mexico. Surg Endosc 21:2304–2307.
31. Polle SW, van BergeHenegouwen MI, Slors JF, et al (2008) Total laparoscopic restorative proctocolectomy: are there advantages compared with the open and hand-assisted approaches? Dis Colon Rectum 51:541–548.
32. Tan YJJ, Tjandra JJ (2006) Laparoscopic surgery for ulcerative colitis – a meta-analysis. Colorectal Dis 8:626–636.
33. Tilney HS, Lovegrove RE, Heriot AG, et al (2007) Comparison of short-term outcomes of laparoscopic vs open approaches to ileal pouch-surgery. Int J Colorectal Dis 22:531–542.
34. Ki AJ, Sonoda T, Milsom JW (2002) One-stage laparoscopic restorative proctocolectomy : an alternative to the conventional approach? Dis Colon Rectum 45:207–210.
35. Sahakitrungruang C, Pattana-arun J, Tantiphlachiva K, et al (2008) Laparoscopic restorative proctocolectomy with small McBurney incision for ileal pouch construction without protective ileostomy. Dis Colon Rectum 51:1137–1138.
36. Casillas S, Delaney CP (2005) Laparoscopic surgery for inflammatory bowel disease. Dig Surg 22:135–142.
37. Kartheuser A, Stangherlin P, Brandt D, et al (2006) Restorative proctocolectomy and ileal pouch-anal anastomosis for familial adenomatous polyposis revisited. Familial Cancer 5:241–260.
38. Heuschen UA, Hinz U, Allemeyer EH (2002) Risk factors for ileoanal J pouch related septic complications in ulcerative colitis and familial adenomatous polyposis. Ann Surg 235:207–216.
39. Polle SW, Dunker MS, Slors JF, et al (2007) Body image, cosmesis, quality of life, and functional out come of hand-assisted laparoscopic versus open ristorative proctocolectomy: long-term results of a randomized trial. Surg Endosc 21:1301–1307.
40. Roses RE, Rombeau JL (2008) Recent trends in the surgical management of inflammatory bowel disease. World J Gastroenterol 14:408–412.
41. Duepree HJ, Senagore AJ, Delaney CP, Fazio VW (2003) Does means of access affect the incidence of small bowel obstruction and ventral hernia after bowel resections? Laparoscopy versus laparotomy. J Am Coll Surg 197:177–181.
42. Wexner SD, Cera SM (2005) Laparoscopic surgery for ulcerative colitis. Surg Clin N Am 85:35–47.
43. Hahnloser D, Pemberton JH, Wolff BG (2004) Pregnancy and delivery before and after ileal pouch-anal anastomosis for inflammatory bowel disease: immediate and long-term consequences and outcomes. Dis Colon Rectum 47:1127–1135.
44. Church J, Simmang C (2003) Standards Task Force of the American Society of Colon and Rectal Surgeons. Practice parameters for treatment of patients with dominantly colorectal cancer (familial adenomatous polyposis and hereditary nonpolyposis colorectal cancer). Dis Colon Rectum 46:1001–1012.
45. Bach SP, MacMortensen NJ (2006) Revolution and evolution: 30 years of ileoanal pouch surgery. Inflamm Bowel Dis 12:131–145.

Rectal Polyps after Ileorectal Anastomosis

17

What is the Future?

Francesco Tonelli and Rosa Valanzano

Abstract Options for prophylactic surgery include colectomy with ileorectal anastomosis (IRA) or proctocolectomy with ileal pouch-anal anastomosis (IPAA). IRA is generally considered to have the advantage of being less invasive, with better functional outcomes than IPAA, but lifelong follow-up and the risk of rectal cancer can be considered. However, the correct selection of patients suitable for IRA on the basis of the number and size of rectal polyps, age of patients, and genetics can assure good results.

Keywords Polyposis • Ileorectal anastomosis • MYH • Subtotal colectomy • Surgery

17.1 Surgical Options in Familial Adenomatous Polyposis

It was not until the 1930s that elective prophylactic surgery for familial adenomatous polyposis (FAP) was introduced, but the first total colectomy with ileorectal anastomosis (IRA) was not performed until 1948, by O.V. Lloyd-Davis at St Mark's Hospital, London [1].

IRA remained almost the only surgical procedure until the 1980s, when proctocolectomy with ileal pouch-anal anastomosis (IPAA) was introduced. Total proctocolectomy with definitive ileostomy was (and still is) reserved for patients with distal rectal cancer or sphincter problems.

IRA is a relatively simple procedure, with a mortality of less than 1% and a low rate of major morbidity (less than 10%) [2]. It avoids temporary ileostomy and assures functional results that are better than IPAA. In some countries, IPAA has gradually become the procedure of choice for FAP, since it avoids the rectal cancer risk and the need for endoscopic follow-up. However, the postoperative complication rate is higher than after IRA, and the long-term postoperative functional results are worse. A recent meta-analysis of 12 comparative studies, including 1002 patients, confirmed these observations: bowel frequency (weighted mean difference 1.62), night defecation (odds ratio 6.64), and use of incontinence pads (odds ratio 2.72) are lower in IRA than in IPAA [3]. However, major disadvantages of IRA are the need for continuous endoscopic follow-up, the appearance and/or the progression of rectal polyps, and the risk of cancer in the rectal stump.

In 2003, a Scandinavian study also demonstrated that female fecundity was reduced by approximately 50% after IPAA, whereas it was unaffected after IRA [4]. Moreover, despite the complete eradication of the rectal mucosa, development of ileal adenomas is becoming increasingly evident, with a prevalence that reaches 13–70% for a median follow-up of 4–6 years after surgery [5–7]. Also, cases of ileal pouch cancer

F. Tonelli (✉)
Department of Clinical Pathophysiology, Surgery Unit,
University of Florence, Florence, Italy

have been reported [8–10]. Therefore endoscopic follow-up is also necessary after IPAA.

In evaluating the risk of cancer in the rectal stump, we also have to consider whether data in the literature are referring to the pre-pouch or the pouch period. A recent study [11] reviewed all 1247 FAP patients submitted to IRA and regular follow-up from 1950 to 2006, and included in the national registry of Denmark, Finland, Sweden, and the Netherland. Rectal cancer developed in the pre-pouch period in 10% of patients, after a median follow-up of 16 years, and in only 2% during the pouch period (mean follow-up 7 years). Also proctectomy for all causes (especially severe rectal polyposis) was 40% in the pre-pouch period and only 30% in the post-pouch period. The same authors had reported a significant reduction in rectal cancer risk from 13% before the introduction of IPAA to none, during the pouch period, and a reduction of proctectomy from 32% to 2% [12]. These results suggest a better indication for IRA in the last period in comparison with the previous period when IRA was the only possible reconstructive procedure.

Since 1965 we have selected FAP patients for IRA on the following criteria: absence of rectal cancer, fewer than ten polyps (size <1 cm), in the last 10 cm of the rectum, and compliance to a regular follow-up. IRA was therefore chosen in 34% of patients. None of them needed re-operation after a mean follow-up of 10 years [13]. Reviewing the patients treated with IRA after a mean follow-up of 12 years, we observed that all patients with fewer than ten rectal polyps at surgery developed a very low number of polyps after IRA (1.62 per year) [14]. A significant difference in the number of recurrent rectal polyps and need for polypectomies was observed in patients with more than ten polyps, including those with diminutive polyps. With respect to genotype, all patients with *APC* mutations at codon 1309 developed carpeting rectal polyposis requiring proctectomy. No cases of cancer in the rectal stump have yet been observed in these patients.

An adequate selection of patients, on the basis of phenotype and genotype, could still allow indication for IRA in a consistent proportion of FAP patients, assuring them safe long-term results.

Therefore, we now focus on the fate of the rectum after IRA, specifying the different factors that correlate with prognosis.

17.2 The Rectal Stump after IRA

17.2.1 The Evolution of Rectal Polyps

In some patients, a spontaneous decrease or disappearance of the polyps present in the rectum can be observed after IRA. The first cases were referred by Hubbard in 1957 [15] and by Dunphy et al [16] two years later. The first Italian observation was in 1970 in two patients aged 38 and 17 years, respectively [17]. The polyps regressed a few months after surgery; they gradually became flatter and completely disappeared within one year. This was observed in 64% of 88 patients of the Mayo Clinic [18]. The regression was complete in 38% of these patients and partial in 26%. The regression does not seem to be strictly correlated to the number of rectal polyps, being equally reported in patients with few (less than 20) or many polyps (more than 100) before surgery. However, in these patients, regression is rarely complete [19]. The regression of the rectal polyps seems to be correlated with modification of the feces after IRA, and direct contact of the ileal feces with rectal mucosa. The role of feces in regression of polyps could be supported by the observation of a lower number of polyps in the right colon (especially in the cecum), and by the fact that the polyps start to disappear near the ileorectal anastomosis that is in direct contact with the ileal fluid. However, no data showed any modification in pH, bacteria, or particular enzymes in the ileal fluid that could explain such an observation [20].

17.2.2 Risk Factors for Polyps and Carcinoma of the Rectal Stump

The cumulative risk reported in the literature is extremely variable. In the Mayo Clinic experience, cancer of the rectal stump develops in 46 out of 178 IRA patients and increases over time, being 5% after 5 years and 59% after 23 years [21]. These data seem even more impressive, since 35 of the patients observed in this study did not show rectal polyps at surgery, and only 126 had a definite diagnosis of FAP. However, in other studies, including data from St. Mark's Hospital [22] and from the Japanese Register [23], the risk seems lower: 3–5% after 5 years, 5–13% after 10 years, 10–25% after 20 years

[12,21,24], and 32% after 40 years [12]. It also increases with age and can be estimated at 30% at 60 years of age [25].

Many factors influence such a risk:

- *the number of rectal polyps*: in a review from Bess et al [21], no patients without rectal polyps at surgery developed rectal cancer, although some of them exhibited polyps at follow-up. However, the total number of colorectal polyps also seems important, so that Slors et al [26] found only three cases of rectal cancer in 44 IRA patients with a *sparse* polyposis. Other authors [23] have reported a high risk (13%) in patients with *diffuse* polyposis. It is extremely important to accurately count the exact number of all rectal (and colonic) polyps to assess the type of polyposis. A colonoscope with magnification, and/or a chromoendoscope, could help to assess the exact number, being able to identify *diminutive* polyps that could cause a misleading evaluation. Since the polyps count is performed on individuals of different ages, one could argue that the difference in the number of polyps might simply reflect different patient ages, instead of different disease phenotype. However, the number of polyps increases only slowly from the time of clinical presentation [27,28], so that the number of polyps assessed at the time of surgery can be considered as defined and can be used in the therapeutic decision
- *the size of polyps*: the risk of cancer is strictly correlated to the size of polyps, being less than 2% in the case of polyps smaller than 5 mm [29], but rising to 47% for polyps of 1 cm [30]
- *age at diagnosis*: in patients below 15 years, the cancer risk is about 1%, but it rises to 23% at the age of 26 years [31,32]. Other authors [33] also report a correlation with age at surgery, with a significantly higher incidence of rectal cancer in subjects operated after 25 years of age, compared to those operated before the age of 25 years. The mean age of onset of rectal stump cancer is about 46–48 years [21,24,34], with a progressively increasing risk that reaches 14–25.7% at the age of 60 years [24,25,34]
- *the length of the rectal stump*: Watne et al [35] observed rectal cancer in 28% and 14% of patients with anastomosis above 14 cm and below 14 cm, respectively. Similar data are reported by Iwama and Mishima [23], who observed rectal cancer in 2 out of 62 patients with a short rectal stump (≤7 cm) compared to 27 out of 161 patients with a rectal stump longer than 7 cm. Also, in an Italian study [13], the incidence of cancer was 6% in cases of ileorectal anastomosis performed at 11–15 cm from the anal verge, and 60% in patients submitted to ileosigmoidostomy
- *the presence of colonic carcinoma at surgery*: data from the literature are not uniform – in the Mayo Clinic experience [21], and in the data of the Italian Register of FAP [36], rectal cancer risk is significantly higher in patients with colonic cancer at surgery. On the other hand, no differences were observed in the Cleveland Clinic [18] and in the Japanese experiences [23] between patients affected or not by colonic cancer
- *the type of mutation*: patients with *APC* mutations located in a region between codons 1251 and 1455 at exon 15 tend to have an aggressive form of disease with development of polyposis, the presence of symptoms, and onset of colorectal cancer at an early age. In particular, mutation in codon 1309 is associated with a severe form of FAP, with a number of colonic polyps that may exceed 5000, and early development of colorectal cancer [34,37,38]. On the other hand, patients carrying germline *APC* mutations in the 3' and 5'-ends show a mild phenotype (AFAP).

It has recently been reported in the literature that some cases of FAP are associated with mutations in the base excision repair gene *MYH*. This gene is associated with an AFAP with 11 to 100 colorectal polyps. The diagnosis occurs at an older age than classic polyposis, and no extraintestinal manifestations are reported. Gastroduodenal polyps are observed, although with a lower frequency compared to the classic FAP [39,40]. The transmission of the disease is compatible with a recessive transmission [41,42]. At present, *MYH* mutations could be involved in approximately 10–20% of AFAP cases [43–45].

17.2.3 Treatment of Rectal Polyps

If we have correctly chosen IRA, we should have no or few small rectal polyps at surgery. It is therefore possible to coagulate those located in the upper colon at surgery, and to treat later those in the lower third of the rectum. In the case of polyps that are larger than 5 mm, it is preferable to remove them preoperatively, in order to assess their histology and possibly modify the surgical choice. Few authors prefer to treat rectal polyps after IRA, because of their possible spontaneous regression after surgery.

Postoperative intensive follow-up is recommended. Sigmoidoscopy is usually scheduled once a year, although it can be altered on the basis of the number and size of polyps and genotype. Endoscopy should be modified with time, becoming more frequent (every six months) after 50 years of age when the risk of rectal stump increases.

17.3 Conclusion

IRA can be considered an appropriate surgical procedure for the treatment of colonic polyposis in 30–40% of FAP patients. Almost all the patients with an AFAP (due to *APC* or *MYH* gene) are suitable for IRA. An accurate endoscopic evaluation of the polyps present in the last 10 cm of the rectum before surgery is mandatory. IRA is indicated only in patients with fewer than ten polyps, independent of their size. A low side-to-end IRA anastomosis (resecting the rectum at the peritoneal reflection and preserving only the distal part) is desirable, since this procedure assures the same functional result as high IRA, but with a lower risk of rectal stump cancer.

References

1. Lockhart-Mummery JP (1934) Causation and treatment of multiple adenomatosis. Ann Surg 99:178–184.
2. Bussey HJR, Eyers AA, Ritchie SM, et al (1985) The rectum in adenomatous polyposis: the St Mark's policy. Br J Surg 72:S29–31.
3. Aziz O, Athanasiou T, Fazio VW, et al (2006) Meta-analysis of observational studies of ileorectal versus ileal pouch-anal anastomosis for familial adenomatous polyposis. Br J Surg 93:2407–2417.
4. Olsen KO, Juul S, Bulow S, et al (2003) Female fecundity before and after operation for familial adenomatous polyposis. Br J Surg 90:227–231.
5. Groves CJ, Beveridge G, Swain DJ, et al (2005) Prevalence and morphology of pouch and ileal adenomas in familial adenomatous polyposis. Dis Colon Rectum 48:816–823.
6. Wu J, Paul P, McGannon E, et al (1998) APC genotype, polyp number and surgical options in familial adenomatous polyposis. Ann Surg 227:57–562.
7. Moussata DN Nancey S, Lapalus MG, et al (2008) Frequency and severity of ileal adenomas in familial adenomatous polyposis after colectomy. Endoscopy 40:120–125.
8. Bassouini MM, Billings PJ (1996) Carcinoma in an ileal pouch after restorative proctocoectomy. Br J Surg 83:506.
9. Church J (2005) Ileoanal pouch neoplasia in familial adenomatous polyposis: an underestimated threat. Dis Colon Rectum 48:1708–1713.
10. Campos FG, Habr-Gama A, Kiss DR, et al (2005) Adenocarcinoma after ileoanal anastomosis for familial adenomatous polyposis: review of risk factors and current surveillance apropose a case. J Gastrointest Surg 9:695–702.
11. Bulow S, Bılow C, Vasen H, et al (2008) Colectomy and ileorectal anastomosis is still an option for selected patients with familial adenomatous polyposis. Dis Colon Rectum 51:1318–1323.
12. Bulow C, Vasen H Jarvinen H, et al (2000) Ileorectal anastomosis is appropriate for a subset of patients with familial adenomatous polyposis. Gastroenterology 119:1454–1460.
13. Tonelli F, Valanzano R, Monaci J, et al (1997) Restorative proctocolectomy or rectum-preserving surgery in patients with familial adenomatous polyposis. World J Surg 21:653–659.
14. Valanzano R, Ficari F, MC Curia, et al (2007) Balance between endoscopic and genetic information in the choice of ileorectal anastomosis for familial adenomatous polyposis. J Surg Oncol 95:28–33.
15. Hubbard TB Jr (1957) Familial adenomatous polyposis of the colon: the fate of the retained rectum after colectomy in children. Am Surg 23:577–586.
16. Dunphy JE, Patterson WB, Legg MA (1959) Etiologic factors in polyposis and carcinoma of the colon. Ann Surg 150:488–495.
17. Tonelli F (1970) Sulla scomparsa dei polipi rettali dopo colectomia totale ed ileorettoanastomosi per poliposi diffusa. [On the passing of rectal polyps after total colectomy and ileorectal anastomosis for polyposis.] Progr Med (Napoli) 26:353–359.
18. Feinberg SM, Jagelman DG, Sarre RG, et al (1988) Spontaneous resolution of rectal polyps in patients with familial polyposis following abdominal colectomy and ileorectal anastomosis. Dis Colon Rectum 31:169–175.
19. Farmer KCR, Plillips RSK (1993) Colectomy with ileorectal anastomosis lowers rectal mucosal cell proliferation in familial adenomatous polyposis. Dis Colon Rectum 36:167–171.
20. Meijers-Severs GI, Cats A, Verschueren RC, et al (1993) Anaerobes and their fermentation products in faeces of patients with familial adenomatous polyposis before and after subtotal colectomy and ileorectal anastomosis. Eur J Clin Invest 23:356–360.
21. Bess MA, Adson MA, Elveback LR (1980) Rectal cancer risk in patients treated for polyposis. Arch Surg 460–467.
22. Nugent KP, Phillips RS (1992) Rectal cancer risk in older patients with familial adenomatous polyposis and ileorectal anastomosis: a cause of concern. Br J Surg 79:1204–1206.
23. Iwama T, Mishima Y (1994) Factors affecting the risk of rectal cancer following rectum-preserving surgery in patients with familial adenomatous polyposis. Dis Colon Rectum 37:1024–1026.
24. De Cosse JJ, Bulow S, Neale K, et al (1992) Rectal cancer risk in patients treated for familial adenomatous polyposis. The Leeds Castle Polyposis Group. Br J Surg 79:1372–1375.
25. Haiskanen I, Jarvinen HJ (1997) Fate of the rectal stump after colectomy and ileorectal anastomosis for familial

adenomatous polyposis. Int J Colectal Dis 12:9–13.
26. Slors JMM, den Hartog-Jager FCA, Trum JW, et al (1989) Long-term follow-up after colectomy and ileorectal anastomosis in familial adenomatous polyposis coli: is there still a place for the procedure? Hepatogastroenterology 36:169–173.
27. Bodmer W (1999) Familial adenomatous polyposis (FAP) and its gene, *APC*. Cytogenet Cell Genet 86:99–104.
28. Cabtree MD, Tomlinson IPM, Hodgson SV, et al (2001) Molecular analysis of the *APC* gene in 205 Duch kindreds with familial adenomatous polyposis: relationship between genotype and phenotype and evidence for modifier genes. Gut 51:420–423.
29. Utsunomiya J (1989) Pathology, genetics, and management of hereditary gastrointestinal polyposis. In: Lynch HT, Hirayama T (eds) Genetic epidemiology and cancer. CRC Press, Boca Raton, FL, pp 219–249.
30. Watne AL (1987) Pattern of inheritance of colonic polyps. Semin Surg Oncol 3:71–76.
31. Mills SJ, Chapman PD, Burn J, Gunn A (1997) Endoscopic screening for familial adenomatous polyposis: dangerous delay. Br J Surg 84:74–77.
32. Iwama T, Mishima Y, Utsunomija J, et al (1993) The impact of familial adenomatous polyposis on the tumorigenesis and mortality at several organs. Ann Surg 217:101–118.
33. Jang YJ, Steinhagen RM, Heimann TN (1997) Colorectal cancer in familial adenomatous polyposis. Dis Colon Rectum 40:312–315.
34. Nugent KP, Philips RKS, Hodgson SV, et al (1994) Phenotypic expression of familial adenomatous polyposis: partial prediction by mutation. Gut 35:1622–1623.
35. Watne AL, Carrer JM, Durham JP, et al (1983) The occurence of carcinoma of the rectum following ileoproctostomy for familial adenomatous polyposis. Ann Surg 197:550–554.
36. Bertario L, Russo A, Radice P, et al (2000) Genotype and phenotype factors as determinants for rectal stump cancer in patients with familial adenomatous polyposis. Ann Surg 231:538–543.
37. Vasen HF, van der Luijt RB, Slors JF, et al (1996) Molecular tests as a guide to surgical management of familial adenomatous polyposis. Lancet 348:433–435.
38. Nagase H, Miyosh Y, Horii A, et al (1992) Correlation between the location of germ-line mutations of *APC* gene and the number of colorectal polyps in familial adenomatous polyposis. Cancer Res 52:4055–4057.
39. Nielsen M, Franken PF, Reinards TH, et al (2005) Multiplicity in polyp count and extacolonic manifestations in 40 Dutch patients with MYH associated polyposis coli (MAP). J Med Genet 42:54–57.
40. Bouguen G, Manfredi S, Blayau M, et al (2007) Colorectal adenomatous polyposis associated with MYH mutations: genotype and phenotype characteristics. Dis Colon Rectum 50:1612–1617.
41. Sieber OM, Lipton L, Cabtree M, et al (2003) Multiple colorectal adenomas, classic adenomatous polyposis, and germ-line mutations in *MYH*. N Engl J Med 348:791–799.
42. Sampson JR, Dolwani S, Jones S, et al (2003) Autosomal recessive colorectal adenomatous polyposis due to inherited mutations of *MYH*. Lancet 362:39–41.
43. Venesio T, Molatore S, Cattaneo F, et al (2004) High frequency of *MYH* gene mutations in a subset of patients with familial adenomatous polyposis. Gastroenterology 126:1681–1685.
44. Lamlum H, Tassan N, Jaeger E, et al (2000) APC variants in patients with multiple colorectal adenomas, with evidence for the particolar importance of E1317Q. Hum Mol Genet 9:2215–2221.
45. Heinimann K, Thompson A, Locher A, et al (2001) Non-truncating APC germ-line mutations and mismatch repair deficiency play a minor role in *APC* mutation-negative polyposis. Cancer Res 61:7616–7622.

Anorectal Polyps and Polypoid Lesions

18

Tomáš Skřička and Pavel Fabian

Abstract Typical histological findings in the most frequent anorectal polyps and polypoid lesions are demonstrated. Less frequent findings are also discussed, and different names for the same polyp explained. Finally, high-grade dysplasia (carcinoma *in situ*) is also discussed.

Keywords Anorectal polyps • Polypoid lesions • Typical pictures

18.1 Neoplastic Polyps

18.1.1 Adenomas

Adenomas are benign neoplastic anorectal polyps. Their classification may also be based upon the gross appearance: pedunculated (stalked), sessile, or flat adenoma. It has been generally taught that tubular adenomas are always pedunculated and that villous adenomas are characteristically sessile. However, tubular adenomas may be sessile and villous adenomas may be pedunculated. The adenomas are often darker or redder than the surrounding mucosa.

Adenomas are important clinically, because they are premalignant lesions that have the potential for developing into cancer. Although this chapter describes anorectal polyps, the reader should be aware that adenomas could, rarely, also occur in upper parts of the gastrointestinal tract. The most common incidence of adenomas is in the colon. The vast majority of what is discussed in this chapter relates to anorectal adenomas. Adenomas of the colon and small intestine will also be briefly discussed.

T. Skřička (✉)
Department of Surgery, Masaryk Memorial Cancer Institute, Brno, Czech Republic

Among asymptomatic individuals undergoing sigmoidoscopy, the prevalence of individuals with distal adenomas is approximately 18% [1]; for multiple adenomas (two or more) it is 36% and 50% respectively [2].

18.1.1.1 Tubular Adenoma

Tubular adenomas are the most common histologic type, accounting for 60–80% of neoplastic mucosal anorectal polypoid lesions (Fig. 18.1). The incidence increases with age, being extremely rare in individuals younger than 20 years. On endoscopic examination they may appear as pedunculated lesions with a stalk, or as sessile lesions with a broad base (Fig. 18.2). Histologically, tubular adenomas have complex branching glands. The likehood of a polyp being malignant is directly related to its size.

Most tubular adenomas feature only mild dysplasia. However, as many as 20% will demonstrate severe atypia, carcinoma *in situ*, or invasive carcinoma. Overall, only 5% of tubular adenomas are malignant [3].

Tubular adenomas tend to be spherical, and have a relatively smooth surface that is often divided into what appears to be "lobules" as a result of intercommunicating clefts in the head of the adenoma.

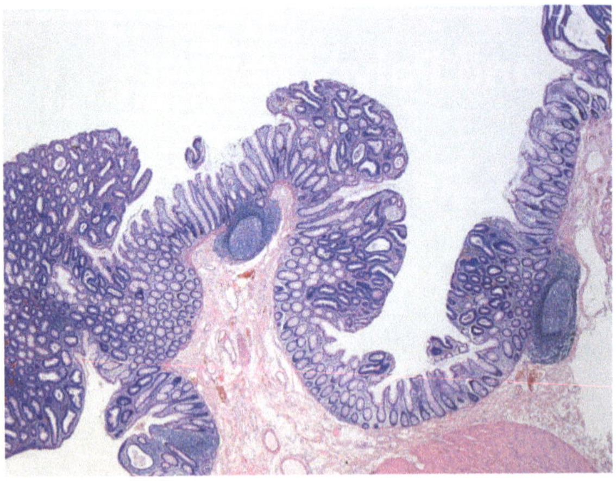

Fig. 18.1 Multiple tubular adenomas in a patient with familial adenomatous polyposis; H&E stain

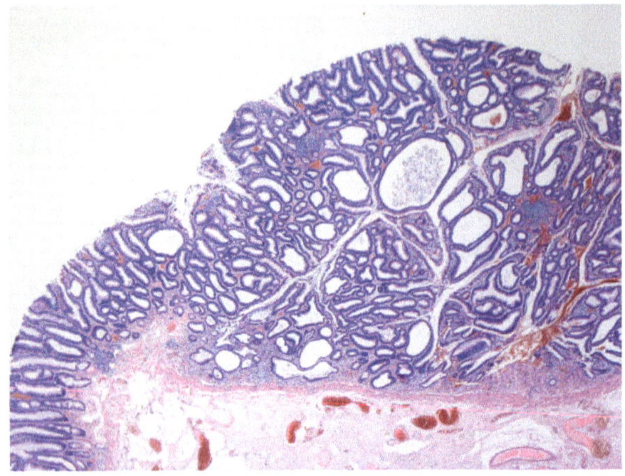

Fig. 18.2 Sessile tubular adenoma with low-grade epithelial dysplasia; H&E stain

In general, adenomas smaller than 1 cm are tubular adenomas.

18.1.1.2 Villous Adenoma

Villous adenomas tend to have a papillary or shaggy-carpet-like appearance. They usually have a "shaggy" surface with obvious papillary fonds.

These polyps have glands arranged in elongated, fingerlike patterns (Fig. 18.3). Most villous adenomas contain some tubular elements. The villous pattern is the one most commonly seen in polyps that are larger than 20 mm, and overall is the most likely to contain malignant foci (Fig. 18.4).

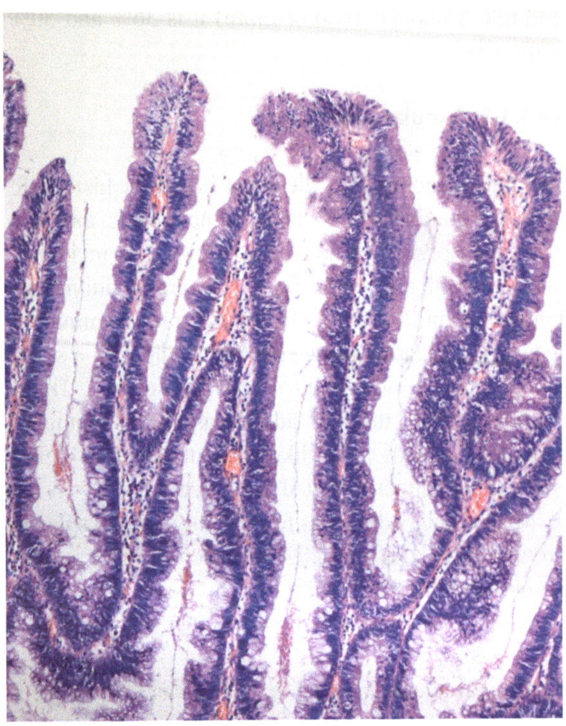

Fig. 18.3 Long, finger-like projections with slim fibrovascular core and low-grade dysplastic epithelia; H&E stain

18.1.1.3 Tubulo-Villous Adenoma (Papillar Adenoma)

These polyps contain the histologic features of both tubular adenomas and villous adenomas. There are both branching glandular patterns, as well as glands arranged in long fingerlike projections. The tubulo-

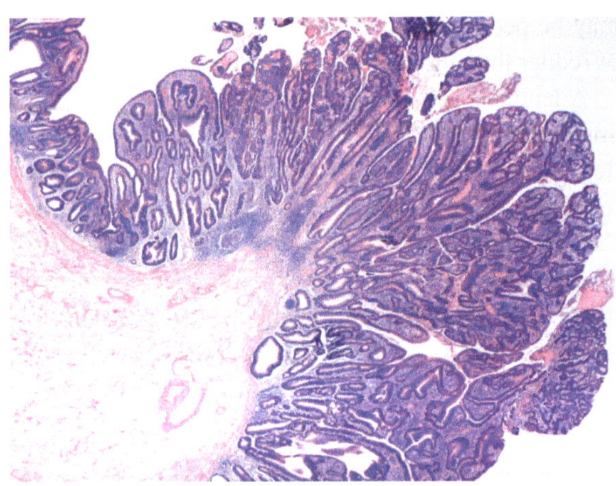

Fig. 18.4 Villous adenoma with superficial carcinomatous transformation (high-grade dysplasia); H&E stain

villous pattern is the most common one seen in polyps measuring 10–20 mm. Tubulo-villous adenomas have an intermediate likehood of being malignant.

18.1.1.4 Serrated Adenoma

Serrated adenoma is a specific type of colorectal adenoma, first described in 1990, and typically occuring in the sigmoid colon and rectum [4]. It usually does not exceed 2 cm in diameter, and can be diminutive, pedunculated, or sessile. Histologically, serrated adenomas have the architecture of a hyperplastic polyp (elongated and dilated crypts with a "saw-tooth" pattern (Fig. 18.5)); however, the lining epithelial cells show nuclear enlargement and other signs of dysplasia (Fig. 18.6) [5]. It is essential to avoid misdiagnosis as a hyperplastic polyp, because serrated adenomas have a similar frequency of high-grade dysplasia and carcinoma as other adenomatous polyps (risk of subsequent carcinoma is 5–6%) [6].

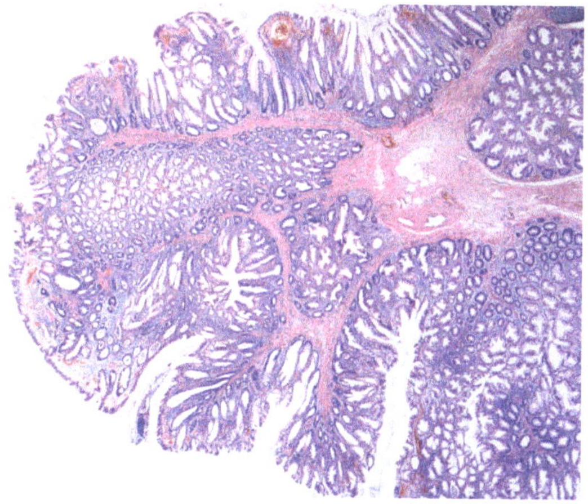

Fig. 18.5 Serrated adenoma: complex branching glands with a "saw-tooth" growth pattern; H&E stain

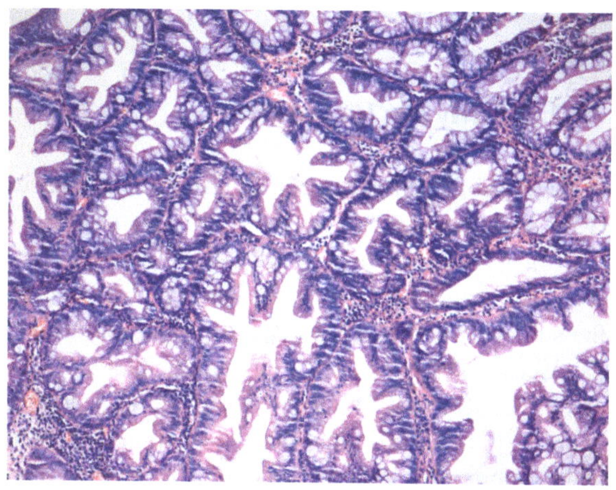

Fig. 18.6 Serrated adenoma: detail of dysplastic epithelial cells; H&E stain

18.1.2 Adenoma with Carcinoma

Anorectal adenomas are precancerous lesions. The incidence of high-grade dysplasia (carcinoma *in situ*) arising in anorectal adenomas is 12.3%, whereas the incidence of invasive cancer arising in anorectal adenomas is about 15% [7]. The term malignant polyp should be restricted to those adenomas (or polypoid carcinomas) in which there is true invasive cancer. The term invasive cancer should be used to describe only those lesions in which cancer has invaded beyond the muscularis mucosae into the submucosa. The lymphatics of the anorectum are closely associated with muscularis mucosae, and only after the cancer has invaded into the submucosa does it have the biological potential for metastasis. Cancer that is limited to the mucosa has been variously termed high-grade dysplasia, carcinoma *in situ*, or intramucosal adenocarcinoma. However, none of these entities has the biological potential for metastases [8].

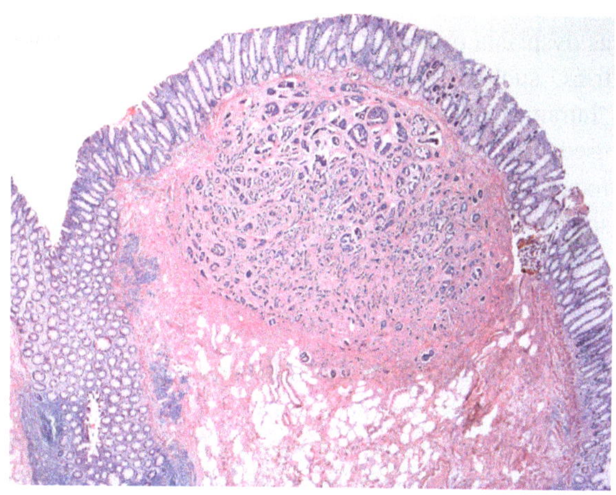

Fig. 18.7 Small, demarcated submucosal carcinoid; covering mucosa remains intact. Note lymphangioinvasion at the bottom of the tumorous infiltrate; H&E stain

18.1.3 Polypoid Carcinoid

This usually occurs in older patients, and commonly presents as a solitary, small nodule covered with intact mucosa (Fig. 18.7). Tumors smaller than 2 cm have a very low risk of metastasis [9]. Histologically they are circumscribed, unencapsulated, and mostly of microacinar architecture [10]. Sertonin production is rare, as well as carcinoid syndrome developement.

18.1.4 Mesenchymal Polyps [Lipomas, Leiomyomas, Hemangiomas, Lymphangiomas, Gastrointestinal Stromal Tumors (GISTs)]

Mesenchymal tumors arising from deeper layers of the rectal wall usually grow as small polypoid or even pedunculated masses covered with normal mucosa. Their histological structure is identical to respective tumors developing in other sites.

18.1.5 Early Rectal Cancer

A macroscopic classification of early rectal cancer (ERC) has been proposed by Kudo and resembles that for gastric cancer [11]. ERC is defined as invasive adenocarcinoma spreading into, but not beyond, the submucosa, that is a T1 tumor in the tumor node metastasis (TNM) classification (International Union against Cancer (UICC)). These tumors have a smaller chance of metastazing to local lymph nodes than adenocarcinoma, because they invade deeper than the submucosa and there is a paucity of lymphatics within colorectal mucosa [12]. Neoplastic cells confined to anorectal mucosa are correctly defined as dysplasia or adenoma (e.g. in UK). In other countries, such as the USA or Japan, the misnomers "intramucosal carcinoma" or "carcinoma *in situ*" are used. ERC may present as a polypoid carcinoma, a focus of malignancy within a large pedunculated or sessile adenoma, or as a small ulcerating adenocarcinoma.

18.1.6 Malignant Lymphomas

Bowel wall infiltration with non-Hodgkin's lymphomas may lead to multiple lymphomatous polyposis. Mantle cell lymphoma can typically present with gastrointestinal mucosa infiltration, either with or without lymph node involvement [13].

18.2 Non-neoplastic Polyps (Tumor-Like Lesions)

8.2.1 Hyperplastic Polyps

Hyperplastic polyps commonly present as sessile polyps that are less than 1 cm in size (Fig. 18.8). Athough traditionally considered non-neoplastic, *ras* mutation [14] and clonality have been demonstrated [15]. However, they are not associated with any risk of malignant transformation. Microscopically, crypt elongation and dilatation with a serrated pattern of bland epithelial proliferation with mucin hypersecretion is observed. Superficial cells lack cytonuclear atypia, which is an important feature in distinguishing serrated adenomas [8,16].

18.2.2 Hamartomatous Polyps

18.2.2.1 Peutz–Jeghers Polyps

These hamartomatous polyps consist of a complex mass of disorganized hyperplastic mucosal glands. Bundles of smooth muscle arborize throughout the polyp, dividing the glandular element into lobules (Fig. 18.9). Dysplasia is rare. Peutz–Jeghers syndrome, carrying an increased risk for cancer (gastrointestinal tract and other sites) development consists of multiple gastrointestinal Peutz–Jeghers polyps associated with brown macules on the skin around the mouth and eyes, and in the buccal mucosa [17].

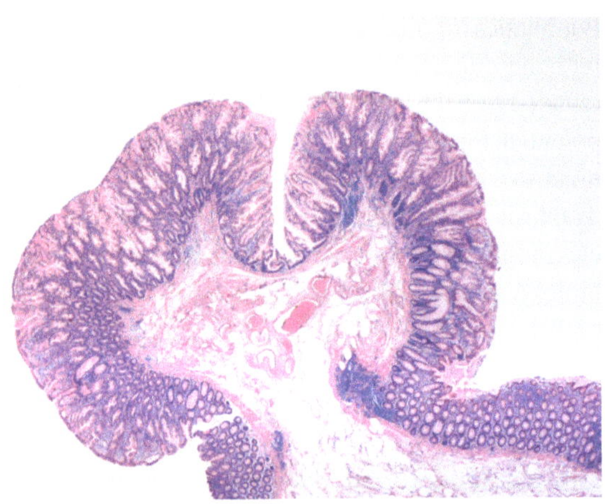

Fig. 18.8 Pedunculated hyperplastic polyp; H&E stain

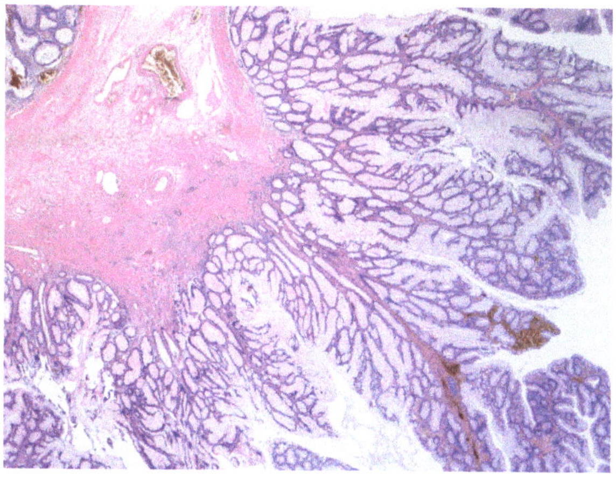

Fig. 18.9 Hamartomatous Peutz–Jeghers polyp. Note the non-dysplastic glands separated by bundles of smooth mucle cells, radiating from the muscularis mucosae layer; H&E stain

18.2.2.2 Juvenile Polyps (Cronkhite–Canada Syndrome)

Sporadic juvenile polyps occur predominantly in children and adolescents, mostly in the rectosigmoid colon. They are typically spherical, pedunculated, and lobulated, and on histology show an abundant edematous stroma vith mixed inflammatory infiltrate and cystically dilated glands containing mucin (Fig. 18.10). Surface erosion is common. The epithelial lining is either normal or exhibits mild reactive changes [18]. Familal juvenile polyposis (multiple juvenile polyps with no other extragastrointestinal lesions) carries a high risk of colon cancer developement. Cronkhite–Canada syndrome is a rare non-hereditary condition characterized by the presence of hundreds of juvenile polyps, brown skin macules, generalized alopecia, and atrophy of the nails, often presenting with diarrhea, abdominal pain, and protein-losing enteropathy [16]. The disease is commonly severe with a fatal outcome.

18.2.3 Benign Lymphoid Polyps

Sometimes, an excessive activation of mucosal and/or submucosal lymphoid tissue may produce macroscopically visible polypoid lesions. These usually do not exceed 15 mm, are symmetric, flat, and covered with normal mucosa. After they have completed initiation of antigenic stimulation, they disappear.

18.2.4 Inflammatory Polyps (Crohn's Disease, Ulcerative Colitis et al)

Inflammatory polyps (also pseudopolyps) consist of epithelium that is regenerating in response to inflammation (Fig. 18.11). They may arise secondary to any inflammatory process, but are most commonly associated with idiopathic ulcerative colitis. They can also develop secondary to an infectious process, such as amebic colitis, chronic schistosomiasis, or bacterial dysentery. Inflammatory polyps have no malignant potential, but in ulcerative colitis they may coexist

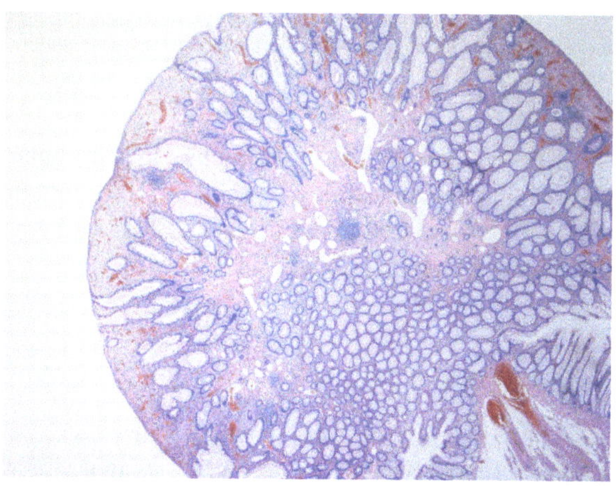

Fig. 18.10 Juvenile polyp: cystically dilated, mucin-containing glands, edematous stroma with imlammatory infiltrate. Superficial erosion (bottom left); H&E stain

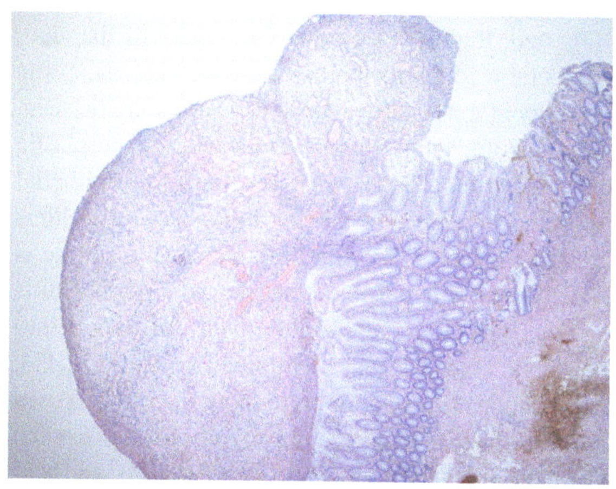

Fig. 18.11 Inflammatory polyp: polypoid lesion consisting mainly of granulation tissue; short stalk with normal mucosa; H&E stain

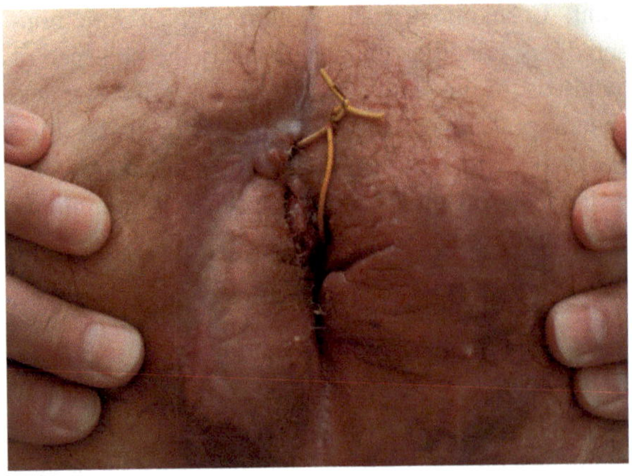

Fig. 18.12 Adenocarcinoma arising in the setting of long-lasting Crohn's proctitis

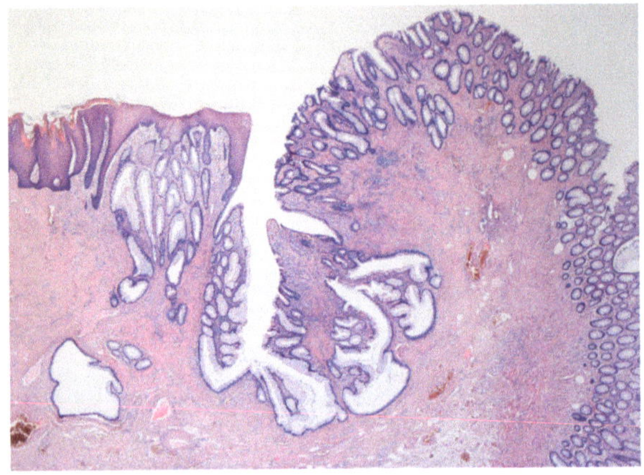

Fig. 18.13 Rectal prolapse syndrome: polypoid mass arising from the anorectal juntion. Distorted glands, partially separated in the deep rectal wall (colitis profunda cystica), inflammatory infiltrate, smooth muscle fibers intervening among the mucosal glands (a feature simulating an invasive cancer); H&E stain

with areas of dysplasia or malignancy. I have personally observed malignancy in Crohn's disease, treated nearly for 20 years by means of semi-invasive methods (Fig. 18.12).

18.2.5 Solitary Rectal Ulcer Syndrome (Localized Colitis Cystica Profunda)

This is a peculiar non-neoplastic condition simulating malignancy either endoscopically or microscopically. Most patients are healthy adults aged between 20 and 40 years. Rectal bleeding, mucus discharge, anorectal pain, and tenesmus are the most common symptoms. Some patients have detectable mucosal rectal prolapse, thus the term "mucosal prolapse syndrome" has been suggested. Endoscopically, solitary ulcer of the anterior rectal wall, erythematous granular mucosa, cystic or polypoid mass, or even villous lesion may be seen. [19]. Histologically, there are distorted, hyperplastic and elongated crypts with regenerative epithelial atypia, accompanied by typical fibromuscular proliferation in the lamina propria. Inflammatory infiltrate is usually scanty. In chronic cases, fibrosis often results in crypt distortion and displacement downward into the submucosa with cystic dilatation (Fig. 18.13). These changes are analogous to those of colitis cystica profunda. This localized form of colitis cystica profunda (hamartomatous inverted polyp) and so-called "inflammatory cloacogenic polyp" have been proposed as a part of solitary rectal ulcer syndrome [8,20].

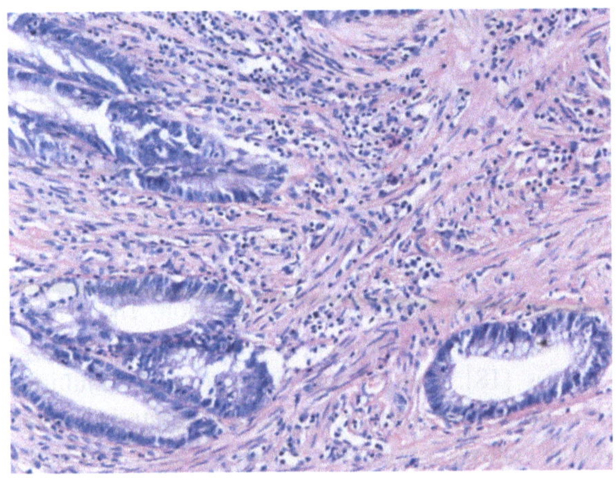

Fig. 18.14 Rectal prolapse syndrome: epithelial atypia, mitoses and surrounding smooth muscle fibers simulate invasive adenocarcinoma; H&E stain

In endoscopic biopsy specimens, the irregular glands with reactive atypia surrounded by smooth muscle fibers can lead to erroneous diagnosis of invasive adenocarcinoma [16,21] (Fig. 18.14).

References

1. Brady PG, Starker RJ, McClave SA, et al (1993) Are hyperplastic rectosigmoid polyps associated with an increased risk of proximal colic neoplasms? Gastrointest Endosc 39:481–485.
2. Konishi F, Morson BC (1982) Pathology of colorectal adenomas: a coloscopic survey. J Clin Pathol 35:830–841.

3. Imbembo AL, Lefor AT (1997) Benign neoplasms of the colon including vascular malformations. In: Sabiston DC Texbook of surgery. The biological basis of modern surgical praxis. W.B. Saunders Company, Philadelphia, pp993–1001.
4. Longacre TA, Fenoglio-Preiser CM (1990) Mixed hyperplastic adenomatous polyps/serrated adenomas. A distinct form of colorectal neoplasia. Am J Surg Pathol 14:524–537.
5. Harvey NT, Ruszkiewicz A (2007): Serrated neoplasia of the colorectum. World J Gastroenterol 13:3792–3798.
6. Song SY, Kim YH, Yu MK, et al (2007) Comparison of malignant potential between serrated adenomas and traditional adenomas. J Gastroenterol Hepatol 22:1786–1790.
7. Neuget AI, Johnsen CM, Forde KA, et al (1985) Recurrence rates for colorectal polyps. Cancer 55:1586–1589.
8. Hamilton SR, Vogelstein B, Kudo S, et al. (2000) Tumours of the colon and rectum. In: Hamilton SR, Aaltonen LA (eds) WHO classification of tumours, pathology & genetics, tumours of the digestive system. IARC Press, Lyon, pp 103–143.
9. Ballantyne GH, Savoca PE, Flannery JT, et al. (1992) Incidence and mortality of carcinoids of the colon. Data from the Connecticut Tumor Registry. Cancer 69:2400–2405.
10. Capella C, Solcia E, Sobin LH, et al. (2000) Endocrine tumours of the colon and rectum. In: Hamilton SR, Aaltonen LA (eds) WHO classification of tumours, pathology & genetics, tumours of the digestive system. IARC Press, Lyon, pp 137–139.
11. Kudo S (1993) Endoscopic mucosal resection of flat and depressed types of early colorectal cancer. Endoscopy 25:455–461.
12. Day DW, Jass JR, Price AB, et al (2003) Epithelial tumours of the large intestine. In: Morson and Dawson's gastrointestinal pathology, 4th edn. Blackwell Science, Oxford, pp 551–609.
13. Weisenburger DD, Armitage JO (1996) Mantle cell lymphoma – an entity comes of age. Blood 87:4483–494.
14. Zauber P, Sabbath-Solitare M, Marotta S, et al (2004) Comparative molecular pathology of sporadic hyperplastic polyps and neoplastic lesions from the same individual. J Clin Pathol 57:1084–1088.
15. Jen J, Powell SM, Papadopoulos N, et al (1994) Molecular determinants of dysplasia in colorectal lesions. Cancer Res 54:5523–5526.
16. Yanling M, Chandrasoma P (1999) Colorectal polyps and polyposis syndromes. In: Chandrasoma P (ed) Gastrointestinal Pathology. Appleton & Lange, Stamford, pp 313–338.
17. Aaltonen LA, Järvinen H, Gruber SB, et al. (2000) Peutz-Jeghers syndrome. In: Hamilton SR, Aaltonen LA (eds) WHO classification of tumours, pathology & genetics, tumours of the digestive system. IARC Press, Lyon, pp 74–76.
18. Soni R, Chandrasoma P (1999) Idiopathic Inflammatory Bowel Disease. In: Chandrasoma P (ed) Gastrointestinal Pathology. Appleton & Lange, Stamford, pp 283–312.
19. Kobayashi G, Chandrasoma P (1999) Nonneoplastic diseases of the colon. In: Chandrasoma P (ed) Gastrointestinal Pathology. Appleton & Lange, Stamford, pp 249–282.
20. du Boulay CE, Fairbrother J, Isaacson PG (1983) Mucosal prolapse syndrome – a unifying concept for solitary ulcer syndrome and related disorders. J Clin Pathol 36:1264–1268.
21. Singh B, Mortensen NJ, Warren BF (2007) Histopathological mimicry in mucosal prolapse. Histopathology 50:97–102.

Polypectomy of Anorectal Polyps and Polypoid Lesions

19

Why, When, and How?

Tomáš Skřička, Lenka Foretová and Pavel Fabian

Abstract The aim of this chapter was to evaluate minimally invasive methods to remove anorectal polyps and polypoid lesions. It seems, that in highly selected cases of early stages of rectal cancer, these can also be treated locally. We show wide-scale use of non-invasive or minimally invasive methods and try to define when and how to use them. Complications are also discussed. Last but not least, the argument for use of these methods argument is a cost–benefit one.

Keywords Anorectal polyps • Cost–benefit • Minimally invasive approach

19.1 Introduction

This chapter presents the rationale for local treatment of different kinds of polyps and small cancer of the rectum, as well as the methods that were or are still currently in use.

Many features of local treatment are attractive. These include removal of the need for major resection with its associated morbidity and mortality, including pain and prolonged recovery time. This is extremely important in the elderly and infirm. Major incentives for patients include maintenance of normal anal sphincter function, and male sexual function, and no need for a stoma. From the cost–benefit viewpoint, local treatment is cheaper, with a shorter hospital stay, fast return to social life, and relatively few complications [1].

T. Skřička (✉)
Department of Surgery, Masaryk Memorial Cancer Institute, Brno, Czech Republic

19.2 Methods of Semi-Invasive and Surgical Therapy of Anorectal Polyps

It is extremely rare for laparotomy or laparoscopy to be indicated for resection of anorectal polyps. Anorectal lesions may be excised locally, using one of a number of transanal techniques, with the goal of complete local excision with clear lateral and deep margins. Preoperatively, patients undergo a standard anorectal preparation with mechanical cleansing, Yal or low enema is sufficient. Some polyps may be succesfully excised in several pieces, using multiple passes with a snare. Patients with sessile polyps that are incompletely removed should undergo radical resection, but it is thought that coagulation of small amounts of residual tissue from a benign tumor is adequate if close follow-up is anticipated.

Large polyps of the rectum can be reached mostly transanally. Low-lying sessile lesions (within 6–8 cm from the anal verge) may be removed using traditional transanal excisional techniques. Various methods have been used for treating higher lesions (8 to 15 or 20 cm), including trans-sacral or trans-sphincteric

G.G. Delaini, T. Skřička, G. Colucci (eds.), *Intestinal Polyps and Polyposis*,

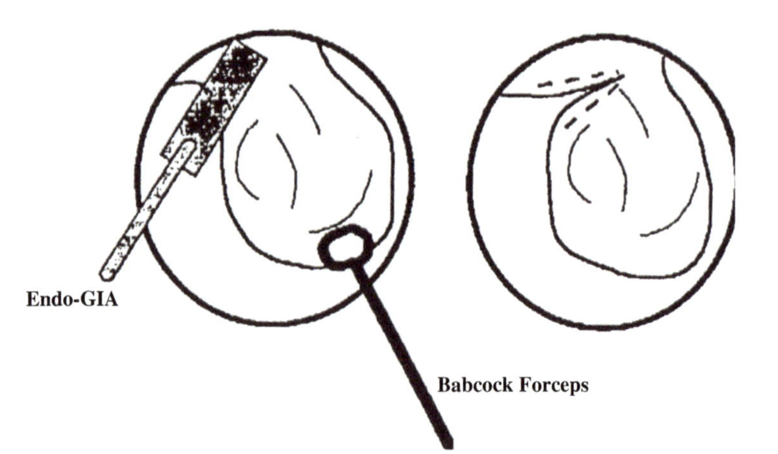

Fig. 19.1 Endo-GIA instrument and Babcock forceps. Reproduced from [4]

resection, electrocautery, repeated attempts at snare excision, transanal excision with a urologic resectoscope, and transanal microscosurgery. These methods are discssed in detail later in this chapter. Recurrence of adenomatous polyps may be treated by further local measures [1].

19.2.1 Direct Transanal Local Excision

The local excision of anorectal polyps is often wrongly considered to be a minor surgical procedure. In reality, the malignant potential of adenomas and the not-infrequent presence of cancer in larger polyps, require, for their removal, an oncologically correct operation with strict indication and accurate execution. Despite an increasing tendency to extend the indications for endoscopic polypectomies to polyps of larger size and villous configuration, the local surgical approach remains the preferred treatment in most cases [2].

Local excision is essentially a wide excisional biopsy. The major advantage of this method is that the entire lesion is available for pathologic examination. Although an occasional benign adenomatous lesion may be removed in the submucosal plane, most are excised through the thickness of the rectal wall. A clear margin is critical, and 1 cm of normal tissue is acceptable. Direct transanal excision of rectal neoplasms may be relatively easy when removing small, low lesions, but it may be a very technically demanding procedure with larger, higher lesions that are within 6–8 cm from the anal verge, that are less than 3–4 cm in diameter, and that are limited to one, or at most two, quadrants of rectal circumference.

Broad-based rectal polyps in the mid-rectum frequently present a management problem because of difficulty of access for local resection. Pai and Morgan [3] used endoloop, and Qureshi and coworkers [4] introduced the use of an Endo GIA instrument (Fig. 19.1), which has originally been removed through a rigid sigmoidoscope. The technique is very simple. The polyp is visualized using a rigid sigmoidoscope (we use the wider tube of transanal endoscopic microsurgery (TEM)), and the ability to prolapse the lesion into the lower third of rectum is assesed by the use of a grasping forceps. The rigid sigmoidoscope is withdrawn, and a bivalve speculum is inserted into the anal canal. The polyp is grasped with Babcock forceps (Fig. 19.1), and traction is applied to move the polyp to the right side of the rectum. A Multifire Endo GIA is introduced into the rectum and the base of the polyp is firmly grasped. The instrument is fired and the procedure repeated for the opposite side. This provides a hemostatic excision of the rectal polyp. According to the authors [4] as well as to our experience, this technique is advised for broad-based rectal lesions of 8–12 cm without any complications.

A similar technique was reported by Allison and coworkers [5] to be succesful in three patients.

Positioning a patient so that the lesion is centered in the dependent or inferior aspect of the operative field is very helpful. However, we prefer the jack-knife position in all cases, where there are no contraindications from the anestetist's point of view.

A recent publication [6] on the long-term results of this technique (30% of the polyps were 8–12 cm from the anal margin and 68% had a diameter larger than 4 cm) reports a recurrence rate of adenomas of 2% and 1.5% for carcinomas, with a complication rate similar to that for the classic Parks' approach.

Transanal local excision, when correctly performed, therefore remains a valuable treatment for rectal polyps of any dimension sited in the low and middle rectum. In experienced hands the excision can be also extended to polyps in the upper rectum.

Local excision of selected rectal tumors using the transanal or trans-sacral approach is an acceptable alternative to more extensive procedures such as abdominoperineal resection (APR) or low anterior resection (LAR). The inherent advantgages of local excision include lower patient morbidity and mortality, shorter hospital stay, and reduced hospital expense. The disadvantages include the inability to accurately select appropriate patients preoperatively who lack lymph node metastases, a slightly increased rate of local recurrence, and the lack of randomized prospective studies to compare the results of APR versus local excision for similarly staged tumours. In a study by Hoth and coworkers [7], the results of patients undergoing local excision of both adenomas and rectal malignances were reviewed to compare morbidity, mortality, and rates of recurrence (local/distant) in the malignant group with those of the historic controls for APR The study showed that results after local excision are comparable in selected cases.

19.2.2 Local Excision Using the Urologic Resectoscope

Many reports from the UK have described the use of a urologic resectoscope for transanal excision of large rectal adenomas and for palliation of rectal carcinomas [8]. This method allows treatment of lesions that are significantly higher than the 6–8 cm reached using standard transanal techniques. No significant anal dilatation is required for access, minimizing sphincter damage. The procedure is said to be simple, inexpensive, and well tolerated. The technique is adapted from those used by urologists and so may be relatively unfamiliar to most surgeons. A major criticism is that pathologic evaluation is severely limited because of the many small samples of tissue submitted. Little may be discerned about the depth of penetration or completeness of resection. This is determined grossly by the surgeon. Thus, when cancer is discovered, it is difficult to decide whether to observe or to pursue more aggressive surgery. This method almost completely substituted now by TEM, but it could play a role in the resection of large adenomas and for palliation of some cases of rectal cancer. It seems not to be good for curative treatment of early rectal malignances.

19.2.3 Transanal Endoscopic Microsurgery

An innovative and important contribution to the local treatment of rectal polyps has been a technique presented by Buess in 1984 and named transanal endoscopic microsurgery (TEM) [9]. This technique, originally presented as an alternative to the posterior approach, uses special equipment applied through the transanal route. I started to use it in 1996 [10]. After initial scepticism expressed by by experienced colorectal surgeons [11], in the last few years the technique has gained popularity such that, in specialized centers, it is adopted extensively for the local treatment of sessile polyps of the low, middle, and high rectum (up to 20 cm from the anal margin). Undoubtedly the increased adoption of the technique when extended also to palliative treatment of very advanced cancers amplifies its advantages and contains the disadvantages.

Advantages are:

- optimal field of vision
- more accurate dissection
- minimally invasive approach
- full-thickness excision
- excision of high polyps
- low morbidity.

Disadvantages are:

- complex and expensive instruments
- long training and learning curve.

TEM is also an adequate method for the local full-thickness excision of large rectum polyps and pT1 "low-risk" rectal carcinomas. In 2006, Schaefer studied prospectively the relevance of this surgical technique concerning complete tumor excision after R1–R2-polypectomy of malignant rectal polyps [12]; 16 patients with pT1 "low-risk" rectal carcinoma and macroscopic (R2) or microscopic (R1) incomplete endoscopic polypectomy were locally resected by TEM. In 12 patients (75%), no residual tumor was found. In the remaining four cases (25%), one adenoma with high-grade atypia, two pT1 "low-risk" carcinomas, and one tumor infiltration in the mesorectal fat were diagnosed. The patient with the mesorectal infiltration was immediately operated on with radical resection. No further tumor cells were found in this specimen. The median follow-up was 21 months. One patient with a pT1 "low-risk" carcinoma developed a local recurrence and a single hepatic metastasis in the left liver lobe after TEM. Both were completely resected. Currently, all patients are alive, with no evidence of tumor recurrence. TEM is a suitable method for the treatment of pT1 "low-risk" rectal carcinomas after incomplete endoscopic polypectomy. In cases of a "high-risk" tumor or deeper tumor infiltration (pT2 and more), radical resection must be carried out after TEM.

19.2.4 Posterior Approaches to the Rectum

The idea of locally removing a neoplasm in the middle rectum from a posterior approach, with the patient lying prone in the jack-knife position or in the left lateral decubitus goes back a long time. Kraske (1885) was the first to propose the excision of the coccyx and the last sacral vertebrae to remove the rectum with the cancer [13]. In 1970, Mason proposed incision of the posterior sphincteric structures and levator muscle to gain access to the middle/lower rectum for removing it completely or for opening its posterior wall and removing polypoid lesions. This approach is still sometimes used [14].

In particular, the posterior approach (fequently represented by a combination of coccygeal removal and partial sphincteric division) seemed to be indicated for the excision of polyps of average dimensions (diameterer 4–5 cm) situated in the anterior wall at a distance of 8–10 cm from the anal verge. Undoubtedly, the visualization of the tumor obtained with such an approach is much better than that obtained using the transanal route. Despite this disadvantage, the role of the posterior route has never been fully established, even though I personally have carried out more than 40 such interventions. Certainly it is not indicated for removal of polyps of the posterior wall or with circumferential extension, due to the risk of cutting into the tumor and implanting neoplastic cells. In the medical literature only 360 posterior approaches have been reported [15]. They included different lesions treated in different ways (submucosal excision, full-thickness excision, pararectal tube resection). It is not surprising that the results regarding postoperative complications and recurrences rates varied greatly between the different experiences. Morbidity from fecal fistula formation, which is the most common postoperative complication, has been repoted to occur in 5–70% of cases, and fecal incontinence in 5–25%. There is a need for a stoma in 20–70% of patients and recurrence of adenoma in 3–33% of the cases. This variability has obviously created some concern about the appropriateness of the posterior approach. In my own limited experience of 13 patients with polyps of the anterior rectal wall, I had three fecal fistulas (one permanent) and three wound infections.

In conclusion, the posterior approach for local excision of a rectal polyp has been progressively abandoned and recent reports of its use are anecdotal.

However, situations are also found in the literature where combined transanal excision of a large rectal polyp could be assisted by trans-sacral manipulation of the rectum [16]. It is recognized that transanal excision of rectal polyps is curative and less invasive than trans-sacral resection of low anterior resection, but it is difficult to resect tumors that are distant from the anal verge. Moreover, in the case of large polyps, the risk of complications, such as hemorrhage or perforation, increases because exposure on the oral side of the tumor is poor. Otsuji and colleagues have described trans-sacral manual assistance to achieve transanal resection of a large tubulo-villous adenoma of the rectum that was hard to resect using the traditional transanal approach [16]. They were not able to assess the proximal extent of the tumor. The tumor could also not be prolapsed because of its large size and the distance from the anal verge to the proximal margin of the tumor. To avoid incision or resection of the rectum in this patient, who was at high risk for postoperative complications, trans-sacral manual assistance was used to achieve transanal resection.

19.2.5 Thermal Destructions and Lasers

19.2.5.1 Electrocoagulation, Fulguration

This method has been popular in the USSR, where it

has been widely used [17]. Results of treatment in 230 patients with malignant transformation of polyps and villous adenomas, carcinoid, and cancer of the rectum and sigmoid have been reported [17]. Cancer patients were not operated radically, because of their general physical condition, or they refused stoma. In malignant polyps, villous tumors, and rectal carcinoids, electroexcision or electrocoagulation seemed to be an adequate procedure. The patients had to be examined every month during the fisrt year postoperatively, and later once every six months. Recently, this procedure has been limited mostly to a palliative one. Another report has shown that the five-year survival rate among selected patients who received electrocoagulation therapy may be similar to that of patients who undergo abdominoperineal resection [18]. This has led to renewed interest in electrocoagulation and its potential curative and palliative benefit in selected patients with rectal cancer.

It has been estimated that approximately 10–80% of patients with rectal cancer do not benefit from a curative surgical procedure secondary to advanced local or regional disease, metastatic disease, or significant concurrent medical illness [19].

19.2.5.2 Laser Vaporization

Lasers have been used in a number of ways in surgery in the past 15–20 years. They may be used to cut tissue like a knife, they may be used to vaporize tissue, they may be used for coagulation, or they may be used as an exogenous energy source to activate substances in tissue cells (photodynamic therapy). There are different types of lasers, including carbon dioxide, neodymium:yttrium–albuminum–garnet, argon, and dye lasers.

In colorectal surgery, the laser offers no advantage over conventional blades, electrocautery, scissors, or a harmonic scalpel for cutting. Indeed, the expense and time required would offset any but the greatest improvements. Laser vaporization is useful in the palliation of obstructing tumors, particularly of the rectum and rectosigmoid junction [20]. Laser vaporization may be used in the same way as electrocautery for local destruction of rectal polyps of known origin (previous histology) or small cancers. Recently this method has been used less and less.

Lasers could be also used for the treatment of rectal bleeding.

19.2.5.3 Photodynamic Therapy

Lasers may be used to stimulate compounds contained within tissue cells. After intravenous injection, certain sensitizing agents (e.g. porphyrin derivates) are selectively retained by malignant cells. With laser light of a specific wavelength and in the presence of oxygen, these compounds are excited and cause cell destruction. This approach has limitations, because the optimal laser wavelength for hematoporphyrin derivative excitation has very little tissue-penetrating capability. Effective tissue destruction is limited to about 2–3 mm in depth. Several groups are now combining endoscopic snare excision followed by phototherapy of the residual tissue. Treatment may need to be repeated several times at 2-week intervals. The proponents of this combined approach believe that it reduces the number of treatments that are necessary [21].

19.2.5.4 Cryotherapy

This method is limited mostly to palliative care. As a radical reatment it is only used ocasionally. When there is a local recurrence of rectal cancer, in most cases it is unresectable and incurable. Relief of symptoms through palliative therapy is the mainstay of treatment. Cryosurgery is an available option that warrants discussion. The advent of the insulated liquid nitrogen cryoprobe in 1964 has made cryosurgery an available resource for palliative treatment of unresectable rectal cancer. In 1978, Osborne reported results of cryosurgery in ten patients who had rectal carcinoma with distant metastases [22]. Good palliation without local complications was achieved without general anesthesia. The cryotherapy temperature was set at 1-180°C, and the interventional field included 2–3 mm of normal mucosa to ensure adequate destruction. During this process, gross tissue fluid and electrolyte changes occur during the freezing process, which in turn causes protein denaturation, enzyme inhibition, capillary thrombosis, and venous stasis. Recent trials of cryosurgery are rare and most of our current knowledge of the concept stems from studies performed more than two decades ago. In summary, cryotherapy is a relatively safe means of palliation for recurrent rectal cancer, offering effective relief of symptoms to approximately 50% of patients with rectal cancer that is not amenable to surgical resection.

19.3 Complications

19.3.1 Anorectal Perforation

Anorectal perforations are rare but serious complications of intestinal endoscopy. Nelson (1982) reported only three perforations in 16,325 proctoscopic examinations [23]. Colorectal perforation during barium contrast studies occurs in approximately 0.02–0.04% of patients, usually in the rectum above the peritoneal reflection [24]. Other iatrogenic anorectal injuries result mostly from surgical procedures. Their incidence is unknown, because the data are rarely published. Significant rectal trauma with rectal perforation requires surgical treatment. Most authors believe that delayed treatment increases mortality from 8% to 20%. Major morbidity and mortality are associated with non-mechanical cleansing of the proximal colon. Treatment involves resection of associated disease, repair of the perforation site, rectal washout, pelvic drainage, and diverting stoma (ileostomy or colostomy).

19.3.2 Hemorrhage

Hemorrhage in the recovery room is the result of an error of technique and is invariably caused by either a missed or a slipped ligature. Occasionally, bleeding may continue undetected, with blood accummulation in a capacious rectum. The first sign of this complication may be pallor, tachycardia, and hypotension, which may require resuscitation with intravenous fluid and blood transfusion. The patient must be returned to the operating room and the bleeding arrested with either diathermy coagulation or suture of the bleeding site. Historically, some have advocated methods such as wound packing and direct pressure at the bleeding site [25,26]. It must be remembered however, that packing alone may provide a false sense of security because of little evidence of external bleeding, while progressive retrograde accummulation of blood leads to deterioration in the patient's condition.

19.3.3 Urinary Retention

The cause of urinary retention after anorectal polypectomy is multifactorial. Postoperative pain is the chief cause of urethral sphincter spasm. Anal packing has been associated with the urinary retention, probably from direct pressure on the bladder outflow track. Spinal anesthesia, injudicious use of intravenous fluids, and opiate analgesics could all lead to urinary retention. Occasionally in elderly men, obstructive uropathy may progress to acute retention of urine after anorectal polypectomy.

19.3.4 Postoperative Pain

Pain may arise chiefly from involment of anoderm below the dentate line, so mostly occurs after anal polypectomy. Polypectomy higher than the dentate line rarely causes postoperative pain.

19.4 Conclusion

In view of recent knowledge, the current trend is toward a minimally invasive approach for anorectal polyps and early stages of cancer. The benefits are a shorter hospital stay, and possibly a lowered prevalence of complications. In addition, patients must be informed of the possibility of recurrence and potential complications of all procedures.

References

1. Orkin BA (1995) Local treatment of rectal neoplasms. In: Mazier WP, et al (eds) Surgery of the colon, rectum and anus. W.B.Saunders Company, Philadelphia, pp 470–489.
2. Fucini C, Segre D, Trompetto M (2004) Local excision of rectal polyp: indications and techniques. Tech Coloproctol 8:300–304.
3. Pai KP, Morgan RH (1995) Endoloop: a helpful new device for transanal rectal polypectomy. Br J Surg 82:784.
4. Qureshi MA, Monson JR, Lee PW (1997) Transanal MULTIFIRE ENDO GIA technique for rectal polypectomy. Dis Colon Rectum 40:116.
5. Allison SI, Adedeji OA, Varma JS (2001) Per anal excision of large rectal adenomas using endoscopic stapler. JR Coll Surg Edinb 46:290–291.
6. Pigot F, Bouchard D, Mortaji M, et al (2003) Local excision of large rectal villous adenomas. Dis Colon rectum 46:1345–1350.
7. Hoth JJ, Waters GS, Penell TC (2000) Results of local excision of benign and malignant rectal lesions. Am Surg 66:1099–1103.
8. Stephenson BB, Shandall AA, Ng KJ, et al (1992) Endoscopic transanal resection of large villous tumors of the

rectum. Ann R Coll Surg Engl 74:54–58.
9. Buess G, Hutterer F, Thiess J, et al (1984) A system for a transanal endoscopic rectum operation. Chirurg 55:677–80.
10. Skřička T (2005) Local excision of rectal cancer: TEM. In: Delaini GG (ed.) Rectal cancer. New frontiers in diagnosis, treatment and rehabilitation. Springer-Verlag Italia, Milan, pp 107–114.
11. Corman M (1991) Principles of surgical technique in the treatment of carcinoma of the large bowel. World J Surg 15:592–596.
12. Schaefer HH, Vivaldi C, Hoelscher AH (2006) Local excision with transanal endoscopis miscosurgery (TEM) after endoscopic R1–R2 polypectomy of pT1 "low-risk" carcinomas of the rectum. Z Gastroenterol 44:647–650.
13. Kraske P (1885) Zur Exstirpation hochsitzender Mastdarmkrebse. Verh Chir 14:464–466.
14. Qiu HZ, Lin GL, Xiao Y, Wu B (2008) Surgery of the rectum: a Chinese 16-year experience. World J Surg 323:1776–1782.
15. Groebli Y, Tschantz P (1994) Should the posterior approach to the rectum be forgotten? Helv Chir Acta 60:599–604.
16. Otsuji E, Fujiyama J, Takagi T, et al (2004) Transanal excision of a large rectal polyp assisted by transsacral manipulation of the rectum. Dis Colon Rectum 47:1420–1422.
17. Fedorov VD, Brusilkovskij MI, Safina FM, Sadovniãij VA (1975) Ekonomnyje operacii pri zlokačestvennych opucholjach prjamoj i sigmovidnoj kišok. Voprosy Onkologii 22:41–45.
18. Eisenstat TE, Oliver GC (1992) Electrocoagulation for adenocarcinoma of the lower rectum. World J Surg 16:458–462.
19. Faintuch JS (1988) Better palliation of colorectgal carcinoma with laser therapy. Oncology 2:33–38.
20. Brunetaud J, Maunoury V, Durcotte P (1987) Palliative treatment of rectosigmoid carcinoma by laser endoscopic photocoagulation. Gastroenterology 92:663–667.
21. Aubert A, Meduri B, Fritsch J, et al (1991) Endoscopic treatment by snare electrocoagulation prior to Nd:YAG laser photocoagulation in 85 voluminous colorectal villous adenomas. Dis Colon Rectum 34:372–377.
22. Osborne DR, Higgins AF, Hobbs KEF (1978) Cyosurgery in the management of rectal tumours. Br J Surg 65:859–861.
23. Nelson RL, Abcarian H, Prasad ML (1982) Iatrogenic perforation of the colon and rectum. Dis Colon Rectum 25:305–309.
24. Williams SM, Harned RK (1991) Recognition and prevention of barium enema examinations. Curr Probl Diagn Radiol 20:123–151.
25. Walker GL, Nogro ND (1959) Postoperative anorectal hemorrhage Surg Clin North Am 39:1655–1660.
26. Goligher JC (1975) Surgery of anus, rectum and colon. Baillière-Tindall, London, pp 158–162.

Quality of Life and Familial Adenomatous Polyposis Patients

20

Gian Gaetano Delaini, Andrea Chimetto,
Marco Lo Muzio, Filippo Nifosì, Maurizio Mainente and
Gianluca Colucci

Abstract Qualty of life (QoL) has become one of the main goals of surgery for familial adenomatous polyposis (FAP). Patients with FAP are usually young and asymptomatic, so the ideal surgery should combine the lowest risk of cancer with the minimum impact to lifestyle and activity. For these reasons restorative proctocolectomy (RPC) with ileo-anal pouch anastomosis (IPAA) has been the gold standard for FAP patients in the last two decades. Many studies report good overall rates of QoL after IPAA. Nevertheless, this surgery is complicated by functional problems, particularly in terms of high frequency of bowel movements, episodes of incontinence (especially mild or soiling in the night time), diet limitations, and, although it usually does not affect social and work life, IPAA can potentially lead to sexual problems; furthermore, there is a significant decrease of fertility in women after IPAA. These considerations lead many authors to reconsider the possibility of FAP patients undergoing ileorectal anastomosis (IRA); this surgery has better functional results and can almost always be converted into secondary IPAA. Nevertheless, IRA necessitates a strict follow-up that can detect but not prevent neoplastic degeneration. Moreover, there is still not a clear superiority of IRA over IPAA in terms of QoL.

Since QoL is highly dependent on the patient's participation and wishes, in combination with surgical outcomes and need for further follow-up, surgery must be discussed and planned with the patients themselves.

Keywords Funtional outcome • Health-related quality of life (HRQOL) • Ileorectal anastomosis • Pouch • Quality of life (QoL) • SF-36

20.1 Introduction

Since patients with familial adenomatous polyposis (FAP) are usually young and asymptomatic, one of the main goals of surgery for polyposis must be to obtain a quality of life (QoL) that is compatible with the lifestyle and personality of these patients.

The surgical procedure (and eventual follow-up) must be planned to obtain the most favorable situation for the desires, aspirations, and activity of the patient, while at the same time guaranteeing the lowest risk of developing cancer.

Consequently, it is fundamental to opt not only for

G.G. Delaini (✉)
Department of Surgery and Gastroenterology, University of Verona, Verona, Italy

the best type of surgery, but also for the best timing.

Until about 20 years ago surgical options for FAP were basically:

- total abdominal colectomy and ileorectal anastomosis (IRA) that would permit maintenance of almost normal bowel habits and continence, but requiring a lifelong follow-up of the remaining rectal mucosa
- proctocolectomy and definitive ileostomy that at least eradicate the risk of cancer but also carry the discomfort of an ileostomy.

At the end of the 1980s, the new technique of restorative proctocolectomy (RPC) with ileo-anal pouch anastomosis (IPAA) initiated a revolution in surgical management of FAP, and soon became the gold standard for FAP and also for ulcerative colitis (UC). In fact this option allows for virtual eradication of the risk of cancer, while still leaving the patient with normal physiological continence. RPC has, nevertheless, been shown to be complicated by a series of functional problems and surgical complications that could potentially affect the QoL after surgery.

When it became evident that the type of surgery for FAP can affect patients' QoL, as long as there was a decrease in mortality and morbidity, many authors began to analyze the results of the different surgical procedures using the indicator of health-related quality of life (HRQOL). Studies reported in the literature about QOL after IPAA (and comparing IPAA with others types of interventions) have multiplied in the last decade. Actually, few studies concerning only FAP patients have used standardized and validated instruments to examine HRQOL as a principal outcome; from the data regarding RPC with IPAA many discussions have arisen about the opportunity to favor this technique over the old IRA or ileostomy in order to obtain the best QoL.

20.2 Quality of Life and Health-Related Quality of Life

In recent decades, constant progress in terms of reduced morbidity and mortality have brought an increasing recognition of QoL as an outcome and indicator in clinical medicine.

In the 1994 the World Health Organization (WHO) gave a definition of QoL [1]:

> The individual's perception of his or her position in life, within the cultural context and value system he or she lives in, and in relation to his or her goals, expectations, parameters and social relations.

The first consideration is that QoL is always focused on the singular person [2]; this centrality of the person or patient implies that any consideration based on QoL must start with the patient's opinions and values in relation to the reality that we want to describe. Another important consideration is that health is just one of the many aspects of QoL; in effect, QoL is a wider concept that depends especially on individual factors such as aspirations and personal values, and environmental factors like social relationships and economic factors.

The multidimensionality of the QoL is outlined in Table 20.1, which lists the areas of value of the WHOQOL, the instrument of measure of global QoL introduced by the WHO. Here, the assessment of QoL depends on the combination of life domains like physical health, psychological state, independence level, social relationships, environment, and spirituality [3].

At this point, it is evident that health has substantial importance in the determination of QoL. It is necessary to develop a definition of health that includes not just the presence or absence of illness, but also the ability to perform activities of daily living with or without limitations [2].

The HRQOL is the patient's own assessment of physical and mental health, social interactions, and general wellbeing related to their state of health and the functional results of therapy [4,5].

Since the HRQOL became one of the main endpoints and indicators of clinical outcomes, many instruments to assess QoL have been introduced into clinical practice [6].

In general there are two types of instruments used to measure QoL:

- generic to measure the overall QoL, considering all the aspects influencing QoL; these can be used to compare subgroups with general populations
- disease-specific instruments designed to measure QoL in specific patient populations.

To be validated, these instruments need to be exhaustive to reflect the centrality of the patient, to be

Table 20.1 WQOL domains for assessment of quality of life. Reproduced from [3], with permission from Elsevier

Areas	Sections
Physical health	Energy and tiredness
	Pain and discomfort
	Sleep and rest
Psychological health	Body image and physical aspect
	Positive emotions
	Negative emotions
	Self-esteem
	Ability to think, concentrate, and memorize
Independence level	Ability to move
	Independence in daily life activities
	Dependence on medications and other medical treatments
	Capacity to work
Social relationships	Interpersonal relationships
	Social support
	Sexual activity
Environment	Financial resources
	Freedom, security and physical integrity
	Health and Social assistance: accessibility and quality
	Domestic environment
	Opportunity to gain new knowledge and ability
	Participation and opportunity for recreation and leisure
	Physical environment (pollution, noise, traffic, climate)
	Transportation
Spirituality and religion	Spirituality/religion/personal convinctions

psycometricrally "valid and reproducible", and to make sense from both the scientific and clinical point of view [2].

One of the most used instruments in clinical research for QoL (and probably the most used for generic HRQOL assessment in colorectal surgery) is the Short Form 36 (SF-36); introduced in the 1980s by a group of searchers of Rand Corporation [4]. Compared with other generic health indices, the SF-36 has been shown to discriminate better between populations with varying QoL [7,8].

The SF-36 contains 36 health-related questions that, when scored according to specific guidelines, define eight domains of HRQOL: general health perception, physical function, physical and emotional role limitations, social functions, mental health, bodily pain, and vitality.

As in other fields of medicine, QoL became one of the main goals and indicators of outcome in surgery. When comparing the HRQOL before and after different types of surgery, this can lead to making better decisions relating to surgical treatment [9]. Different QoL questionnaires have been introduced in clinical practice to measure results in colorectal surgery (e.g. EORT QLQ-CR38, IBDQL, etc), and more specifically after IPAA (e.g. the Cleveland Global Quality of life score, Pemberton score, Oresland score, etc).

20.3 Quality of Life after Restorative Proctocolectomy and Ileo-Anal Pouch

Interest in QoL after surgery for colorectal disease, especially for UC and FAP, increased after the introduction of RPC with IPAA.

This type of surgery preserves anal function, but it has a high rate of early or late complications [10]. The long-term problems of an ileo-anal pouch are due to surgical complications, but also to a series of functional problems of the pouch, especially in relation to continence and bowel movements, which are thought to influence lifestyle after this surgery (diet, sexuality, social, work, activity, etc). For this reason, HRQOL and not just objective functional outcomes have become a point of discussion in deciding which is the best surgical option for FAP and UC.

The data available are mostly concerned with UC patients, as this is the main indication for IPAA (UC 87%, FAP 8.9%, others 3.6%) [10]. At present, these data do not always show a clear superiority of a certain type of surgery.

In 2005, Hueting et al [10] published an extremely interesting a meta-analysis that included data from 9317 patients. The data from this meta-analysis allowed us to form a picture of complications of early and long-term complications after IPAA. Therefore we will next analyze which aspects and complications can influence and affect QoL.

20.3.1 Long-term Complications of IPAA

Complications after IPAA occur in about 13% to 62% of patients [11,12]; it is interesting that FAP patients are usually reported to develop fewer significant complications in the postoperative course (10–25%) [11,13], probably due to the fact that these patients are usually in good general health and are not using medications like corticosteroids or immunosuppressants.

The principal long-term complications are represented in Table 20.2 [10]. As can be seen, the creation of an ileo-anal pouch is complicated by pelvic sepsis in 9.5%, formation of a pouch/anal vaginal fistula in 5.5%, stricture of the anastomosis in 9.2% (sometimes with the necessity to perform pneumatic or digital dilation), and sexual dysfunction in 3.6% of cases. The most frequent non-pouch-related complication is small bowel obstruction (a situation that not infrequently results in further intervention), reported in 13.1% in this meta-analysis but in a higher percentage in many single institution series – from 15% to 25% [12,14].

20.3.2 Pouchitis

Pouchitis is defined as an idiophatic and aspecific inflammation of the mucosa of the pouch; it is one of the most common long-term complications. The diagnosis is based on the combination of clinical, endoscopic, and histological criteria. Clinically, pouchitis is characterized by increased frequency of loose, watery, and sometimes bloody stool; urgency; lower abdominal cramping; and fever.

The pooled incidence of patients that refer to at least one episode of pouchitis is 18.8% in the meta-analysis by Hueting et al [10], but the real incidence is difficult to determine. The cumulative probability of pouchitis reported by the principal groups is higher and reaches 59% [15]. Pouchitis can be classified according to etiology (idiopathic versus secondary), disease activity (remission versus active), duration (acute versus chronic with a 4-week cut-off), or disease course (infrequent, relapsing, or continuous), or according to response to therapy (responsive versus refractory) [16].

The QoL seems to be affected significantly only in patients with chronic refractory pouchitis [17,18], which fortunately is not common, occurring in less than 5% of cases [19]. Pouchitis is rarely seen in patients with FAP (3–14 %) [12–20]; UC patients are 9.3 times more likely to develop pouchitis than FAP patients [11]. Globally, pouch failure, defined as the necessity for excision or permanent defunctionalization, is reported in up to 8.5% of these patients [10]. Pouch failure is largely due to pelvic sepsis and mostly

Table 20.2 Pooled incidences of pouch related complications. Adapted from [10], with permission from S. Karger AG, Basel

Complication	Number of studies	Number of patients	Pooled %	95% CI
Pelvic sepsis	41	9082	9.5	8.2–10.9
Fistula	30	5120	5.5	4.3–7.0
Stricture	28	5185	9.2	6.8–12.4
Pouchitis	33	7289	18.8	15.7–22.4
Sexual dysfunction	21	5112	3.6	2.7–4.7
Pouch failure	39	8877	6.8	5.4–8.4
Small bowel obstruction	27	5853	13.1	11.0–15.7
Other	22	3441	3.4	2.4–4.8

occurs in the first year after surgery [19]; the other main cause for pouch failure is chronic refractory pouchitis.

20.3.3 Cancer in Transitional Columnar Mucosa and in the Pouch

The criticism of the double-stapled ileo-anal pouch is that by leaving a cuff of 1–3 cm of transitional columnar mucosa (ATZ), there is a potential risk of developing dysplasia and cancer. In fact, it has been demonstrated that this risk cannot be completely erased, even after endoanal total mucosectomy, since it has been demonstrated that small islets of columnar epithelial cells can be left after mucosectomy [21,22]. The incidence of dysplastic polyps in the residual ATZ after double-stapled IPAA has been estimated to be up to 28–31%, as compared with 10–14% after endoanal mucosectomy [21,22]. Until now, only eight cases of cancer from the pouch-anal anastomosis have been reported, four of these after mucosectomy [13].

FAP patients may develop adenomas in the ileo-anal reservoir; the risk of developing adenomas has been calculated at 7%, 35%, and 75% respectively at 5, 10, and 15 years [23]. Most of these adenomas are mild dysplasia and unlikely to develop into cancer. Regular endoscopic monitoring of FAP pouches is mandatory. The cumulative risk of developing rectal cancer after colectomy and ileorectal anastomosis for FAP is 3.9%, 10.4%, 12.1%, and 25.8% respectively after 10, 15, 20, and 25 years of follow-up [13].

20.3.4 Functional Results of RPC

With the term "functional results of IPAA" (and more widely of the entire colorectal surgery) we consider all aspects of life that concern bowel activity. In particular this includes stool frequency; the presence of any grade of incontinence, soiling or urgency; and peri-anal irritation.

We can measure the functional results of an ileo-anal pouch (or of an ileorectal anastomosis or ileostomy) directly or indirectly, by considering a spectrum of situations such as the need for drugs to reduce the number of bowel movements, limitations of diet, use of pads, and other similar factors.

Again, the data of the meta-analysis by Hueting et al [10] show that the mean defecation frequency of the ileo-anal pouch population is 5.2 per 24 hours, with a mean night-time frequency of 1.0. Patients suffer mild incontinence in 17% of cases, and urgency (that is inability to defer defecation for less than 15 minutes) in 7.3%; however, severe incontinence is reported by only 3.7% of patients.

The ability to discriminate between gas and feces is perfect in more than 75% of patients [19]; up to 17% of patients in the daytime and 43% of patients in the night time experience soiling (leakage of a small quantity of mucus from the anus) occasionally [24,25]. An important observation about continence after IPAA is that after the double-stapled technique there is a significant decrease of the ratio of patients reporting night-time incontinence/soiling when compared to mucosectomy and manual anastomosis [26,27].

Peri-anal skin irritation is reported occasionally in about half of patients; the use of pads, especially at night time, is reported in 18% to 68% of these patients [19,24,25].

From these data, the observation that functional results of an ileo-anal pouch may influence the QoL becomes obvious.

The reality is that there is no definitive correlation between QoL and functional results. In fact, a real and definitive answer does not exist and if some authors have found a correlation between functional results and QoL [24,28,29], there are others that have found the opposite [30,31], demonstrating that even when alteration of continence exists this may not diminish quality of life [32].

20.3.5 Diet Limitations

Another important factor in determining functionality of the pouch is the need to impose dietary limits to reduce defecation frequency. The rate of patients who need to modify their diet is quite high, up to 62% [33]. In particular, patients usually tend to avoid food with high fiber content (especially green vegetables) and alcohol, and also they tend to have dinner quite early to prevent night-time bowel movements; some authors have found that these modifications can influence QoL negatively [33]. The regular use of antidiarrheic drugs can reach up to 60%, is significant in all the main series [19,24,25], and is an indirect index of diet limitations and high frequency of bowel movements.

20.4 Fertility and Pregnancy

Young patients with FAP are generally fit, and many are childless at the time of operation. It is obvious that changes in fertility and pregnancy have to be considered before surgery for FAP [34,35].

Pregnancy and delivery are not jeopardized in patients with an ileo-anal pouch. Although soon after the introduction of IPAA some authors recommended Caesarean delivery after RPC [36], successive studies widely demonstrated that safe pregnancy and normal delivery are possible [37–39]. The function of the pouch is not affected by pregnancy or by the type of delivery [13].

Other factors regarding fertility must be considered. Olsen et al [34] reported that fertility decreases to 54% after construction of a pouch for FAP. The reasons for this decrease are not yet well understood, but are probably related to the formation of pelvic adhesions [34,37,40]. This reduction in fecundity in women with FAP (which is even higher in UC patients) undergoing IPAA must be discussed with patients before planning surgery [34,41].

20.5 Sexual and Urinary Functions

The risk of injury to the presacral nerves (sympathetic plexus) and the nervi erigentes (parasympathetic plexus) should focus attention on the possible sequelae such as urinary retention, erectile dysfunction, or retrograde ejaculation. After the introduction of the nerve-sparing technique (or "close rectal wall") [42], these kinds of complications can be almost completely avoided [43–45].

The meta-analysis of Hueting et al [10] revealed urinary and sexual complications in 3.6% of patients. Colwell and Gray [46] reported rates of sexual dysfunction in males of 0.5% to 1.5% for erectile impotence, and of 3% to 4% for ejaculatory dysfunction after IPAA.

Transient postoperative dysuria and urinary retention have been reported in less than 2% of patients [13,43]. Again, Colwell and Gray [46] showed that from 3% to 22% of women with IPAA could experience dyspareunia (probably due to the mechanical changes in the pelvic space) [47]; in addition, 3% of female patients may report inhibited sexual relations because of fear of stool leakage.

20.6 Social and Work Restrictions

Another important issue is the impact of IPAA on work and social life. This is especially important considering that the average age at surgery in patients with polyposis is around the third decade of life.

In 1999, a group at the Cleveland Clinic [48] analyzed the follow-up of almost 1000 patients who underwent IPAA (3.8% for FAP). They used the SF-36 and the Cleveland Global Quality of Life (CGQL) instrument (Fazio score), which is a validated QoL indicator for pouch patients. They found that even in long-term follow-up, the rate of patients who reported no restrictions in work and social life reached about 90%. This is in agreement with data from Hanloser et al at the Mayo Clinic [19] on 1885 patients, who found that of 92% of patients who had remained in the same employment, 83% were not affected by surgery.

20.7 Measure of Quality of Life After IPAA and After Surgey for FAP

In the last 20 years, IPAA has been deemed the gold standard worldwide for surgery for FAP; therefore, if we want to consider QoL after surgery for FAP we must consider this procedure. In general, most of the studies considering HRQOL after IPAA do not distinguish or group patient data according to the diagnosis. Data related to FAP patients are only available in a few studies, particularly in relation to the index of QoL.

Different groups using validated instruments, mainly the generic SF-36, have demonstrated that global QoL after IPAA is good and comparable to that of a normal healthy population [11,28–30,49] in almost all domains of the scores.

These same papers showed that despite a comparable QoL with a normal population, IPAA patients are found to have a decreased health perception [28,49], and Barton et al [11] noted a worse SF-36 score in role limitation due to emotional problems in patients with FAP. The HRQOL seems to be stable, even at the long-term follow-up [19,48].

Nevertheless, there are some obscure points that emerge from discussion in the literature. For example, as already stated, functional results are not always directly related to QoL, and therefore they do not interfere with HRQOL scores. We have conflicting results regarding this [24,28–30,32]. Other aspects that have been assumed to affect QoL after IPAA are diet limitation [33], pouchitis [33,50], and age at time of surgery [29].

Nevertheless, the general scores of IPAA patients still are quite high, and most patients report that they are able to maintain work [19,25,48] and social activity, or even improve their abilities following surgery [49].

20.7.1 Quality of Life: IPAA Versus Other Surgical Options

In 1991, Kohler, Pemberton and others from the Mayo Clinic [51], in one of the first papers on QoL after IPAA (even without the support of validated instruments), compared the QoL after conventional and continent-Koch ileostomy and after IPAA; they found that most patients were satisfied, although 39% with conventional and 14% with continent ileostomy desired a change, in contrast to only 4% of those with ileal pouch-anal anastomoses; in effect, even though the majority of all the categories of patients were able to return to normal social and work activities, the IPAA patients scored significantly better results in sexual and sport activities.

Also McLeod and Baxter [6], in 1998, analyzed the literature and found high and similar levels in the global QoL after ileostomy, continent ileostomy, and IPAA, with the same observation that body image and sexuality were markedly improved after intervention that was able to preserve continence, such as Koch ileostomy and more IPAA.

In 2000, Seidel et al [52] compared results after IPAA and after ileostomy and noted that, despite a significantly higher rate of complications after IPAA than after ileostomy (53% versus 16%), both groups reported favorable responses in relation to QoL domains.

20.7.2 Results in FAP Patients

Considering FAP patients in particular, there are some observations that can be made: since the beginning of these QoL studies, many authors have found that FAP patients experience a lower QoL after surgery; most attribute this to the fact that before surgery patients with FAP are young, in good health and asymptomatic, so restrictions on lifestyle after surgery may be perceived as more severe by this subgroup of patients [33,52,53,54]. In particular, Seidel registered a difference in the fact that UC patients define their QoL as better than before surgery in 95% of cases (both IPAA and ileostomy), while FAP patients only say it is better in 57% of cases [52]. Fujita et al observed that patients with FAP are less satisfied, even when functional results and stool frequency are similar [54].

Kohler et al, from the Mayo Clinic registered that FAP patients reported more sexual and sports limitations after surgery, for both ileostomy and IPAA [51].

All these observations lead back to the discussion of whether primary IPAA should still be considered the gold standard, especially in patients with FAP.

In recent years, some papers have been published that have compared the results after IPAA and after IRA in patients with a diagnosis of FAP. In these studies, the HRQOL has always been considered one of the main outcomes; these studies also tried to find a correlation between QoL and functional results.

In 2000, Ko et al [30] studied functional results and QoL of two subgroups of patients that underwent IPAA and IRA for FAP. The functionality of the bowel in terms of bowel frequencies, soiling, pad usage, peri-anal skin problems, diet limitations, and inability to distinguish gas were worse after ileo-anal pouch. In final analysis, data on HRQOL based on the SF-36 were not significantly different in any of the eight dimension of SF-36.

Van Duijvendijk et al [55] performed a similar analysis, selecting patients with FAP from the Netherlands registry, comparing HRQOL after IPAA and IRA. The SF-36 and EORTC QLQ CR-38 showed no differences between patients who underwent the two types of anastomoses; however, the comparison between a healthy population and FAP patients after surgery showed poorer scores after either IPAA or IRA. Hassan et al, from the Mayo Clinic [5], also found no difference in HRQOL based on SF-36 between IPAA and IRA patients in polyposis.

In 2006, Aziz et al [56] published a meta-analysis of the observational studies of ileorectal versus ileoanal pouch anastomosis for FAP. The data extrapolated showed undoubtedly better results after IRA, as shown in Table 20.3, even though, curiously, they found a worse rate of urgency after IRA compared to IPAA. In addition, Gunther et al [57] found better

Table 20.3 Comparisons of functional results after ileorectal and ileopouch-anal anastomosis. Adapted from [56]

	IRA	IPAA
Bowel frequency per day (%)	2–6.1	3.8–8
Need for night defecation (%)	8.2	44.1
Day (24h) incontinence (%)	30	50.5
Night incontinence (%)	3.8	21
Pad use (%)	5.4	14.5
Faecal urgency (%)	39.2	14.2

functionality after IRA; in their analysis they also found that this superiority, especially in terms of continence, leads to a superior QoL score after IRA compared to after IPAA.

20.8 Conclusion

FAP patients nowadays are usually young and asymptomatic when they undergo surgery. QoL is an important issue when deciding which kind of surgery to choose. Even if the last word is far from being said, it seems that IRA can offer a better functional outcome and probably a better QoL than IPAA [57]. On the other hand, Vasen et al [58] have demonstrated that the life expectancy after IPAA is 1.8 years longer than after IRA. Aside from this, 47 out of 659 patients after IRA developed rectal cancer, with 75% of them having had a clear endoscopy within a year from the diagnosis [58].

It is also important to underline that IPAA can be accomplished after IRA, with good functional results [59,60].

If genetics can facilitate the decision-making process, by defining mutations that are linked to the less aggressive form of FAP [61], it could be difficult to decide which is the best surgical option with respect to QoL.

The need for a strict follow-up, the risk of cancer, and the threat of a re-operation can be an unacceptable source of stress and lead to reduced QoL.

Expectations and previous health status are also important factors when judging the results. A clear example of this is the lower scores in terms of QoL after IPAA in patients with FAP in comparison to patients with UC. This is probably due to the fact that these types of patients, unlike symptomatic colitis patients, have a normal HRQOL before surgery, and the impact of altered bowel function after IPAA may be more important.

The surgeon plays a key role in providing the patient with all possible options; therefore, the patient should receive a tailored surgery through a careful process of informed consent. This could be facilitated in the future if further studies are able to refine the measurements of QoL and provide a precise interpretation of the data.

References

1. The WHOQOL Group (1994) The development of the World Health Organization quality of life assessment instrument (The WHOQOL). In: Orley J, Kuyen W (eds) Quality of life assessment: international perspectives. Springer-Verlag, Heidelberg, pp 41–57.
2. Apolone G, Mosconi P, Ware JE (1997) Questionario sullo stato di salute SF-36. Manuale d-uso e guida all'interpretazione dei risultati. Guerini & Associati Editore, Milan.
3. The WHOQOL Group (1995) The World Health Organization quality of life assessment (WHOQOL): position paper from the World Health Organisation. Soc Sci Med 41:1403–1409.
4. Ware JE, Sherbourne CD (1992) The MOS 36-item short form health survey (SF-36). I. Conceptual framework and item selection. Med Care 30:473–483.
5. Hassan I, Chua HK, Wolff BG, et al (2005) Quality of life after ileal pouch-anal anastomosis and ileorectal anastomosis in patients with familial adenomatous polyposis. Dis Colon Rectum 48:2032–2037.
6. Mc Leod RS, Baxter NN (1998) Quality of life of patients with inflammatory bowel disease after surgery. World J Surg 22:375–381.
7. Chrispin PS, Scotton H, Rogers J, Lloyd D, Ridley SA (1997) Short form 36 in the intensive care unit: assessment of acceptability, reliability and validity of the questionnaire. Anaesthesia 52:15–23.
8. Heyland DK, Guyatt G, Cook DJ, et al (1998) Frequency and methodologic rigor of quality of life assessments in the critical care literature. Crit Care Med 26:591–598.
9. Thirlby RC, Land JC, Fenster FL, Lonborg R (1998) Effect of surgery on health –related quality of life in patients with inflammatory bowel disease. Arch Surg 133:826–832.

10. Hueting W, Buskens E, van der Tweel I, et al (2005) Results and complications after ileal pouch anal anastomosis: a meta-analysis of 43 observational studies comprising 9317 patients. Dig Surg 22:69–79.
11. Barton JG, Paden MA, Lane M, Postier RG (2001) Comparison of postoperative outcomes in ulcerative colitis and familial polyposis patients after ileoanal pouch operations. Am J Surg 182:616–620.
12. Fazio VW, Ziv Y, Church JM et al (1995) Ileal pouch-anal anastomoses complications and function in 1005 patients. Ann Surg 222:120–127.
13. Kartheuser A, Stangherlin P, Brandt D, Christophe R, Sempoux C (2006) Restorative proctocolectomy and ileal pouch-anal anastomosis for familial adenomatous polyposis revisited. Fam Cancer 5:241–260.
14. Meagher AP, Farouk R, Dozois RR, Kelly KA, Pemberton JH (1998) J ileal pouch-anal anastomosis for chronic ulcerative colitis: complications and long term outcome in 1310 patients. Br J Surg 85: 800–803.
15. Simchuk EJ, Thirlby RC (2000) Risk factors and true incidence of pouchitis in patients after ileal pouch-anal anastomosis. World J Surg 24:851–856.
16. Mahadevan U, Sandborn WJ (2003) Diagnosis and management of pouchitis. Gastroenterology 124:1636–1650.
17. Abdelrazeq AS, Lund JN, Levenson SH (2005) Implications of pouchitis on the functional results following stapled restorative proctocolectomy. Dis Colon Rectum 48:1700–1707.
18. Lepisto A, Lukkonen P, Jarvinen HJ (2002) Cumulative failure rate of ileal pouch-anal anastomosis and quality of life after failure. Dis Colon Rectum 45:1289–1294.
19. Hanloser D, Pemberton JH, Wolff BG, et al (2007) Results at up to 20 years after ileal pouch-anal anastomosis for chronic ulcerative colitis. Br J Surg 94:333–340.
20. Hulten L (1998) Proctocolectomy and ileostomy to pouch surgery for ulcerative colitis. World J Surg 22:335–341.
21. van Duijvendijk P, Vasen HF, Bertario L, et al. (1999) Cumulative risk of developing polyps or malignaci at the ileal pouch-anal anastomosis in patients with familial adenomatous polyposis. J Gastrointest Surg 3:325–330.
22. Thompson-Fawcett MW, Warren BF, Mortensen NJ (1998) A new look at the anal transitional zone with reference to restorative proctocolectomy and stapled anastomosis. Br J Surg 85:1517–1521.
23. Parc YR, Olschwang S, Desaint B, et al (2001) Familial adenomatous: prevalence of adenomas in the ileal pouch after restorative proctocolectomy. Ann Surg 233:360–364.
24. Berndtsson I, Lindholm E, Oresland T, Borjesson L (2007) Long-term outcome after ileal pouch-anal anastomosis: function and health related quality of life. Dis Colon Rectum 50:1545–1552.
25. Wheeler JMD, Banerjee A, Ahuja N, et al (2005) Long-term function after restorative proctocolectomy. Dis Colon Rectum 48:946–995.
26. Reilly TW, Pemberton JH, Wolff BG, et al (1997) Randomized prospective trial comparing ileal pouch-anal anastomosis performed by excising the anal mucosato ileal pouch-anal anastomosis performed by preserving the anal mucosa. Ann Surg 225:666–677.
27. Hallgren TA, Fasth SB, Oresland MO, Hulten L (1995) Ileal pouch-anal function after endoanal mucosectomy and handsewn ileoanal anastomosis compared with stapled anastomosis without mucosectomy. Eur J Surg 161:915–921.
28. Tianen J, Matikainen M (1999) Health-related quality of life after ileal J-pouch-anal anastomosis for ulcerative colitis: long term results. Scand J Gastroenterol 34:601–605.
29. Carmon E, Keidar A, Ravid A, Goldman G, Rabau M (2003) The correlation between quality of life and functional outcome in ulcerative colitis patients after proctocolectomy ileal pouch anal anastomosis. Colorectal Dis 5:228–232.
30. Ko CY, Rusin LC, Schoetz DJ Jr, et al (2000) Does better functional result equate with better quality of life? Implications for surgical treatment in familial adenomatous polyposis. Dis Colon Rectum 43:829–835; discussion 835–837.
31. Steen J, Meijerink WJ, Masclee A, et al (2000) Limited influence of pouch function on quality of life after ileal pouch-anal anastomosis. Hepatogastroenterology 47:746–750.
32. Holubar S, Hyman N (2003) Continence alterations after ileal pouch-anal anastomosis do not diminish quality of life. Dis Colon Rectum 46:1489–1491.
33. Coffey JC, Winter DC, Neary P, et al (2002) Quality of life after ileal pouch-anal anastomosis: an evaluation of diet and other factors using the Cleveland Global Quality of Life instrument. Dis Colon Rectum 45:30–38.
34. Olsen KO, Juul S, Bulow S, et al (2003) Female fecundity before and after operation for familial adenomatous polyposis. Br J Surg 90:227–231.
35. Gallagher MC, Sturt NJH, Phillipq RKS (2003) Female fecundity before and after operation for familial adenomatous polyposis. Br J Surg 90:759–762.
36. Pezim ME (1984) Successful childbirth after restorative proctocolectomy with pelvic ileal reservoir. Br J Surg 71:292.
37. Hanloser D, Pemberton JH, Wolff BG (2004) Pregnancy and delivery before and after ileal pouch-anal anastomosis for inflammatory bowel disease: immediate and long term consequences and outcomes. Dis Colon Rectum 47:1127–1135.
38. Metcalff A, Dozois RR, Beart RW, Wolff BG (1985) Pregnancy following ileal pouch-anal anastomosis. Dis Colon Rectum 28:859–861.
39. Nelson H, Dozois RR, Kelly KA, et al (1989) The effect of pregnancy and delivery on the ileal pouch anal-anastomosis functions. Dis Colon Rectum 32:384–388.
40. Olsen KO, Jull S, Berndtsson I (2002) Ulcerative colitis: female fecundity before diagnosis, during disease and after surgery compared with a population sample. Gastroenterology 122:15–19.
41. Soravia C, Cohen Z (2004) Familial adenomatous polyposis. In: Fazio VN, Church JM, Delaney CP (eds) Current therapy in colon and rectal surgery, 2nd edn. Philadelphia: Elsevier-Mosby, Philadelphia, pp 349–353.
42. Nicholls JR (1993) Controversies and practical problem solving In: Nicholls J, Bartolo D, McMortensen N (eds) Restorative proctocolectomy. Blackwell Scientific Publications, Oxford, pp 53–81.
43. Kartheuser A, Parc R, Penna C, et al (1996) Ileal pouch-anal anastomosis as the first choice operation in patients with familial adenomatous polyposis. A ten-year experience. Surgery 119:615–623.
44. Kartheuser A, Charre L, Kayser J, et al (2003) Laparoscopic restorative proctocolectomy without ileostomy. Colorectal

Dis 5(suppl 2):170.
45. Scaglia M, Oresland T, Delaini GG, Hulten L (1994) Functional results following restorative proctocolectomy. In: Serio G, Delaini GG, Hulten L, Nicholls J, Vestweber KH (eds) Inflammatory bowel disease. Graffham Press, Edinburgh, pp 63–73.
46. Colwell JC, Gray M (2001) What functional outcomes and complications should be taught to the patients with ulcerative colitis or familial adenomatous polyposis who undergo ileal pouch anal anastomosis. J Wound Ostomy Continence Nurs 28: 184–189.
47. Metcalff A, Dozois RR, Kelly KA (1986) Sexual functions in women after proctocolectomy. Ann Surg 204:624–627.
48. Fazio VW, O'Riordan MG, Lavery IC, et al (1999) Long-term functional outcome and quality of life after stapled restorative proctocolectomy. Ann Surg 230:575–586.
49. Robb B, Pritts T, Gang G, Warner B, et al (2002) Quality of life in patients undergoing ileal pouch-anal anastomosis at the University of Cincinnati. Am J Surg 183:353–360.
50. Karlbom U, Raab Y, Ejerblad S, et al (2000) Factors influencing the functional outcome of restorative proctocolectomy in ulcerative colitis. Br J Surg 87:1401–1408.
51. Kohler LW, Pemberton JH, Zinmeister AR, Kelly KA (1991) Quality of life after proctocolectomy: a comparison of brooke ileostomy, Kock pouch and ileal pouch-anal anastomosis. Gastroenterology 101:679–684.
52. Seidel SA, Newman M, Sharp KW (2000) Ileoanal pouch versus ileostomy: is there a difference in quality of life? Am Surg 66:540–547.
53. Heuschen UA, Heuschen G, Lucas M, et al (1998) Pre-and postoperative quality of life of patients with ulcerative colitis and familial adenomatous polyposis with ileoanal pouch operation. Chirurg 69:1329–1333.
54. Fujita S, Kusukanoki M, Shoji Y, et al (1992) Quality of life after proctocolectomy and ileal J pouch-anal anastomosis. Dis Colon Rectum 35:1030–1039.
55. van Duijvendijk P, Slors JF, Taat CW, et al (2000) Quality of life after total colectomy with ileorectal anastomosis or proctocolectomy and ileal pouch-anal anastomosis for familial adenomatous polyposis. Br J Surg 87:590–596.
56. Aziz O, Athanasiou T, Fazio VW, et al. (2006) Meta-analysis of observational studies of ileorectal versus ileal pouch-anal anastomosis for familial adenomatous polyposis. Br J Surg 93:407–417.
57. Gunther K, Braunrieder G, Bittorf BR, Hohenberger W, Matzel KE (2003) Patients with familial adenomatous polyposis experience better bowel function and quality of life after ileorectal anastomosis than after ileoanal pouch. Colorectal Dis 5:38–44.
58. Vasen HF, van Duijvendijk P, Buskens E, et al (2001) Decision analysis in the surgical treatment of patients with familial adenomatous polyposis: a Dutch-Scandinavian collaborative study including 659 patients. Gut 49:231–235.
59. von Roon AC, Tekkis PP, Lovegrove RE, et al (2008) Comparison of outcomes of ileal pouch-anal anastomosis for familial adenomatous polyposis with and without previous ileorectal anastomosis. Br J Surg 95:494–498.
60. Penna C, Kartheuser A, Parc R, et al (1993) Secondary proctectomy and ileal pouch-anal anastomosis after ileorectal anastomosis for familial adenomatous polyposis. Br J Surg 80:1621–1623.
61. Vasen HF, van der Luijt RB, Slors JF, et al (1996) Molecular genetic tests as a guide to surgical management of familial adenomatous polyposis. Lancet 348:433–435.

Psychological and Medico-legal Aspects of Genetic Counseling in Familial Adenomatous Polyposis

21

Lenka Foretová and Tomáš Skřička

Abstract Familial adenomatous polyposis is a severe autosomal dominant disease with complex clinical symptoms and a high risk of colorectal and other cancers. Genetic counseling and testing is recommended and is important for probands but also for other relatives at risk. A germline mutation in the *APC* gene is the cause of the disease in most families. Genetic counseling is a process that includes several sessions with the patient and family members, and should help the family in process of making decisions about testing and cancer prevention. The primary and secondary prevention is complex and should be provided by several professionals with in-depth knowledge of the syndrome. Many psychological and social problems may be anticipated and should not be overlooked.

Keywords Familial adenomatous polyposis • Genetic counseling • Genetic testing • Prevention • Psychological and social issues

21.1 Introduction to the Genetics of Polyposis Syndromes

Familial adenomatous polyposis (FAP) is one of the most important hereditary causes of colorectal cancer, with a prevalence of about 1:8000 [1], and autosomal dominant inheritance. In about 30% of cases no polyposis was found in a previous generation and the disease is caused by *de novo* mutation. Also, somatic mosaicism of the *APC* mutation in one of the parents can cause sporadic occurrence of the disease, or polyposis of two siblings with no affected father or mother. Non-paternity may be another cause of sporadic disease, but should only be discussed if necessary.

L. Foretová (✉)
Department of Cancer Epidemiology and Genetics,
Department of Surgery, Masaryk Memorial Cancer Institute,
Brno, Czech Republic

FAP is a complex disease leading to more than 100 colonic polyps, which may increase in size and number with age. The first clinical signs of the disease are mostly detected in adolescence, but in attenuated form may develop much later. Colorectal cancer occurs at an earlier age than in sporadic cases but usually after the age of 20 years. There is a high risk of extracolonic gastrointestinal manifestations, mostly gastroduodenal polyps (gastric non-adenomatous fundic polyps, small bowel adenomas) that may cause a high risk of malignancy and some mechanical complications. Duodenal adenomas are frequently seen around the papilla of Vater [2]. Among extraintestinal manifestations, subcutaneous benign tumours such as epidermoid cysts of the scalp, osteomas, dental abnormalities, and desmoid tumours of soft tissue are frequently seen. These manifestations are called Gardner's syndrome as a variant of FAP [3,4]. Congenital hypertrophy of the retinal epithelium (CHRPE) manifests as pigmented ocular lesions, which are seen in about 60–90% of FAP patients [2]. Thyroid, brain (Turcot's syndrome with medulloblastoma), and adrenal gland tumors or hepatoblastomas in children are seen less frequently in patients with FAP.

There are other genes related to multiple polyposis and this should be taken into account especially in less serious polyposis and other histological types of polyps [5]. *MYH*-associated polyposis (MAP) is an autosomal recessive disease with usually 5–100 adenomas (rarely more), in some cases with duodenal polyps and other extracolonic features (dental abnormalities, dermal cysts, osteomas, or CHRPE). The mean age of clinical diagnosis is about 50 years, but in some cases the disease is diagnosed very early [6,7]. The clinical differentiation of MAP from FAP is problematic, and the condition is often sporadic or with two or more affected siblings. In families with vertical transmission of the disease, phenocopies (sporadic cases in a family with hereditary disease) or colorectal cancer in a heterozygote of *MYH* mutation can be the cause. So far, only a small effect on cancer risk is expected in *MYH* heterozygotes [8]. MAP polyps are usually tubular or tubulo-villous adenomas or hyperplastic polyps, and MAP cancers are predominantly left sided (71%), and in 27% they are metachronous or synchronous [9].

Hereditary mixed polyposis syndrome (HMPS) may present with atypical juvenile polyps, hyperplastic polyps, serrated adenomas, or classical adenomas and a high risk of cancer and no extracolonic manifestations. The search for the gene is still ongoing [10,11].

Peutz–Jeghers syndrome is an autosomal dominant inherited disease with hamartomatous polyps and characteristic mucocutaneous pigmentation, which is mostly around the lips and buccal mucosa. The skin pigmentation may fade but the buccal manifestation continues into adulthood. Hamartomatous polyps may be of different size and may cause mechanical complications. The risk of cancer is increased (39% for colorectal cancer by the age of 65 years), but the risk of other cancers is also increased (pancreatic, stomach, breast, ovarian, small bowel, endometrial, oesophagus, lung, and sex cord tumors). In about half of tested probands, *STK11/LKB1* germline mutation is found [12].

Juvenile polyposis syndrome is diagnosed in children with more than ten juvenile polyps (hamartomas with overgrowth of the lamina propria, and mucin retention cysts). One-third of cases are familial with autosomal dominant inheritance. The disease-causing mutation in the *SMAD4* gene may be detected in about one-third of the tested cases, and mutation in the *BMPR1A* gene in another one-third of cases. There is an increased risk of colorectal, but also gastric, duodenal, and pancreatic cancer.

21.2 The role of Genetic Counseling in FAP Diagnoses

Rectal bleeding, pain, and diarrhea can be the first symptoms of the disease. In some cases, other symptoms like large desmoids, osteomas, and cysts may lead to the diagnosis of FAP. Colorectal cancer or other cancers may complicate the disease in those of a young age. When the clinical diagnosis of adenomatous polyposis is clear, or there is a suspicion of FAP, genetic counseling and testing should be offered to the patient and the whole family. A geneticist, or in some countries a genetic counselor, invites the patient or other family member for the first visit. Genetic counseling is an approximately one-hour session when personal history and detailed family history are discussed. The interviewer asks about all the diseases (especially cancerous) in siblings and children of the proband, and the mother's and father's family, with at least a four-generation pedigree. Information about the diagnoses and the age of diagnoses, about the treatment, the lifestyle and the risk factors should be included. When possible, the information should be confirmed in medical records, pathology records (with written permission), or through death certificates. By recording the pedigree, a better understanding of possible inheritance, distribution and spectrum of diseases, and ages of diagnoses is possible. The geneticist should decide if *APC* or other genes (for related syndromes) should be tested in the family. Clinical symptoms of FAP, heredity, the risk for other family members, primary and secondary prevention, and also laboratory testing are discussed with the patient. In the case of clear or suspicious FAP, genetic testing of the *APC* gene is offered to the patient. Informed consent for the DNA testing should be signed by the patient and the physician, stored in the medical records, and sent to the laboratory with the DNA sample. In some cases physicians avoid genetic counseling before genetic testing and send blood for genetic testing directly to the genetic laboratory. This is not the recommended way of genetic testing.

In families with detected causal mutation, predictive testing is offered to relatives at risk starting at the age of 10 years. The age when the testing should be offered depends on many factors, mostly related to the clinical relevance and the maturity of the child. In cases of classical FAP the first symptoms can be seen at a very young age and may cause some complications like rectal bleeding, anemia, pain, diarrhea, etc. All predictive testing can only be done after genetic counseling and with signed informed consent (parent consent in the case of testing children).

21.3 The Possibilities of Genetic Testing in FAP

The *APC* gene (chromosome 5q21) was found by linkage analysis and positional cloning in 1991. This gene is part of the Wnt signalling pathway and plays an important role in the degradation of β-catenin in the cell cytoplasm. It has 15 exons coding a protein of 2843 amino acids, which exists mostly in the cytoplasm and interacts with other molecules such as β-catenin, α-catenin, axin, tubulin, and other proteins. Truncating germline mutations in the *APC* gene are responsible for about 70–90% of FAP cases [5].

The position of germline mutations is important for the prediction of clinical symptoms of the disease. Severe polyposis is seen in mutations between codons 1290 and 1400, CHRPE in mutations between codons 457 and 1444, and osteomas and desmoids in mutations in codons 1400 and more. Attenuated FAP is seen in mutations between codons 78–167, 1581–2843, and in exon 9 [2]. However, there is a wide variability of clinical symptoms even within the family, which may be caused by modifying genes, environmental factors, lifestyle, and diet.

Genetic testing is offered to probands with FAP, after genetic counseling. The patient needs to sign informed consent for testing of the *APC* gene, and decide what should be done in the laboratory with his/her DNA after the testing is finished. It can be very helpful if they are informed by geneticists about the importance of DNA storage for the purpose of additional genetic testing in the family in the future, about the possibilities of testing other genes and modifying factors, and about the possibilities of being involved anonymously in future research. Any additional clinical genetic testing should again be consulted with the patient, and a new informed consent should be signed.

Testing of the first person in the family, the patient with FAP, is usually a laborious process, which takes several months. It is necessary to screen all coding exons, and intron and exon boundaries for possible mutations. The use of testing methods varies in different laboratories. There are new, more-sensitive and less-laborious methods available, and new equipment is developed every year. There is no real hot spot in the *APC* gene, and mutations can be found in any part of the gene. Each family can have a unique mutation. In most families conventional mutations like small insertions, deletions, or substitutions are detected, but in some families large gene rearrangements are found. Thus the clinical testing should be extended for the detection of these large changes in the *APC* gene. In many families only a variant with unknown clinical significance is detected, and the testing results are uninformative with no causal pathogenic mutations found. The predictive testing of these unknown variants is not provided to the other family members.

In families with known causal mutation, predictive testing is offered to family members who are at risk. The predictive testing is less complicated and usually takes a short time since the location and the type of mutation are already known.

21.4 The Disclosure of Genetic Testing Results to the Patient and Family Members

When the genetic testing is completed, the patient or the family members are invited for another session with the geneticists. The results and other related issues are discussed, and the patient receives a detailed genetic report. If the pathogenic mutation in the family has been found, predictive testing of the same mutation is offered to other relatives. This information should be disseminated in the family by the proband.

Those family members who test negative in predictive testing do not need any special surveillance if any other risk factors are not seen in the family pedigree and their personal history. Their children do not need to be tested. In those who test positive, preventive surveillance is recommended starting at the age of finding the result, or from 10–12 years of age. In families where genetic testing was not informative or where the patient with FAP is not available for testing, preventive follow-up should be offered to all relatives who are at risk.

The geneticist discusses the need for genetic testing in family members with the patient. In some families the dissemination of information is not a problem, but in others the patient may refuse to provide the information to relatives. There may be no contact with distant family members, and the relationship even with close relatives may not be good enough to talk about the disease, or there may be other reasons for non-disclosure. The geneticist should explain (sometimes repeatedly) the importance of genetic testing of relatives and support the process of dissemination, but the final decision depends only on the proband.

An information brochure should be prepared in familial cancer clinics and given to the tested person together with the genetic report. It is also important to suggest another possible meeting with a geneticist or genetic counselor when needed.

21.5 Coping with FAP Syndrome

Different types of acceptance of the information can be expected, depending on the type of person, and their age, and personal, familial, and social situation. The revelation of a genetic disorder is usually devastating for the patient and family members. In many cases nothing serious was expected and suddenly they are informed about the severity of the disease, chronic illness, surgical procedures needed, and, in some cases, about the cancer diagnosis or the high risk of cancer. Patients and families need a sympathetic person, a geneticist or other clinician, who understands the psychological aspects of the disease and who is able to discuss all the important issues with them [13,14]. The psychological consequences of the revelation of genetic test results are different and may include anxiety, cancer phobia, denial, or blame of other people. Interfamilial relationships may be changed, and some explanation of the situation should also be available for spouses (if asked for). The geneticist should offer the help of other professionals, especially a psychologist, who may be consulted on all related problematic issues.

The heredity of FAP needs to be explained. For many patients it is difficult to understand that the disease is hereditary when nobody in their family has had the same problems. They have to be informed that their children have a 50% risk of inheriting the same mutation. In patients who will be planning a family, the explanation of possible prenatal or preimplantation diagnostics should be involved. Preimplantation embryo selection is offered in many centers, and for some couples is more acceptable than prenatal diagnosis from amniotic fluid.

21.6 Preventive Measures in FAP

Secondary prevention should be explained in detail by the geneticist and by the physician-specialist. It should also be explained to children who need to be followed up. Endoscopic examination can be painful, and if done without any previous preparation and explanation it may traumatize the child and lead to a reduction in compliance. Adequate sedation before the examination, but also friendly clinicians, nurses, and environment may be helpful for cooperation. Anxiety and fear of the surgery and cancer are very frequent [2]. It is important to help family members to cope with the disease by giving them all the information they need about the disease and possible clinical symptoms, about the lifestyle, risk factors, preventive care, examinations, schedules, prophylactic surgery, and the possible influence on their quality of life.

It should be stressed that primary prevention is very important. All the risk factors for cancer development should be explained, and a healthy lifestyle with no smoking or alcohol consumption, and a healthy diet should be strictly supported. In many cases these people are young, they engage in many sporting activities and the disease can interfere with this. It should be explained that physical activity is healthy, but should be done with caution because of the possible complications related to FAP. This is particularly important in individuals with classical FAP and a mutation in the risk area of the gene for desmoid tumours, in whom sports with a high risk of body damage are not recommended. Trauma, female sex, and estrogens are recognized as risk factors for desmoid tumours. The prevalence of desmoids among FAP patients is about 10–25% [15,16], and it is one of the major causes of morbidity and mortality in FAP (besides metastasis of unknown primary and duodenal cancer) [17].

Secondary prevention starts early and is complex. It is recommended to start sigmoidoscopy at the age of early adolescence, or even from 10–12 years of age, at 1–2-year intervals. If polyps are found, total colonoscopy should be provided every 6 months or each year. Prophylactic colectomy is planned according to the severity of clinical symptoms. Detailed discussion with a surgeon about the operation and the type of surgical procedure is needed. In the case of polyps, or at 25 years of age, regular gastroduodenoscopy should be started. Since the disease is systematic, annual physical examination with evaluation of soft tissue and bone lesions, abdominal masses, the thyroid gland, and other neurological symptoms is necessary. Abdominal ultrasound should be done regularly, even in childhood until the age of 10 years, because of the risk of hepatoblastoma. Regular eye examination should be conducted yearly, especially when CHRPE is diagnosed. Dental controls are recommended twice a year.

It might be helpful to have specialized nurse who discusses with the family all the preventive examinations, terms and conditions, and other family-related problems like planning the examination for the most suitable time.

In most cases the timing of prophylactic surgery should take into account other FAP-related problems like age, school, and family planning, job related issues, or vacations. It can be delayed if the polyps are rare and not large, but in most FAP cases the surgery has to be done in the late teens or early twenties [18–20]. Follow-up continues even after the surgery, with regular physical examination, abdominal ultrasound and/or magnetic resonance imaging (MRI), upper endoscopy, and other controls.

In the case of polyps (especially duodenal), chemoprevention may be started in order to slow the growth of the polyps. The risk of malignancy is usually not influenced, but the risk of mechanical or other complications may be decreased. The selective COX-2 inhibitor celecoxib is approved for preventive treatment at the dosage of 800 mg/day. There may be increased risk of cardiovascular complication (heart failure or stroke), perforation, or bleeding from the stomach or intestine after long-term use, and a careful clinical follow-up is needed [20]. Celecoxib, sulindac, or the anti-estrogenic drug tamoxifen can be used in the prevention or treatment of desmoids (Fig. 21.1).

21.7 Social and Psychological Issues Related to FAP

FAP is a hereditary disease where clinical symptoms may occur in childhood, and predictive genetic testing is frequently offered to children. Children are very vulnerable and need a lot of psychological support. Childhood may be affected by many symptoms of the disease, frequent controls, and very unpleasant examinations like sigmoidoscopy or colonoscopy. If prophylactic surgery needs to be done it may be a traumatic and mutilating procedure. Children may miss a lot of school, and may be disqualified from some activities; contact with their friends can be disrupted for some time. It is important to speak to the child and explain that this is temporary and they will get back to normal activities after some time. It is good if they explain the disease to their friends and prepare them for some changes. The psychologist may help the child in coping with all of these problems and improve the quality of their life. Geneticists should also consult all the issues carefully with children and young adults and be prepared to answer their questions.

In young adults, planning of children is frequently discussed. There are issues of transmitting the disease to their children, whether to have children, when to have children, whether to use prenatal or preimplantation diagnoses, when to plan the prophylactic surgery, and whether to expect problems with pregnancy after the surgery. In women with sparse polyps, an ileorectal anastomosis with biannual endoscopic examination is preferred, especially before childbearing. After pouch surgery, reduced fertility may occur [21].

Genetic testing of FAP may change intrafamilial relationships. The negatively tested family members may feel guilty when their close relatives have to deal with such a serious disease. Those who are positively tested may feel angry, disappointed, or anxious; they may blame other family members for having the disease. Some family members do not want to consider the information and do not want to be tested or have colonoscopy. There are different scenarios in families, and possible problems should be asked about and counseled by a geneticist.

Medical information should be confidential, but the genetic disease is a family problem and the genetic testing is done for the patient and his/her family members. There is an ethical dilemma if the patient does not want to let other relatives know about the

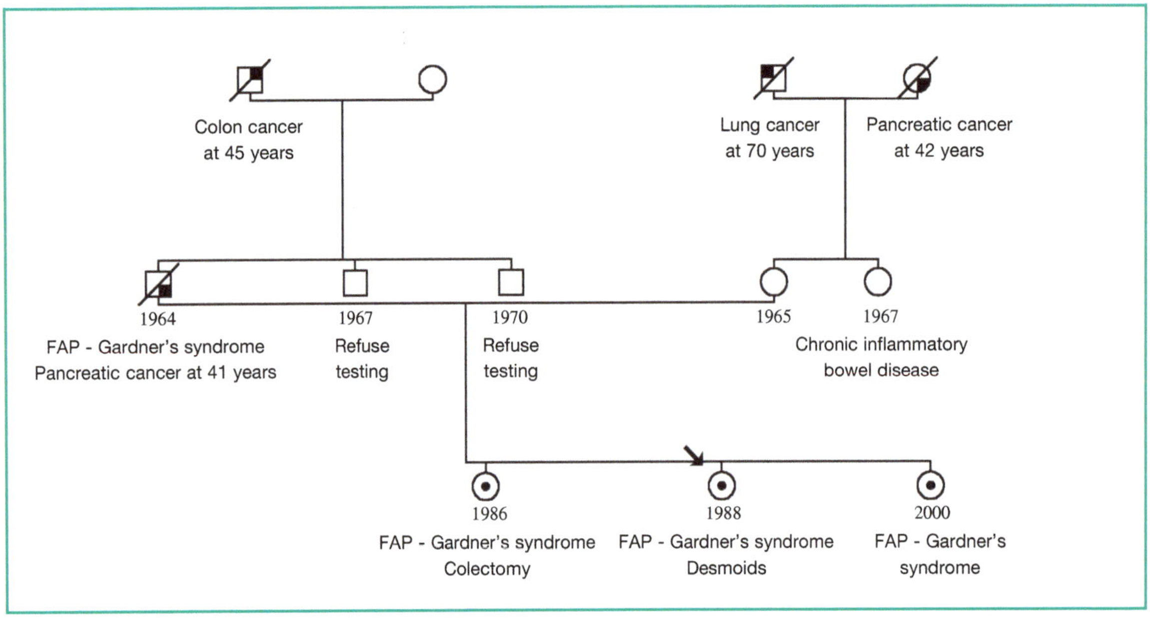

Fig. 21.1 Family with severe FAP and a large desmoid complication. The proband (born 1988) was diagnosed with a large desmoid growing from the external part of the chest inside and causing collapsing of the left lung and deviation of the mediastinum. She did not have any major problems, just anemia, but started to feel short of breath and tired at school when exercising, and detected an enlarging mass on the left side of her chest. She had computed tomography (CT) scan of her abdomen before the surgical procedure of her chest, and multiple polyps of the colon were detected. She was sent to the geneticist for genetic counseling for suspicion of FAP. In her personal history she had some scalp atheromas repeatedly removed, osteoma of the mandible resected, and problems with dental development. The surgery for a large desmoid was done, but complete resection was not possible. A germline mutation in the *APC* gene, exon 15, c.4666dupA, pThr1556AsnfsX3, was detected. This is the part of the *APC* gene where mutations cause the classical FAP and the risk of desmoids may be increased. After the chest surgery she had colonoscopy, which detected multiple polyposis, but also gastroduodeno-intestinal polyposis. She had to have a surgical procedure because of double intestinal invagination. Colectomy was not recommended because of severe polyposis of the whole gastrointestinal tract, and a massive and life-threatening desmoid. The oncologist also saw her, and treatment with tamoxifen and celebrex was started because of the desmoid progression. In her family history her father died of pancreatic cancer at 41 years. The medical records were examined and information about colonic polyposis and polypectomy at 31 years was found in the report, but the family had not been informed and examined. He also had multiple gastroduodenal polyps and atheromas. His father died of cancer of the rectum at 45 years.
Two sisters of the proband were tested; both were carriers of the same mutation. The sister born in 1986 had colectomy for multiple polyposis, atheromas, and dental abnormalities. The youngest sister born in 2000 was referred to the children's hospital for an examination. There are also a lot of familial problems, starting after the death of the father. Communication between the proband and the mother has ceased, and there is no communication with other relatives (father's brothers). In this family the clinical symptoms of the disease are very severe, especially in the proband. Her prognosis is poor and the colectomy was finally not indicated. Both her sisters are carriers. The family was not informed about the father's hereditary disease, FAP, by the clinicians of the local hospital, where he was treated for pancreatic cancer and died

genetic problem [22]. In many cases it is caused by inadequate understanding of the risks of the disease and the necessity of early prevention. Other sessions with the geneticist should be scheduled and the information should be provided repeatedly.

Some patients are afraid of discrimination from their employers, losing their job, having their health insurance denied (in countries where health insurance is not mandatory), or having a problem with life and disability insurance [23,24]. It is important that the geneticist also counsels about these issues and advises individuals to take out their life insurance before the predictive testing. Job-related issues are critical for some patients and they should be advised by social workers. Some working processes may carry a risk for FAP patients and they should be able to discuss this issue with a specialist. It is helpful if the geneticist can provide contacts to other specialists.

21.8 Cooperation between the Geneticist and Other Specialists

As a systematic disease, FAP needs complex follow-up and care from different specialists. This is usually provided in cancer centers at high-risk clinics. There are different organizations of care and the cooperation of many professions is needed. A gastroenterologist sees the patient regularly, doing all the examinations of the gastrointestinal tract. A surgeon is involved in the care, especially if prophylactic surgery is planned. An oncologist is an important integral part of the team, carrying out physical examination, planning all additional examinations, and evaluating the results. In case of cancer, duodenal multiple polyps, desmoids, and other complications, the oncologist should offer chemoprevention or therapy. A specialized nurse is also a part of the team, planning examinations, contacting the patient with time schedules, asking about possible problems, and discussing some issues, which may be important for the patient. The nurse should inform the physician about any problems or complications that the patient has. The nurse is usually easily contacted by patients and has time to consult them.

Other specialists are regularly involved in patient care, such as psychologists, ophthalmologists, dentists, radiotherapists, special surgeons, and geneticists.

There are many psychological issues related to FAP. Little research on decision making or psychological outcomes of the surgical procedures has been carried out [25]. The role of a psychologist is fundamental in many families and should not be omitted.

The geneticist has to provide all the information about the testing, testing results, heredity, and prevention, and has to consult all the family members for predictive testing. Even if the results of the genetic testing are not informative (negative results of testing in the proband), the family members should be advised about the need for colonoscopy for clinical diagnosis of FAP. Additional sessions should be offered to patients or relatives, even after the results are disclosed.

21.9 Conclusion

FAP is a complicated systematic disease with a high risk of cancer and a requirement of lifelong medical follow-up, treatment, or surgical procedures. For many people it is hard to cope with the diagnoses and they need specialized help. It is essential that all the professionals in the team taking care of patients with FAP are able to cooperate and address all the problematic issues related to the disease.

References

1. Alm T, Licznerski G (1973) The intestinal polyposes. Clin Gastroenterol 2:577–602.
2. Rozen P, Macrae F (2006) Familial adenomatous polyposis: the practical application of clinical and molecular screening. Fam Cancer 5:227–235.
3. Gardner EJ (1951) A genetic and clinical study of intestinal polyposis, a predisposing factor for carcinoma of the colon and rectum. Am J Hum Genet 3:167–176.
4. Büllow S, Berk T, Neale K (2006) The history of familial adenomatous polyposis. Fam Cancer 5:213–220.
5. Lipton L, Tomlinson I (2006) The genetics of FAP and FAP-like syndromes. Fam Cancer 5:221–226.
6. Sampson JR, Dolwani S, Jones S, et al (2003) Autosomal recessive colorectal adenomatous polyposis due to inherited mutations in MYH. Lancet 362:39–41.
7. Gismondi V, Meta M, Bonelli L, et al (2004) Prevalence of *Y165C*, *G382D* and *1395delGGA* germline mutations on Italian patients with adenomatous polyposis coli and colorectal adenomas. Int J Cancer 109:680–684.
8. Kambara T, Whitehall VL, Spring KJ, et al (2004) Role of inherited defects of MYH in the development of sporadic colorectal cancer. Genes Chromosomes Cancer 40:1–9.
9. Lipton L, Halford SE (2003) Carcinogenesis in MYH-associated polyposis follows a distinct genetic pathway. Cancer Res 63:7595–7599.
10. Jaeger EE, Woodford-Richens KL, Lockett M, et al (2003) An ancestral Ashkenazi haplotype at *HMPS/CRAC1* locus on 15q13-q14 is associated with hereditary mixed polyposis syndrome. Am J Hum Genet 72:1261–1267.
11. Farrington SM, Tenesa A, Barnetson R, et al (2005) Germline susceptibility to colorectal cancer due to base-excision repair gene defects. Am J Hum Genet 77:112–119.
12. Wang ZJ, Churchman M, Avizienyte E, et al (1999) Germline mutations of the LKB1 (STK11) gene in Peutz-Jeghers patients. J Med Genet 36:365–368.
13. Lynch HT, Watson P, Shaw TG, et al (1999) Clinical impact of molecular genetic diagnosis, genetic counselling, and management of hereditary cancer. Part I: Studies of cancer in families. Cancer 86:2449–2456.
14. Chapman PD, Burn J (1999) Genetic predictive testing for bowel cancer predisposition: the impact on the individual. Cytogenet Cell Genet 86:118–124.
15. Lofti AM, Dozois RR, Gordon H, et al (1989) Mesenteric fibromatosis complicating familial adenomatous polyposis: predisposing factors and results of treatment. Int J Colorectal Dis 4:30–36.
16. Bertario L, Russo A, Sala P, et al (2001) Genotype and phenotype factors as determinants of desmoid tumours in patients with familial adenomatous polyposis. Int J Cancer 95:102–107.

17. Sturt NJH, Clark SK (2006) Current ideas in desmoid tumours. Fam Cancer 5:275–285.
18. Lynch PM (1999) Clinical challenges in management of familial adenomatous polyposis and hereditary non-polyposis colorectal cancer. Cancer 86(suppl):2533–2539.
19. Phillips RKS, Spigelman AG (1996) Can we safely delay or avoid prophylactic colectomy in familial adenomatous polyposis? Br J Surg 83:769–770.
20. Phillips RKS, Wallace MH, Lynch PM, et al (2002) A randomized double blind, placebo controlled study of celecoxib, a selective cyclo-oxygenase 2 inhibitor on duodenal adenomatosis in familial adenomatous polyposis. Gut 50:857–860.
21. Olsen KO, Juul S, Bulow S, et al (2003) Female fecundity before and after operation for familial adenomatous polyposis. Br J Surg 90:227–231.
22. Wilcke JTR, Seersholm N, Kok-Jensen A, et al (1999) Transmitting genetic risk information in families: attitudes about disclosing the identity of relatives. Am J Hum Genet 65:902–909.
23. McEwen JE, McCarty K, Relly PR (1993) A survey of medical directors of life insurance companies concerning use of genetic information. Am J Hum Genet 53:33–45.
24. Rodrigez-Bigas M, Vasen HFA, O´Malley L, et al (1998) Health, life and disability insurance and heredity nonpolyposis colorectal cancer. Am J Hum Genet 62:736–737.
25. Patenaude AF (2005) Genetic testing for cancer. Psychological approaches for helping patients and families. American Psychological Association, Washington, DC.

The Role of a Registry in Familial Adenomatous Polyposis

22

Monica Mazzucato, Silvia Manea, Oliviana Gelasio, Cinzia Minichiello and Paola Facchin

Abstract The revolution in information technology has transformed the processes of collection, transmission, analysis, and storage of data, leading to the improvement of many large-scale systematic data-collection systems of health-related events. Familial adenomatous polyposis (FAP) represents a paradigmatic example of how the establishment of a data-collection system, starting from family tracing and evolving to more complex registration processes, increases knowledge on different aspects of the disease and ultimately can influence patients' clinical management and professionals' attitudes. Many years have passed since the establishment of the first polyposis registry at St Mark's Hospital in London. An historical overview of FAP registries is presented, discussing potentials and limitations of population-based registries versus clinical registries. In the context of recently developed health policies on rare diseases, the experience of a registry monitoring FAP as well as other rare conditions, population-based, and at the same time with features of a clinical one, is presented. The value of FAP registries relies on their ability to provide a snapshot of aspects of disease as well as of its management, producing evidence and translating this into clinical practice. Our belief is that such registries, really patient-centered, should be developed as an integral part of the healthcare network, becoming connecting informative tools between vertical networks, namely centers of expertise, horizontal care networks, and other institutions and persons involved in the care of FAP patients.

Keywords Familial adenomatous polyposis • Rare diseases • Registry

22.1 Background

The basis of cancer-prevention activities is the early identification and appropriate management of premalignant conditions. Familial adenomatous polyposis (FAP) is one of the diseases that most benefit from effective prevention interventions. These interventions are addressed to all potentially at-risk individuals and are based on appropriate follow-up and treatment programs.

For these reasons, the complex reconstruction of family pedigrees has always been part of the clinical activities of centers dealing with FAP. At the same time it strongly encouraged the creation of polyposis registries. These registries have progressively evolved from simple cases' repositories to complex informa-

M. Mazzucato (✉)
Coordinating Centre for Rare Diseases, Veneto Region, Padua, Italy

G.G. Delaini, T. Skřička, G. Colucci (eds.), *Intestinal Polyps and Polyposis*,

tive tools, structured to follow FAP patients and, in general, all susceptible individuals.

FAP registries represent a paradigmatic experience in the field of systematic data collection on individuals affected by the same rare condition, with the aim of making the available information less dispersed. In fact, when dealing with rare conditions such as FAP, one of the key problems faced not only by patients but also by clinicians is the scarce knowledge of the relevant aspects of epidemiology and natural history, as well as specific issues related to diagnosis and therapy. FAP registries represent an excellent opportunity to tackle all these questions.

22.2 Disease Registries

In epidemiology, the term "registry" is applied to a file of data concerning all cases of a particular disease or other health-relevant conditions in a defined population, such that the cases can be related to a population base [1].

Registration is the process of systematic and continuous data collection of all the events or case of interest using multiple sources of information.

The continuity requisite distinguishes a registry from *ad hoc* studies. In registries the cases collection is continuous, providing a wide spatio-temporal coverage.

Since their beginning, health data collections have been implemented for very practical reasons. Their aims have been closely related to the field of public health, since the information collected usually has a more or less direct return on the community. Over the years, registries, initially static in their structure and function, have evolved towards more complex forms of data collection, in which the starting point of all actions is the individual, promoting a more active participation and involvement of patients.

Disease registries represent systems organized for the collection, storage, analysis, and interpretation of data concerning individuals affected by a particular disease or a group of related diseases. At the beginning, this kind of registry collected synthetic information: anagraphic and few other data, namely diagnosis and outcomes, with the collection being either prospective or retrospective. Besides representing lists of patients, useful for research activities, these registries promoted the gathering of important information on the epidemiology and natural course of the monitored diseases.

Among disease registries, an important distinction has to be made between patients' registries, maintained by referral centers, and population-based registries.

A population-based registry collects information on all the cases of a particular disease arising from a population residing in a defined area, namely a city, a region, a country, etc.

The aim is to obtain epidemiological measures, useful for health interventions planning. In these registries, implying a wide population monitoring activity, a major limit is the sustainability over time of a long-lasting registration. This can be particularly difficult if the registration is completely separated from any clinical activity.

An evolution of the described registries is surveillance systems. In these information systems, the process of data collection is characterized by a strict link with action. Examples are the congenital malformation registries and the injury surveillance systems. These systems are organized in order to activate specific interventions, if collected measures, namely incidence, exceed a predefined cut-off. The interventions aim to identify the possible causes underlying the monitored phenomenon as well as at reducing the number of events at population level, when possible.

These systems are focused on the event; cases are expressions of the event and have to be monitored exactly when the event takes place. These systems are not suitable for the monitoring of events experienced by the same individual during a long lifetime period.

The monitoring of chronic conditions or risk conditions during the entire life course requires attention to be focused on the patient experiencing the event rather than on the event itself.

In this context, the first experiences of clinical data collection were developed, involving a group of individuals affected by the same condition and/or followed by the same clinical center.

Since the process of collecting information is long-lasting, involving different persons over time, predefined and common data-collection registration systems have been developed.

These systems, initially implemented with follow-up purposes, have been progressively standardized in their content. The process of data collection is strictly linked to the clinical dimension, leading to the fact that information has to be collected when it is generated and by whom it is generated.

In the field of rare disorders, many clinical centers, especially referral centers, set up specific databases, initially created with administrative purposes or with the aim of better organizing follow-up programmes.

The inclusion of adjunctive information has transformed these databases into more clinical databases, which have progressively become essential sources of information for the conduction of research studies. Over the years, important forms of collaborative activities have been developed by centers involved in the same clinical activities, promoting the use of unique and common systems for data collection.

Thanks to the scientific enthusiasm of motivated clinicians and the availability of relevant patient collections, the first disease registries were established, initially on a local and then on an international basis. The centralization of the information in a center, identified as the coordinator, was initially based on non-electronic forms filled in by the participating centers.

During the last decades the impressive progress in the field of information technology has facilitated the collection, transmission, processing, and storage of data, while improvements in systematic data-collection systems for health-related events havc allowed the development of increasingly complete and reliable databases. This favors the collection of information when it is generated, and its rapid accessibility.

22.3 FAP Registries: a Historical Overview

Menzel described the first case of multiple colorectal polypoid lesions in an article published in 1721 in a German journal [2]. In 1925, the surgeon Lockhart-Mummery of the St Mark's Hospital in London described three families with familial polyposis in his article "Cancer and heredity" in the *Lancet* [3]. He first stated the existence of a separate clinical entity from inflammatory polyps, that he called "multiple adenomata of the colon", highlighting the inherited aspect of the condition.

He demonstrated that the follow-up of FAP families, characterized by a high incidence of colorectal cancer, remarkably decreased the occurrence of this type of neoplasm. In 1925 at St Mark's Hospital in London, Lockhart-Mummery, his colleague, the pathologist Dukes, and Bussey founded the first familial polyposis registry in the world. This experience derives from the natural phenomenon that characterizes the care of patients with rare diseases, with FAP patients among them: aggregation of cases in centers of expertise, which, as a consequence of patients' referral, progressively acquire more experience and knowledge. Following a notable number of cases, these centers have been pioneers in developing clinical data-collection systems. This is the case for St Mark's Hospital Polyposis Registry, an experience which is still ongoing.

Since its foundation, this hospital-based registry has allowed, the development of both clinical activities and relevant research studies. Multiple aspects of the disease have been explored starting from the collected data. It was worth the effort to increase knowledge of the disease, as well as of the treatments and preventive actions needed [4,5]. In 1952, Dukes, one of the registry's founders, acknowledged the value of cases registration in rare diseases, stating:

> "... polyposis is a rare disease and no one surgeon can expect to see more than a few cases in a lifetime. For this reason, it is necessary when studying this disease to collect information from different sources, to collate it, and to keep a register of affected individuals and their relatives" [6].

One of the peculiar features of St Mark's Hospital Polyposis Registry is the fact that, from its outset, it involved different health professional figures, highlighting the importance, when dealing with complex pathologies, of a multidisciplinary approach in both health care and the research field.

Starting from this first experience, many FAP registries have been established worldwide, with differences in organization, aims, and covered population.

Church and McGannonn, analyzing these experiences, distinguished three main types of FAP registries: national registries, regional registries, and registries developed in the context of tertiary referral centers for the disease [7].

In fact, the distinction between national and regional registries is not so definite, as the size of the population in some countries can be compared to that monitored by some regional registries. What really differentiates registries is whether they are population based or not.

In the following sections, some examples of FAP registries are presented.

22.4 FAP Population-based Registries

Population-based registries are usually based on a centralized surveillance system that is able to guarantee a national or regional coverage. Each new FAP case is stored in a unique database, the core of the entire system. The prerequisite is a well-defined catchment area, covering the entire monitored country or region. In this way, the collected cases refer to a defined population, used as denominator for each epidemiological indicator calculated using the registry's data.

Registry staff are not necessarily involved in the patients' clinical care, focusing their activities mainly in data collection and analysis.

Prerequisites of these registries are: a homogeneous population, a complete and long-term registration of patients, an exhaustive, or at least high, coverage, and a healthcare system that is based on equal access for citizens.

This type of registry is ongoing in the Scandinavian countries. The Swedish Polyposis Registry, founded in 1957, represents the first population-based registry in Europe [8]. Later, the Finnish Registry (1961) [9,10], the Danish Registry [11,12], with a national coverage since 1975, and the Dutch Registry (1984) [13] were established.

These registries are based on a referral system of cases, which is integrated with the information coming from other multiple sources of information. In these countries, a unique personal identification code, used in different information flows, both health- and non-health-related, is available. Through record-linkage processes, this allows identification of nearly all the present cases and enables their healthcare pathways to be traced.

The main sources of data are family histories, church registries, medical records, death certificates, autopsy reports, and data from central offices of civil registration. Moreover, data collected by these registries can be integrated and compared to those of other registries, particularly national cancer registries.

One of the most interesting developments of these registries is the promotion of collaborative studies. The aim is to enrol a relevant number of patients, who are followed for long periods, in order to obtain information on long-term outcomes, to compare the efficacy of different surgical options, and to elaborate shared guidelines for disease management, based on the evidence produced [14–17].

While national registries are common in north European countries, in other European and extra-European countries registries have mainly been established on a regional basis. They essentially differ in the reduced size of the monitored population. The objective is always to achieve an exhaustive monitoring. These registries have been more easily developed where the healthcare system is mainly organized on a regional basis.

In the United Kingdom, regional registries are widespread. Historically, they were established in the context of clinical genetics services, serving a population from one to five millions inhabitants. They have rapidly been acknowledged as an essential part of prevention programs for colorectal cancer in subgroups of at-risk populations, such as FAP families [18].

Because of their strict link with genetic services, research topics have dealt with the identification of FAP families and the distribution of their mutations. A relatively recent aspect of these registries' activity is the development of collaborations, aiming to recruit a considerable number of cases, allowing study of genotype–phenotype correlations, with significant consequences for family management [19].

In Italy, FAP registries were developed in the context of wider registration systems, such as cancer registries [20]. An example is the colorectal cancer registry of Modena Province, North Italy, one of the first specialized cancer registries in Italy, founded in 1984 [21,22].

Outside Europe, an example of a regional FAP registry is the Canadian FAP Registry [23], founded in 1980 at Toronto General Hospital, with the corresponding catchment area being the Ontario Province. This registry has features of both a regional registry and a registry developed in a tertiary referral centre. Since its foundation, it has faithfully followed the structure of the St Mark's Polyposis Registry. For its initial aims, the Canadian Registry focused on knowledge of the pathology's natural history and on assessment of the efficacy of colorectal cancer-prevention activities. Afterwards, the identification of markers for detection of the disease became another relevant objective. Great attention has been paid to psychosocial aspects of the disease and to patients' support and advocacy [24]. The discovery of genes causing other hereditary colorectal cancer syndromes determined the birth of registries monitoring all hereditary gastrointestinal cancers. Data derived from these registries, and those coming from death certificates, allowed identification of areas with a high colorectal cancer incidence, which became the bases of specific surveillance systems [25].

Other non-European FAP registries, established in the context of clinical centers, but with defined catchment areas, are the Japan Polyposis Registry [26] and the Singapore Polyposis Registry [27].

In the history and activities of the above-mentioned registries, even though specific characteristics are present, some common elements can be identified. The activities have been carried out in different phases. They usually start with collecting data on probands, using multiple sources of information and then focusing on the reconstruction of family pedigrees. This shifting represents a peculiar feature of FAP registries, as the collected data refer both to the already diagnosed cases, and to the so-called "call-up cases".

At the beginning, these population-based registries provided relevant information on FAP epidemiology, namely on prevalence and incidence and their trend over time [28]. Furthermore, they increased knowledge on natural history and the efficacy of treatments and follow-up programs. A natural consequence of the registration has been the organization of preventive activities, first addressed to probands' relatives. There is strong scientific evidence about the efficacy of these interventions in decreasing morbidity and mortality due to colorectal cancer in FAP patients [13].

In many cases, peventive activities involved other health professionals working in specialized centers, as well as in the primary care network, such as general practitioners, with the aim of recruiting the highest number of at-risk individuals [29]. In this way, the registries increased knowledge of the disease among physicians, playing a relevant role in spreading information, and positively influencing both clinical decisions and the healthcare pathways of the patients [30,31].

Due to new progress in molecular diagnosis, the role of the registry has undergone a dramatic evolution. Besides the previously described aims of the registries, new ones emerged: study of the frequency and distribution of predisposing *APC* mutations in the monitored kindred, genotype phenotype correlation analysis, and identification of additional susceptibility genes. The development of the genetic component of these registries has allowed identification of which mutations cause the most severe phenotypes of disease, and therefore need strict monitoring of carriers. Survival data show that colorectal cancer mortality notably decreased in these years, mainly as a consequence of registration and follow-up programs. The same positive trend was not observed for the extracolonic manifestations, particularly for dermoid and duodenal cancers.

Despite the importance of the data derived from these registries, a major criticism concerns the link with the clinical dimension, which is not always evident. Another issue is the difficulty of using registries' data to make assumptions on the whole population. Data can be incomplete, as the determinants of a patient's presence in the registry depend on a complex set of variables, such as the geographical area where they live or the knowledge they have of the registry's existence.

Furthermore, in population-based registries a general problem of organization exists. Their functioning needs specific and dedicated competences and resources, for example to maintain the system, to elaborate data using different information sources, and to spread the information obtained both to those who generated it and to patients and their families. All these activities require a specific interest and involvement of the institutions in order to keep the registration systems active and of good quality.

22.5 FAP Clinical Registries

Besides population-based registries, other different forms of registration have been developed, especially in the context of centers that specialize in the management of inherited colorectal cancer syndromes. Such registries are particularly common in the United States. One of their main features is the lack of a defined catchment area, so that they can enrol patients residing in different areas from the one in which the center is located.

These registries have been created by individual clinicians demonstrating specific interest in the management and study of these conditions. They have promoted, organized, and maintained lists of patients and recorded information on them as part of their clinical and research activities.

In 2004, Church et al carried out a survey aiming to establish how many registries linked to centers of expertise for inherited colorectal cancer syndromes existed in the United States [32]. There were 30 centers and 18 registries. Registries were, on average, recently instituted, having a median age of 5.5 years. Nevertheless, some of them have been operating for many years, among them the Johns Hopkins Hospital Registry, the Cleveland Clinic Registry, the Creighton University Registry, and the MD Anderson Cancer Centre.

The median number of patients enrolled in the registries was low: 45 in the case of FAP. A considerable variability in aspects of dimension, function, and tools used for data collection emerged, making collaborative studies potentially difficult to conduct.

Another major limitation highlighted in the survey was incomplete coverage of the population. Considering data in the literature compared to that of registries, it was estimated that only between 19% and 23% of all theoretically existing families were enrolled in a registry, implying undermonitoring of large areas of the country. This is of particular concern from a public health point of view, as it affects patients' opportunities to access the most effective treatments and follow-up programs available.

Furthermore, registries established in the context of centers of expertise can operate a preselection in the process of patient enrolment, leading to the presence in the registry of non-homogenous subsets of patients, despite their having the same clinical diagnosis.

In some cases, the patients enrolled are those who require a more complex management – with more severe forms of the disease, or less responsive to treatments or inteventions. In other cases, enrolment of less severely affected patients is possible, since it is easier for them to contact the center and adhere to follow-up programs.

In general, clinical registries are less useful than population-based registries, such as the Scandinavian ones, in producing reliable epidemiological data. On the other hand, other aspects such as influence on healthcare activities, elaboration of guidelines, patients' education and support, spreading of information, and collaboration with patients' associations appear to be more developed.

22.6 Collaborative Registries' Networks

FAP registries represent great opportunities to promote collaborative research activities. In recent years, collaborative networks connecting FAP reference centers have been strongly encouraged, identifying powerful tools in registries for pooling data and sharing information, on the basis not only of personal experiences, but also of common systematic data-collection processes.

The aim is, above all, the identification of which are the best practices for the appropriate and effective management of patients and FAP families, in order to increase survival and ameliorate quality of life.

Considerable variability in data-collection systems, and the lack of common diagnostic criteria, represent the main obstacles for these kind of collaborations.

Despite these problems, in recent years in the field of colorectal cancer syndromes, great efforts have been made to promote international collaboration.

In 1985, experts from 45 different countries gathered together in London to celebrate St Mark's Hospital's 150th anniversary. Efforts in pooling and sharing knowledge about polyposis were strongly promoted. One of the first activities of the group was to conduct a survey of the existing polyposis registries around the world, in order to find out what information was already available on the disease. In 2003, the Leeds Castle Polyposis Group and the International Collaborative Group on HNPCC merged together, becoming the International Society for Gastrointestinal Hereditary Tumors (InSiGHT) [33].

Over time, collaborative groups of experts have placed more emphasis on the fact that care of patients and families has to be based on a multidimensional approach, highlighting the importance of registries in organizing care, and promoting research and education. The birth of these collaborative experiences has given great impulse to the creation of new registries worldwide and to the development of joint actions in the field of research.

22.7 FAP Registries: Potential and Limitations

Considering the different types of registries described, one might ask which is the best form of registration in familial polyposis.

The value of a registry must be considered with respect to the aims that lead to its establishment, and depends on the quality and usefulness of the collected information.

Considering the FAP registries that are already established, the great distinction is between population-based registries, whether regional or national, and clinical registries, whether annexed to a single center or to a network of centers.

Undoubtedly, population-based registries present some advantages, namely an exhaustive registration of cases, unique opportunity to obtain reliable epidemiologial data, and the standardization of data collection – a prerequisite for conducting collaborative studies.

Epidemiological data produced by population-based registries are essential from a health-planning point of view, and consequently the relationship between these registries and national health authorities is more strict. Due to their role of support in planning health policies, it is more likely that these registries receive adequate funding and can be directly involved in the organization of preventive programs. Furthermore, every planned intervention can theoretically reach all the individuals to whom it is addressed, due to the exhaustive character of the registration, representing an important example of equitable health policy.

Despite these aspects that are of great value, at the same time population-based registries present some major limitations. Commonly, these registries can be completely distinct from the patient care dimension, being maintained by other professionals who are not directly involved in clinical activities.

This can be less pronounced in some experiences, for example some regional population-based registries, where the link with clinical centers is more evident. Even if population-based registries represent unique and valuable sources of data, the manner in which this information can be accessed and used by both clinicians and patients is crucial.

Based on a complex organization, the registries need considerable professional and technical resources. Furthermore, their role must be well recognized so that their existence and sustainability over time can be guaranteed.

A long period of registration is needed in order to obtain reliable survival data. This is one of the main problems of every registration system: the maintenance of a good coverage and long-term high quality of collected data – results that cannot be achieved without motivation and resources.

The relationship with the patient care dimension characterizes registries established as part of the activities of FAP referral centers. In these situations, as well as in population-based registries, the advantages of the registration system are not evident in the short term, especially when monitoring rare diseases. Even these registries can be affected by sustainability problems, as they are often mantained by the special interest and strong motivation of individual clinicians. Despite initial enthusiam, data collection can become an onerous task that is difficult to maintain over time. Another problem is the extreme variability in the contents of the registration, in registries' organization, and in the information programs used in data collection. As these registries are usually not funded by national or regional authorities, they are generally supported by the contributions of patients and their associations, which can give rise to problems of transparency on the one hand, and researchers' autonomy on the other. Due to their strong link with the care dimension, great attention has been paid to the aspects of active involvement and participation of patients. This can increase competition between centers for enrolling a large number of patients, making the development of collaborative experiences potentially more difficult.

Clinical registries are of limited usefulness from a public health point of view, especially if compared to population-based registries. One of the main issues is accessibility of information for stakeholders. Furthermore, attention is usually focused on very specific topics, which can differ over time as new knowledge is progressively achieved.

Patients' mobility between centers and ill-defined catchment areas affect the possibility of using these registries to obtain information on the epidemiology of the monitored condition and its burden at a population level.

22.8 Rare Diseases Registries

According to the prevalence criteria adopted by the European Commission, a disease is defined as rare when it affects less than one person per 2000 inhabitants in Europe [34], while in the USA a disease is rare when it affects no more than one in 1250 inhabitants

[35]. Data derived from several different FAP population-based registries put the prevalence of this condition under the cut-off that defines a disease as rare. As well as other rare diseases, FAP is challenging from the point of view of patient care, as it involves different health professionals, operating both in the so-called "vertical networks", constituted by centers of expertise, and in the horizontal care networks, namely the primary care setting.

New interest and attitudes towards the care of these complex patients have been accompanied by the development and use of technical solutions aiming to facilitate the collection of large amounts of data, while at the same time favoring interaction between all the different health professionals involved in a patient's care. Recent progress in this field is demonstrated by the development of registries combining aspects of population-based registries, useful as epidemiological sources of data and supporting health planning, as well as aspects of more clinical registries, collecting data that are useful in the clinical decision-making process.

In these systems, which can be defined as informative as they go beyond mere registration purposes, collected data are those that help and orientate clinical decision processes regarding individual patients. At the same time, through the stratification of the information generated by different centers, it is possible to achieve a high level of standardization in diagnostic and therapaeutic approaches. Furthermore, these systems, developed as part of specific health policies, represent the basis for the provision of services and benefits for patients, an aspect that guarantees their quality and maintenance in the long term.

As an example, the Italian legislation on rare diseases represented a valuable opportunity to implement this kind of information system.

In 2001 a specific law was issued, containing a list of 331 diseases or groups of disorders defined as rare, and divided into 13 nosological categories. Familial polyposis is included in the neoplasms group.

A national network based on referral centers for diagnosis, treatment, and prevention of rare diseases was established, together with area-based registers monitoring all patients. An exemption from specific healthcare costs was introduced, leading patients affected by rare disorders to join the surveillance system in order to achieve the benefits. The experience of the Veneto Region, in the north east of Italy, with nearly 4,700,000 inhabitants, is presented. In enforcing the rare disorders' national law, specific health policies and integrated planning strategies were developed in this area.

An inter-regional network of centers of excellence for each group of rare disorders was established, and a unique computerized monitoring system was implemented, allowing diagnosis recording, exemption leading to benefit entitlement, and case enrolment in the register. This system connects all the identified centers of excellence, all the healthcare districts, and the local pharmaceutical services. In this way a patient refers, or can be referred by the general practitioner for instance, to a specific center of the network in order to have a complete assessment and a timely diagnosis. The patient enters the surveillance system after a specific diagnosis of rare disease has been made, and this is followed by the exemption issued by the local health districts. Only in this way can the patient receive the benefits they are legally entitled to, such as specific drugs or interventions listed in a personal therapeutic plan defined by the referral center. At the same time, the patient's clinical history can be collected and enriched with the information coming from other current health data flows, such as the hospital discharge records, death records, etc.

The information collected by the registry is useful for epidemiological studies, but also represents the starting point for supplying services to the patients. Moreover, it is essential to support specific research programs. For instance, it is possible to have homogenous subgroups of patients who can be more strictly monitored.

The registry has enrolled nearly 14,000 patients with rare diseases. At present, 51 FAP families are monitored.

From a health-planning point of view, an exhaustive monitoring system is undoubtedly a very useful resource. At the same time, the information system described, based on the registry for rare disease, clearly differs from a disease registry or from a registry set up to produce mere epidemiological data. The added value relies on the fact that it represents an essential part of the healthcare network, connecting vertical and horizontal care networks, through a common information tool.

The information system is the basis for the provision of services for patients, and is designed, in both content and logic structure, to simplify patients' care pathways. In this system, information becomes the glue connecting the patient and all the health professionals who, in different contexts and phases of the disease, participate in the care process.

22.9 Future Perspectives

The future and the role of FAP registries cannot be defined without taking into account their influence in two contexts, which are distinct and at the same time interdependent: the context of information and that of culture and attitudes.

With regard to the first, relatively recent progress in technology has given great impetus to the birth and development of registries, giving rise to important open questions concerning the use of the information collected and generated by these systems. A crucial aspect is represented by security, which involves technical, legal, and ethical issues. It is essential to have knowledge and respect for existing regulations in the field of data management, and it is also highly recommended that the correct choices are made in relation to complex technology solutions that ensure a higher level of protection than the minimum required by law.

Despite technical aspects, major issues in establishing every registration system are related to how the information should be organized, with respect to which priorities, logic framework, and rules. These three aspects vary in different types of data collection, as the aims according to which information is produced can differ. Furthermore, the detail of the collected information differs, as the purposes of the registration can be administrative, clinical, or research ones. According to the experience of the rare diseases registry, the collected information is that which is useful in clinical practice and in the care of patients.

Starting from the gathering of basic information, more complex modules, facilitating patient management, for example prescription of drugs, and formulation of care plans, have been progressively added.

A future development will be the organization of the information using a hierarchical logic, which can differ according to the phase of disease and type of patient.

Considering the aspects of culture and attitudes, FAP registries have represented a unique source of information on epidemiology and natural course of the disease. At the same time, they have been, and still are, essential to define best clinical practices.

While producing new knowledge, they have strongly influenced attitudes and decisions. The development of collaborative networks between clinical centers, promoting shared methods and common instruments of data collection, has led to the elaboration of management guidelines. This process, effectively combining experience and research, has made the best practices, which are already performed in some centers, available and easily accessible to a broader audience, with great advantage for patients and their families.

FAP registries are also key instruments of education and spreading of knowledge. The collaboration between health professionals from different backgrounds, while maintaining their respective competences, has promoted the development of a multidisciplinary approach in patient care. A future perspective and urgent need is the development and spreading of knowledge regarding tools and methods that not only define the clinical diagnosis, but also specifically assess the healthcare needs of affected individuals. The healthcare plan, tailored to the patient and designed according to his/her health needs, is the key instrument to define the impairment profile and residual resources in the patient, considered within a precise context and phase of disease.

Modern registration systems should not only be tools to be used for designing and monitoring these healthcare plans, but should also represent the core of the network connecting centers of expertise – vertical care networks and horizontal networks, namely the primary care setting, and the other institutions and persons involved in the care of individuals with FAP.

References

1. Last JM (2001) A dictionary of epidemiology. Oxford University Press, New York.
2. Menzel D (1721) De excrescentiis verrucoso cristosis copiose in intestinis crassis dysenteriam passi observatis. Acta Med Berol IX:68–71.
3. Lockhart-Mummery JP (1925) Cancer and heredity. Lancet 1:427–429.
4. Lockhart-Mummery HE, Dukes CE, Bussey HJR (1956) The surgical treatment of familial polyposis of the colon. Br J Surg 43:476–481.
5. Dukes CE (1958) Cancer control in familial polyposis of the colon. Dis Col Rect 1:413–423.
6. Dukes CE (1952) Familial intestinal polyposis. Ann R Coll Surg Eng. 10:293–304.
7. Church JM, McGannon E (1995) A polyposis registry: how

to set one up and make it work. Semin Colon Rectal Surg 6:48–54.
8. Kanter-Smoler G, Fritzell K, Rohlin A, et al (2008) Clinical characterization and the mutation spectrum in Swedish adenomatous polyposis families. BMC Med 6:10.
9. Moisio AL, Järvinen H, Peltomäki P (2002) Genetic and clinical characterisation of familial adenomatous polyposis: a population based study. Gut 50:845–850.
10. Renkonen ET, Nieminen P, Abdel-Rahman WM, et al (2005) Adenomatous polyposis families that screen APC mutation-negative by conventional methods are genetically heterogeneous. J Clin Oncol 23:5651–5659.
11. Bülow S (1984) The Danish Polyposis Register. Description of the methods of detection and evaluation of completeness. Dis Colon Rectum 27:351–355.
12. Bülow S, Burn J, Neale K, et al (1993) The establishment of a polyposis register. Int J Colorectal Dis 8:34–38.
13. Vasen HF, Griffioen G, Offerhaus GJ, et al (1990) The value of screening and central registration of families with familial adenomatous polyposis. A study of 82 families in The Netherlands. Dis Colon Rectum 33:227–230.
14. Bülow S, Alm T, Fausa O, et al (1995) Duodenal adenomatosis in familial adenomatous polyposis. DAF Project Group. Int J Colorectal Dis 10:43–46.
15. Vasen HF, van Duijvendijk P, Buskens E, et al (2001) Decision analysis in the surgical treatment of patients with familial adenomatous polyposis: a Dutch-Scandinavian collaborative study including 659 patients. Gut 49:231–235.
16. Bülow S, Bülow C, Vasen H, et al (2008) Colectomy and ileorectal anastomosis is still an option for selected patients with familial adenomatous polyposis. Dis Colon Rectum 51:1318–1323.
17. Vasen HF, Möslein G, Alonso A, et al (2008) Guidelines for the clinical management of familial adenomatous polyposis (FAP). Gut 57:704–713.
18. Morton DG, Macdonald F, Haydon J, et al (1993) Screening practice for familial adenomatous polyposis: the potential for regional registers. Br J Surg 80:255–258.
19. Sampson JR, Dolwani S, Jones S, et al (2003) Autosomal recessive colorectal adenomatous polyposis due to inherited mutations of MYH. Lancet 362(9377):39–41.
20. Modica S, Roncucci L, Benatti P, et al (1995) Familial aggregation of tumors and detection of hereditary non-polyposis colorectal cancer in 3-year experience of 2 population-based colorectal cancer registries. Int J Cancer 62:685–690.
21. Ponz de Leon M, Sassatelli R, Scalmati A, et al (1993) Descriptive epidemiology of colorectal cancer in Italy: the 6-year experience of a specialised registry. Eur J Cancer 29A:367–371.
22. Ponz de Leon M, Rossi G, di Gregorio C, et al (2007) Epidemiology of colorectal cancer: the 21-year experience of a specialised registry. Intern Emerg Med 2:269–279.
23. Stern HS (1996) The Canadian Familial Adenomatous Polyposis Registry: past, present and future. J R Soc Med 89:153P–154P.
24. Esplen MJ, Berk T, Butler K, et al (2004) Quality of life in adults diagnosed with familial adenomatous polyposis and desmoid tumor. Dis Colon Rectum 47:687–695; discussion 695–696.
25. Green RC, Green JS, Buehler SK, et al (2007) Very high incidence of familial colorectal cancer in Newfoundland: a comparison with Ontario and 13 other population-based studies. Fam Cancer 6:53–62.
26. Iwama T, Tamura K, Morita T, et al (2004) A clinical overview of familial adenomatous polyposis derived from the database of the Polyposis Registry of Japan. Int J Clin Oncol 9:308–316.
27. Goh HS, Wong J (1992) The Singapore Polyposis Registry. Ann Acad Med Singapore 21:290–293.
28. Björk J, Akerbrant H, Iselius L, et al (1999) Epidemiology of familial adenomatous polyposis in Sweden: changes over time and differences in phenotype between males and females. Scand J Gastroenterol 34:1230–1235.
29. Bülow S (2003) Results of national registration of familial adenomatous polyposis. Gut 52:742–746.
30. Bülow S, Faurschou Nielsen T, et al (1996) The incidence rate of familial adenomatous polyposis. Results from the Danish Polyposis Register. Int J Colorectal Dis 11:88–91.
31. Bülow S, Bülow C, Nielsen TF, et al (1995) Centralized registration, prophylactic examination, and treatment results in improved prognosis in familial adenomatous polyposis. Results from the Danish Polyposis Register. Scand J Gastroenterol 30:989–993.
32. Church J, Kiringoda R, LaGuardia L (2004) Inherited colorectal cancer registries in the United States. Dis Colon Rectum 47:674–678.
33. Neale K, Bülow S (2003) Origins of the Leeds Castle Polyposis Group. Fam Cancer S1:1–2.
34. US Orphan Drug Act. Public Law 97-414, 4 January 1983.
35. Regulation (EC) No 847/2000 of April 27, 2000. Laying down the provisions for implementation of the criteria for designation of a medicinal product as an orphan medicinal product and definition of the concepts 'similar medicinal product' and 'clinical superiority'. Official Journal of the European Communities 28 April 2000:L103/5.

Coping with FAP

The Role of Patients' Associations

23

John D. Roberts and Mick J. Mason

Abstract This is our personal story of how two senior citizens found out about familial adenomatous polyposis (FAP) late in life. Circumstances gave us the time to investigate and understand FAP from a layman's point of view and also to help others come to terms with the challenge of FAP.

We recognised the need for uncomplicated information for those affected by FAP and for a centralised point for this information without the need to scour the internet. So FAPGene (http://www.fapgene.org.uk) was born. Those searching on the internet for information about the *APC* gene which is linked to FAP are most likely to have found a mass of technical articles often requiring a subscription. We found that a search for FAP is far more productive to a patient's needs. Very early on we found that patients/individuals often made initial contact via email with no further enquiry after the information they needed was passed on. This is reflected in our article where FAP is described as a lonely disease. One of the success points shown is our ability to work with health professionals and not cross that border into providing medical information. Again, many patients/individuals were given contact numbers for these medical questions. The question of whether FAPGene will progress to a true support group or patient association depends on several things including the age of John and Mick and their own state of health. At the moment they are fortunately able to continue improving the scope of FAPGene but the future is uncertain.

Keywords Education • Familial adenomatous polyposis • Forums • Information days • Support • Websites

23.1 Introduction

We have been asked to write about FAP patient associations. Patient associations or support groups, which we will use interchangeably, can take several forms, such as a website, an internet forum or a physical group where people can meet. As Laura Szabo-Cohen wrote 'There is no community in more need of support groups than the FAP population . . . but neither is there a population more scattered, so we must make our own arrangements' [1].

Our own attempt started as a website, and after three years seems to be turning into a patient-led support group. This is a result of organising, through

J.D. Roberts (✉)
FAPGene, Derby, UK

Fig. 23.1 Swarkestone Sailing Club, Derbyshire, UK

the website, an FAP Family and Information day at Swarkestone Sailing Club (Fig. 23.1) in the East Midlands of the UK. We are pleased to report that we have just held the third of these events and each one has attracted an increasing number of attendees.

To demonstrate the need for patient associations we would like to highlight the need that FAP patients have for information, and the lengths they usually have to go to to find it. More especially, we should emphasise how initially they may prefer to be anonymous; the need to meet other people with the same condition develops once they have made contact and come to understand that there are other people with the same worries and problems as themselves.

23.2 John's Story

John's mother died of bowel cancer in 1970, but it was not until 1985 when his brother, Edwin (then 39 years), developed polyps in his large bowel, had a total colectomy and ileorectal anastomosis (IRA), that FAP was mentioned. John then had an initial examination with a sigmoidoscope and was declared clear of polyps. In 1992 his son developed hepatoblastoma, (a liver cancer) at age 6 years. As there is a statistical link between hepatoblastma and FAP, it was suggested that John be checked again, and this time polyposis was found and later confirmed as FAP. Aged 52 years he had a total colectomy and an (internal) ileo-anal pouch was installed. Later, two of Edwin's three children had total colectomies with IRA.

John first started looking for information in 1993 by writing a letter to the journal of the Ileostomy and Internal Pouch Support Association Group (The IA) [2], a small number of whose members were people with FAP looking for support for pouch or ileostomy problems. This provided two contacts in the UK so they were able to support each other over the telephone for several years. Also, a third lady, in Vancouver, British Columbia, sent a long letter of her experience and copies of the Hereditary Colon Cancer newsletter. This enabled John to get on the mailing list for a time and alerted him to the wide range of problems that people with FAP can suffer from.

Later in the quest there was contact with Kay Neale at the St Mark's Hospital FAP Registry at Harrow, London. This led to his involvement in the production of a newsletter 'Polypost' for a year or so. A little later, at a Red Lion Group meeting (another pouch group) at St Mark's, Kay told him about Mick, who had set up a website about FAP, and his experiences. Fortunately, Mick lived within 40 miles of John and was the first person with FAP that John had met face to face locally. Others were family and the four or five people at the Red Lion Group meeting in London 120 miles away . . .

23.3 Mick's Story

In February 2002, at the age of 59 years, Mick was diagnosed with FAP following colon cancer and a genetic test in June 1999. He and his wife Ann came away from the genetic centre worried and confused. If only they had had something to take home and read, it would have helped so much.

Early in 2003 the thought of an information website on FAP crossed their minds, as very little was available on the internet. They had still met no one else with FAP but felt sure there must be others in the same position, almost crying out for information.

APC is the name of the actual gene, but searching in Google with this only seemed to find rather technical information. Mick was told he had FAP and it was genetic, so FAPGene sounded about right as the name for his new website. Also, he felt patients were more likely to search for FAP rather than APC.

Mick decided only non-medical information would be included, and to work with the health professionals rather than against them. Links to other suitable websites, personal stories and latest news on FAP research would also be included.

In late 2004 the website was ready and FAPGene was launched to the world (http://www.fapgene.org.uk). There were not a lot of pages but Mick had managed to add two personal stories – of his own. These were articles he had written for the IA journal. *The fashion show*, his account of his debut as a male model at an event organised by Stoma Care Nurses at The Leicester Royal Infirmary, was followed by the booklet *My genetic journey* (http://www.fapgene.org.uk/booklet.html), which speaks for itself.

There were four critical moments which really helped to boost the profile of FAPGene. A small snippet in a magazine by CancerBackup stated that having a genetic cancer gene did not mean you would get cancer. With FAP, cancer is almost certain to develop without major preventative surgery. A phone call to CancerBackup was well received and resulted in Mick's name being forwarded for involvement in a joint Department of Health and Macmillan Cancer Support Genetics Partnership Programme [3]. This was the second critical point which eventually led to a very receptive audience for promoting awareness of FAP and the FAPGene website.

In 2005 the website was having visitors in a steady, if not spectacular, stream. Details had been forwarded to all and sundry with little in return. This was until the third critical moment came, when Mick received a call from Kay Neale at the Polyposis Registry at St Marks Hospital in Harrow.

Kay had seen the website and invited Mick to meet her staff. His idea of eventual family information days was something the registry had thought about in the past. However, a lack of time and resources had been a major stumbling block. Mick left with high hopes that the Polyposis Registry would soon have their first event, while he felt FAPGene's was still a few years away.

The final critical moment for Mick was when Kay put him in touch with John Roberts who lived in Derby. They had exactly the same views about the need to raise awareness of FAP and organise family/information days.

It sounds so simple, but Mick was a little sceptical at first when John mentioned holding the first event the following March of 2006 – not only that, but with quite high-profile speakers.

The rest is history as they say, and is well documented in other parts of the article and perhaps best left to John. Mick added that a combination of things had helped to make FAPGene successful rather than anything major. To hold their first family/information day in 2006 just months after the first one at St Marks was a great achievement. They felt that they had finally arrived.

23.4 Forums

The following explains why FAPGene does not have its own internet forum. Initially, there were security concerns, and also if there is a forum in a country with a small FAP population it is best to share one rather than compete.

23.4.1 FAPGene and the IA

With many more people now on the internet, it is not surprising that discussion forums are so popular. A forum at FAPGene seemed an obvious move.

Mick decided to use a ready-made product from the website host. Within days, literally hundreds of postings appeared with almost every one giving direct links to questionable websites. This was totally unacceptable and the venture had a premature ending.

Mick enquired of the IA if a section on their forum could be made for FAP patients. Several members of The IA had FAP, but at first the administrator thought FAPGene was better providing its own. He said he would help in any way possible in setting up a more-secure forum. In September 2004, however, a young lady in Scotland also expressed an interest, which triggered a change of heart and saw an FAP section provided by the IA. Within weeks it blossomed, and to avoid any duplication it was decided to provide a direct link from and to FAPGene. This also helped to promote the FAPGene website itself.

There were many topics raised but a small number of replies in comparison to the number of views. This seems to prove Mick and John's own theory that many patients with FAP are content to find information anonymously and without committing themselves to any form of membership.

There are also other forums available, generally American based but with an international user list. Several of our members have also used Facebook on the internet (Table 23.1).

Table 23.1 Forum statistics

Topic name	Replies	Views
Desmoid tumours	221	5577
Duodenum polyps	80	2784
General chat	33	1204
Information days/support	67	1717
Cancer after surgery	23	623
26 June 2008, total topics 103	854	21,000

23.5 Organisation of FAPGene

How is http://www.fapgene.org.uk organised?

Basically Mick and John find information about FAP which they think other FAP sufferers may find interesting, and add it to the website. Sometimes this is a link to a particular website, or a reproduced article if permission is obtained.

The knowledge they gained from doing this and their experience on medical support group information days (IA and Red Lion Group) gave them the confidence to form a support group. In this, they were supported by Kay Neale, of the St Mark's Hospital Polyposis Registry, and other professionals, without whom they perhaps would have not been successful.

Prior to the first 2006 information day, a small committee with Mick as treasurer/secretary (Fig. 23.2), John as chairman (Fig. 23.3), and Mick's wife Ann as a committee member, was formed in order to open a bank account; this allowed them greater flexibility. However after listening to Wolfram Nolte, Vice President of Familienhilfe Polyposis coli e.V. at Swarkestone in 2008, who spoke of their 250 members throughout Germany, they realised that with FAPGene there is no formal review of what they are doing as there are no actual members. (One could say they must be doing something right as people do return to their yearly meetings.)

Therefore, they aim to formalise an idea that Mick had whereby payment for attendance at the information day also includes a year's membership to the group. An AGM could be held at the start of the information day, which would allow the members to increase participation in the running of FAPGene. They also feel that they have to consider what will happen to FAPGene in its present form if they find that they cannot continue to run it themselves. They are both retired and although committed to continue for the foreseeable future, when the inevitable happens they would both wish that FAPGene carries on and continues to develop.

FAPGene has been fortunate to receive several donations from individuals over the past two years. These donations, with the registration fees (7.50//£6.00) from the Swarkestone family/information days, have given them an annual budget of around 375€ (£300). The funds now provide for a buffet

Fig. 23.2 Mick Mason, Founder

Fig. 23.3 John Roberts, Chairman

lunch and a donation to Swarkestone Sailing Club in lieu of a hire charge.

Whereas the family/information days are mainly John's domain, Mick provides the website expertise on a voluntary basis, and internet costs are 35€ (£28.00) per annum.

The extra income this year already ensures they have the funds in place to support all their costs for 2008/2009.

23.5.1 Education/Research

Mick's contacts with CancerBackup and Macmillan led to invitations to speak to groups of professionals and people in training. Recently, Mick and John gave a presentation to a group of nurse students on a masters course, which was well received. Tony Farine, who organised the occasion at Nottingham Medical School, sent the following comments, 'Thank you so much for your time and also the presentation. I think that the talk went very well'. FAPGene has also been active in promoting the 'Family Talk' project at the School of Health Sciences at Birmingham University [4]. 'Family Talk' is a research project that hopes to find out more about families' experiences of living and coping with certain types of illness or disability, and how it is discussed amongst family members. Dr Alison Metcalfe, the project lead, gave an overview of its progress at Swarkestone 2008. Also at the same Information Day Dr Emma Tonkin gave a talk on the NHS National Genetics Education and Development Centre's resource, 'Telling Stories: Understanding Real Life Genetics' [5]. This has been developed to illustrate the impact and utility of genetics on real-life health care. Later, she emailed to say, 'My thanks go to Mick and John for allowing me the opportunity to talk. I found the day really interesting and learnt a lot'.

FAPGene finds it interesting that patient stories can be used in different ways, first to explain the dynamics of communication within the family, and second to record patient experience and use it in a teaching experience for professionals using a website.

Research is a subject that most people affected by FAP are interested in, one of the main concerns being the ability to have children who are free of FAP. This is now possible with current UK laws on preimplantation genetic diagnosis [6].

In 2006 SLA Pharma, a pharmaceutical company, gave a talk at Swarkestone on Alfa Capsules containing omega-3 The company also then printed 500 copies of Mick's *My genetic journey* booklet for FAPGene, which has been used to raise awareness of FAP around the UK.

Fig. 23.4 Family day, Swarkestone 2008

23.5.2 Family/Information Days

These days are arranged around four main speakers. There are local speakers in the morning with the first giving a general outline of FAP and its history; this is followed by a specialty speaker.

In the afternoon there are speakers from St Mark's Hospital, Harrow and London. One of these is generally Kay Neale, Nurse Specialist in Polyposis and manager of the Polyposis Registry talking about genetics or various complications of FAP. This year Miss Sue Clark, a consultant surgeon, spoke about which operation to carry out for FAP patients. She also, answered questions about desmoid tumours as several of those present had this problem (Fig. 23.4).

The simple design of the building, which accommodates 50 people, allows attendees to mingle and exchange information and also to encourage each other. This is enhanced by a structured programme, which also includes free time for extra discussions.

23.6 Conclusion

There is a need for another channel for FAP information other than the usual medical one. The size of the problem is more than can be comfortably dealt with in the consulting room; one could say it is almost a degree-level course in its own right. While the website is a good first move, information days have provided an environment where clinicians and patients join together to gain knowledge, while allowing patients the chance for face-to-face meetings with their peers.

The strengths of FAPGene include the respectability it has gained from many professionals and professional bodies. Also, despite having no formal membership as yet, an increasing number of people are interested in adding their own personal stories.

Another strength of FAPGene is that it is mainly promoted though its website, and as such is available to almost everyone worldwide. It is an ideal place to advertise both the FAPGene and St Mark's Hospital annual information days for those people who wish to meet face to face. Before John and Mick helped to generate enough confidence that an information day would be well attended, there were none; now there are two a year in the UK – one in London and one in the Midlands. Prior to 2005, contacting FAP patients directly in the UK had to be through the medical profession. Now there are two patient groups, who work together to inform, educate and support people with FAP.

23.6.1 Personal Note – Mick

I feel that perhaps FAPgene has progressed as far as is possible with John and myself. Our website and family information days are a success and have undoubtedly helped many people over the past three years. I have no doubts that this will continue for a number of years but with the uncertainty over other effects of FAP I feel it is wise that we do not try and expand into something that will detract from what we do successfully.

My own future health is governed by any changes in the many duodenum polyps I have. There are too many to remove and any surgery which might be needed would certainly restrict my involvement with FAPGene. I do feel fortunate that I have escaped lightly in the past compared to some. This is one reason I have been able to cope with FAP and start FAPGene. To actually search for information while suffering from major problems would have been an impossible task for many people myself included.

It is therefore my hope that others will read this article and build on what John and I have started. The number of people that attend a family/information day is not important. What is important is the number that go away with new information and a realisation that others are in a similar position, and an understanding of how they manage to cope.

23.6.2 A Professional's Comment

This wonderfully informative and friendly website started as a result of a real need for information and support specifically for people whose lives have been touched by FAP. The great success of FAPGene speaks for itself; it has brought together an ever-expanding group of people and enabled them to share experiences and information through both the internet and the family days. Members come from all over the UK and beyond to give and receive support and, perhaps most importantly, take some comfort from the realisation that they are not alone in coping with all that FAP throws at them. I thoroughly recommend this excellent site and all it has to offer, and congratulate Mick and John for developing such a valuable resource (Dr. Andrea Pithers, Trinity Health Innovations Ltd, andrea.pithers@thihealth.co.uk).

23.6.3 Comment from Jo Aston (Patient)

Jo visited the Swarkestone Family/Information Day for the second time in March 2008:

> Hi John & Mick
>
> Once again thank you for a very informative day! I really enjoyed it and know my family did too. I particularly enjoyed chatting to other people and especially people I recognised from last year. I particularly enjoy finding out about other families' circumstances, tips for coping etc.
>
> The idea of paying a yearly membership is a good one – it was a thought I had too when listening to the gentleman from Germany. Newsletters would be good too, it would certainly keep us informed of developments throughout the year. Good ideas!

The amount of work you two put into the day is obvious. I hope you got as much out of it as we did.

Since Saturday we have contacted Gill Plumridge with regard to the Family Talk Project and hope to help out with that. We have also contacted West Midlands Genetics Service with regard to repeat testing for the members of our family who had received a negative result (this was after advice given on Saturday by Kay Neale). We are also looking at information for the children in our family (in response to the information from yourselves with regard to the booklet).

So in all it was a very productive day for our family!

If there is anything at all we can do to help – please email one of us.

Thanks again – take care!

Jo Aston

Further Resources

- **Cancerbackup** (charity providing cancer/genetic information): 3 Bath Place, Rivington Street, London, EC2A 3JR, UK.
  Tel: +44 (0)20 7696 9003; fax: +44 (0)20 7696 9002; website: www.cancerbackup.org.uk
- **Familienhilfe Polyposis Coli e.V.**: Am Rain 3a, 36277, Schenklengsfeld, Germany.
  Tel.: +49 (0)6629 1821; fax: +49 (0) 6629 915193; email: info@familienhilfe-polyposis.de; website: www.familienhilfe-polyposis.de
- **Ileostomy and Internal Pouch Support Group**: Peverill House, 1–5 Mill Road, Ballyclare, County Antrim, Northern Ireland, UK.
  Tel: +44 (0)800 0184724; fax: +44 (0)28 9332 4606; website: www.iasupport.org
- **Macmillan Cancer Support** (cancer support and campaign charity): 89 Albert Embankment, London, SE1 7UQ, United Kingdom.
  Tel: +44 (0)20 7840 7840; fax: +44 (0) 20 7840 7841; website: www.macmillan.org.uk
- **NHS National Genetics Education and Development Centre**: Morris House, Birmingham Women's Hospital, Edgbaston, Birmingham, B15 2TG, UK.
  Tel: +44 (0)121 623 6987; website: www.geneticseducation.nhs.uk
- **Polyposis Registry**: St Mark's Hospital, Northwick Park, Watford Road, Harrow, Middlesex, HA1 3UJ, UK.
  Tel: +44 (0)20 8235 4270; Fax: +44 (0)20 8235 4278; website: www.polyposisregistry.org.uk
- **Red Lion Group** (ileo-anal pouch charity): St Mark's Hospital, Watford Road, Harrow HA1 3UJ, UK. Website: www.redliongroup.org/j/index.php

Acknowledgements
We would like to acknowledge the help of Swarkestone Sailing Club, without whom we would have struggled to find a venue with the funds we have:

- Swarkestone Sailing Club: Swarkestone Lake Swarkestone Derbyshire. Website: www.swarkestonesc.co.uk

We would also like to acknowledge support from SLA Pharma for research work and for printing our booklet *My genetic journey*:

- UK Operating Office SLA Pharma: AG Elite House Hill Farm Industrial Estate, Leavesden, Watford WD25 7SA, UK. Tel: +44 (0)1923 681 001; fax: +44 (0)1923 681 221; website: www.slapharma.com

References

1. Szabo-Cohen L. Finding a support group for polyposis. Familial Colon Cancer Registry at Huntsman Cancer Institute Spring 2001:6. www.huntsmancancer.org/pdf/fccr/FCCR_Spring_2001.pdf (accessed 22 August 2008).
2. The Ileostomy and Internal Pouch Support Group, Peverill House, 1–5 Mill Road, Ballyclare Co Antrim, BT39 9DR, UK. +44 (0)28 9334 4043; fax: +44 (0)28 9332 4606; website: www.the-ia.org.uk
3. MacMillan Cancer Support. Genetics Partneship Programme www.macmillan.org.uk/About_Us/Specialist_healthcare/Primary_care_cancer_leads/Genetics_Partnership_Programme.aspx (accessed 22 August 2008)
4. Gill Plumridge. Family Talk Project. School of Health Sciences, 52 Pritchatts Road, University of Birmingham B15 2TT, UK. Tel: +44 (0)121 415 8740; website: http://healthscinet.bham.ac.uk/FamilyTalk (accessed 22 August 2008).
5. Emma Tonkin. Telling Stories Understanding Real Life Genetics. Faculty of Health Sport and Science, University of Glamorgan, PontypriddCF37 1DL, UK. Tel: +44 (0)1443 483156; website: www.geneticseducation.nhs.uk/tellingstories/index.asp (accessed 22 August 2008).
6. Human Fertilisation and Embryology Authority. HFEA licenses PGD for inherited colon cancer. www.hfea.gov.uk/en/1049.html (accessed 22 August 2008).

Subject Index

G

H

I

J

K

L

M

N

P

MIX
Papier aus verantwortungsvollen Quellen
Paper from responsible sources
FSC® C105338

If you have any concerns about our products,
you can contact us on
ProductSafety@springernature.com

In case Publisher is established outside the EU,
the EU authorized representative is:
Springer Nature Customer Service Center GmbH
Europaplatz 3, 69115 Heidelberg, Germany

Printed by Libri Plureos GmbH
in Hamburg, Germany